AF581849

Innovations in the Food System: Exploring the Future of Food

Innovations in the Food System: Exploring the Future of Food

Editor

Maria Lisa Clodoveo

MDPI • Basel • Beijing • Wuhan • Barcelona • Belgrade • Manchester • Tokyo • Cluj • Tianjin

Editor
Maria Lisa Clodoveo
University of Bari
Italy

Editorial Office
MDPI
St. Alban-Anlage 66
4052 Basel, Switzerland

This is a reprint of articles from the Special Issue published online in the open access journal *Foods* (ISSN 2304-8158) (available at: https://www.mdpi.com/journal/foods/special_issues/Innovations_Food_System_Exploring_Future_Food).

For citation purposes, cite each article independently as indicated on the article page online and as indicated below:

LastName, A.A.; LastName, B.B.; LastName, C.C. Article Title. *Journal Name* **Year**, *Volume Number*, Page Range.

ISBN 978-3-0365-5663-5 (Hbk)
ISBN 978-3-0365-5664-2 (PDF)

Contents

About the Editor . . . . **vii**

Maria Lisa Clodoveo
Special Issue "Innovations in the Food System: Exploring the Future of Food"
Reprinted from: *Foods* **2022**, *11*, 2183, doi:10.3390/foods11152183 . . . . **1**

Francesca Anna Ramires, Stefania De Domenico, Danilo Migoni, Francesco Paolo Fanizzi, Dror L. Angel, Rasa Slizyte, Katja Klun, Gianluca Bleve and Antonella Leone
Optimization of a Calcium-Based Treatment Method for Jellyfish to Design Food for the Future
Reprinted from: *Foods* **2022**, *11*, 2697, doi:10.3390/foods11172697 . . . . **7**

Vincenzo Russo, Marco Bilucaglia, Riccardo Circi, Mara Bellati, Riccardo Valesi, Rita Laureanti, Giuseppe Licitra and Margherita Zito
The Role of the Emotional Sequence in the Communication of the Territorial Cheeses: A Neuromarketing Approach
Reprinted from: *Foods* **2022**, *11*, 2349, doi:10.3390/foods11152349 . . . . **27**

Jing Xu, Jiajia Cai, Guanxin Yao and Panqian Dai
Strategy Optimization of Quality Improvement and Price Subsidy of Agri-Foods Supply Chain
Reprinted from: *Foods* **2022**, *11*, 1761, doi:10.3390/foods11121761 . . . . **43**

Giordano Stella, Biancamaria Torquati, Chiara Paffarini, Giorgia Giordani, Lucio Cecchini and Roberto Poletti
"Food Village": An Innovative Alternative Food Network Based on Human Scale Development Economic Model
Reprinted from: *Foods* **2022**, *11*, 1447, doi:10.3390/foods11101447 . . . . **63**

Porzia Maiorano, Francesca Capezzuto, Angela Carluccio, Crescenza Calculli, Giulia Cipriano, Roberto Carlucci, Pasquale Ricci, Letizia Sion, Angelo Tursi and Gianfranco D'Onghia
Food from the Depths of the Mediterranean: The Role of Habitats, Changes in the Sea-Bottom Temperature and Fishing Pressure
Reprinted from: *Foods* **2022**, *11*, 1420, doi:10.3390/foods11101420 . . . . **85**

Estee Ngew, Wut Hmone Phue, Ziruo Liu and Saji George
Composite of Layered Double Hydroxide with Casein and Carboxymethylcellulose as a White Pigment for Food Application
Reprinted from: *Foods* **2022**, *11*, 1120, doi:10.3390/foods11081120 . . . . **117**

Caterina Palocci, Karl Presser, Agnieszka Kabza, Emilia Pucci and Claudia Zoani
A Search Engine Concept to Improve Food Traceability and Transparency: Preliminary Results
Reprinted from: *Foods* **2022**, *11*, 989, doi:10.3390/foods11070989 . . . . **133**

Umile Gianfranco Spizzirri, Paolino Caputo, Cesare Oliviero Rossi, Pasquale Crupi, Marilena Muraglia, Vittoria Rago, Rocco Malivindi, Maria Lisa Clodoveo, Donatella Restuccia and Francesca Aiello
A Tara Gum/Olive Mill Wastewaters Phytochemicals Conjugate as a New Ingredient for the Formulation of an Antioxidant-Enriched Pudding
Reprinted from: *Foods* **2022**, *11*, 158, doi:10.3390/foods11020158 . . . . **147**

Vincenzo Russo, Margherita Zito, Marco Bilucaglia, Riccardo Circi, Mara Bellati, Laura Emma Milani Marin, Elisabetta Catania and Giuseppe Licitra
Dairy Products with Certification Marks: The Role of Territoriality and Safety Perception on Intention to Buy
Reprinted from: *Foods* **2022**, *10*, 2352, doi:10.3390/foods10102352 . . . . **169**

Qi Tao, Xiaohui Cui, Hongwei Ding and Huixia Wang
Application Research: Big Data in Food Industry
Reprinted from: *Foods* **2021**, *10*, 2203, doi:10.3390/foods10092203 . . . . **179**

Simone Mancini, Giovanni Sogari, Salomon Espinosa Diaz, Davide Menozzi, Gisella Paci and Roberta Moruzzo
Exploring the Future of Edible Insects in Europe
Reprinted from: *Foods* **2022**, *11*, 455, doi:10.3390/foods11030455 . . . . **195**

Maria Lisa Clodoveo, Marilena Muraglia, Vincenzo Fino, Francesca Curci, Giuseppe Fracchiolla and Filomena Faustina Rina Corbo
Overview on Innovative Packaging Methods Aimed to Increase the Shelf-Life of Cook-Chill Foods
Reprinted from: *Foods* **2021**, *10*, 2086, doi:10.3390/foods10092086 . . . . **207**

Maria Lisa Clodoveo, Marilena Muraglia, Pasquale Crupi, Rim Hachicha Hbaieb, Stefania De Santis, Addolorata Desantis and Filomena Corbo
The Tower of Babel of Pharma-Food Study on Extra Virgin Olive Oil Polyphenols
Reprinted from: *Foods* **2022**, *11*, 1915, doi:10.3390/foods11131915 . . . . **219**

About the Editor

Maria Lisa Clodoveo

Maria Lisa Clodoveo is Associate Professor in Food Science and Technology at the Interdisciplinary Department of Medicine, University of Bari, Italy. She is the Director of the Short Master in "Health claims of Extra Virgin Olive Oil as Marketing Tool to Improve the Company's Competitiveness", and Director of the Short Master in "The Olive Oil Sensory Science and the Culinary Art". She is a member of the "Accademia dei Georgofili" and of "Accademia Nazionale dell'Olivo e dell'Olio". She is the founder of the "Research Centre for Olive Growing and Olive Oil Industry" at the University of Bari. She was the winner of the award "Antico Fattore" (2016) assigned by "Accademia dei Georgofili". She was awarded with the Menvra Award 2018 as woman of oil. She is a Principal Investigator of the European Project Horizon 2020—Fast Track for Innovation—Olive Sound. She is the inventor of two patents and the author of more than 100 articles and book chapters in the olive oil sector.

 MDPI

Editorial

Special Issue "Innovations in the Food System: Exploring the Future of Food"

Maria Lisa Clodoveo

Interdisciplinary Department of Medicine, Piazza Giulio Cesare, University of Bari, 70124 Bari, Italy; marialisa.clodoveo@uniba.it

The Food and Agriculture Organization of the United Nations (FAO) in 2018 provided a definition of "food systems" highlighting that they "encompass the entire range of actors and their interlinked value-adding activities involved in the production, aggregation, processing, distribution, consumption, and disposal of food products that originate from agriculture, forestry or fisheries, and food industries, and the broader economic, societal and natural environments in which they are embedded". The COVID-19 pandemic has imposed a stop that has caused the population to rethink their lifestyles, production, and consumption, also accelerating the transformation progress necessary in light of the objectives of the 2030 Agenda for Sustainable Development—adopted by all United Nations Member States in 2015—which provides a shared blueprint for the peace and prosperity of people and the planet, now and in the future. Actual food systems account for nearly one-third of global GHG emissions; consume large amounts of natural resources; result in biodiversity loss and negative health impacts (due to both under- and over-nutrition); and do not allow fair economic returns and livelihoods for all actors, in particular, for primary producers. With regard to these observations, innovations should aim to develop the following food systems:

Inclusive: ensuring economic and social inclusion for all food system actors, especially smallholders, women, and youth;

Sustainable: minimizing negative environmental impacts, conserving scarce natural resources, and strengthening resiliency against future shocks;

Efficient: producing adequate quantities of food for global needs while minimizing post-harvest loss and consumer waste;

Nutritious and healthy: enabling the consumption of a diverse range of healthy, nutritious, and safe foods.

These are ambitious goals that will require multidisciplinary effort—from engineering to life sciences, biotechnology, medical sciences, social sciences, and economic sciences. New technologies and scientific discoveries are the solution to the increasing demand for sufficient, safe, healthy, and sustainable foods influenced by the increased public awareness of their importance. This Special Issue is composed of 11 papers.

Jing Xu and his collaborators [1] focused their work on strategy optimization of quality improvement and price subsidy of the agri-foods supply chain. In this paper, the differences in the quality safety, price, and market demand of agri-foods in the supply chain are compared and analyzed demonstrating that the maximum profit of supply chain participants decreases with the increase of price elasticity of demand. When the quality of agri-foods is upgraded in a producer-led manner, the quality of agri-foods in the supply chain does not undergo substantial improvement, and the maximum profit of agri-foods operators is insensitive to the price elasticity of demand at this time. When the seller-led quality upgrading is launched, the maximum profit of the producer decreases with the increase of the quality elasticity of demand, the maximum profit of the seller increases with the increase of the quality elasticity of demand, and the total profit of the supply chain also increases with the increase of the quality elasticity of demand under the centralized decision

Citation: Clodoveo, M.L. Special Issue "Innovations in the Food System: Exploring the Future of Food". *Foods* **2022**, *11*, 2183. https://doi.org/10.3390/foods11152183

Received: 21 July 2022
Accepted: 21 July 2022
Published: 22 July 2022

situation. The quality and safety of agri-foods as well as the overall profit of the supply chain can be improved most effectively under the centralized control decision with the goal of maximizing the supply chain benefits. The authors concluded that in terms of quality and price, quality improvement actions of agri-foods driven by supply-side producers are less effective than those driven by demand-side consumption. In addition, cost-sharing contracts can significantly improve the quality of agri-foods in the supply chain and make them more "high-quality and low-price" than before the adoption of cost-sharing contracts.

Giordano Stella and co-authors [2] discussed the "Food Village", an innovative alternative food network based on a human scale development economic model. The researchers suggest that although the different alternative food networks (AFNs) have experienced increases worldwide for the last thirty years, they are still unable to provide an alternative capable of spreading on a large scale. They in fact remain niche experiments due to some limitations on their structure and governance. Their study proposes and applies a design method to build a new sustainable food supply chain model capable of realizing a "jumping scale". Based on the theoretical and value framework of the Civil Economy (CE), the Economy for the Common Good (ECG), and the Development on a Human Scale (H-SD), the proposed design model aims to satisfy the needs of all stakeholders in the supply chain. Max-Neef's Needs Matrix and Design Thinking (DT) tools were used to develop the design model. Applying the design method to the food chain has allowed us to develop the concept of the "Food Village", an innovative food supply network far from the current economic mechanisms and based on the community and eco-sustainability.

Maiorano et al. [3] addressed the ambiental issue of Food from the Depths of the Mediterranean and the role of habitats, changes in the sea-bottom temperature, and fishing pressure. This paper reviews studies that highlight a link between deep-sea fishery resources (deep-sea food resources) and vulnerable marine ecosystems, species, and habitats in the Mediterranean Sea, providing new insights into changes in commercial and experimental catches of the deep-sea fishery resources in the central Mediterranean over the last 30 years. About 40% of the total landing of Mediterranean deep-water species is caught in the central basin. Significant changes in the abundance of some of these resources with time, sea-bottom temperature, and fishing effort have been detected, as well as an effect of the Santa Maria di Leuca cold-water coral province on the abundance of the deep-sea commercial crustaceans and fishes. The implications of these findings and the presence of several geomorphological features, sensitive habitats, and vulnerable marine ecosystems in the central Mediterranean are discussed with respect to the objectives of biodiversity conservation combined with those of management of fishery resources.

Estee Ngew and her research team [4] studied the composite of layered double hydroxide with casein and carboxymethylcellulose as a white pigment for food application. The authors studied the composite of layered double hydroxide with casein and carboxymethylcellulose as an alternative to the titanium dioxide (TiO_2). Titanium dioxide is commonly used in food, cosmetic, and pharmaceutical industries as a white pigment due to its extraordinary light scattering properties and high refractive index. However, as evidenced by recent reports, there are overriding concerns about the safety of nanoparticles of TiO_2. As an alternative to TiO_2, Mg-Al layered double hydroxide (LDH) and their composite containing casein and carboxymethyl cellulose (CMC) were synthesized using wet chemistry and compared with currently used materials (food grade TiO_2 (E171), rice starch, and silicon dioxide (E551)) for its potential application as a white pigment. These particles were characterized for their size and shape (Transmission Electron Microscopy), crystallographic structure (X-ray Diffraction), agglomeration behavior and surface charge (Dynamic Light Scattering), surface chemistry (Fourier Transform Infrared Spectroscopy), transmittance (UV–VIS spectroscopy), masking power, and cytotoxicity. their results showed the formation of typical layered double hydroxide with flower-like morphology which was restructured into pseudo-spheres after casein intercalation. Transmittance measurement showed that LDH composites had better performance than pristine LDH, and the aqueous suspension was heat and pH-resistant. While its masking power was not on a par with

E171, the composite of LDH was superior to current alternatives such as rice starch and E551. Sustainability score obtained by MATLAB® based comparison for price, safety, and performance showed that LDH composite was better than any of the compared materials, highlighting its potential as a white pigment for applications in food.

Caterina Palocci and collaborators [5] presented their preliminary results on a search engine concept to improve food traceability and transparency. The authors started from the evidence that in recent years, the digital revolution has involved the agrifood sector. However, the use of the most recent technologies is still limited due to poor data management. The integration, organization, and optimized use of smart data provide the basis for intelligent systems, services, solutions, and applications for the food chain management. With the purpose of integrating data on food quality, safety, traceability, transparency, and authenticity, an EOSC-compatible (European Open Science Cloud) traceability search engine concept for data standardization, interoperability, knowledge extraction, and data reuse, was developed within the framework of the FNS-Cloud project (GA No. 863059). For the developed model, three specific food supply chains were examined (olive oil, milk, and fishery products) in order to collect, integrate, organize, and make available data relating to each step of each chain. For every step of each chain, parameters of interest and parameters of influence—related to nutritional quality, food safety, transparency, and authenticity—were identified together with their monitoring systems. The developed model can be very useful for all actors involved in the food supply chain, both to have a quick graphical visualization of the entire supply chain and for searching, finding, and re-using available food data and information.

Gianfranco Spizzirri [6] and its research group developed a tara gum/olive mill wastewaters phytochemicals conjugate as a new ingredient for the formulation of an antioxidant-enriched pudding. Olive mill wastewater, a high polyphenols agro-food by-product, was successfully exploited in an eco-friendly radical process to synthesize an antioxidant macromolecule, usefully engaged as a functional ingredient to prepare functional puddings. The chemical composition of lyophilized olive mill wastewaters was investigated by HPLC-MS/MS and 1H-NMR analyses, while the antioxidant profile was in vitro evaluated by colorimetric assays. Oleuropein aglycone (5.8 μg mL^{-1}) appeared as the main compound, although relevant amounts of an isomer of the 3-hydroxytyrosol glucoside (4.3 μg mL^{-1}) and quinic acid (4.1 μg mL^{-1}) were also detected. LOMW was able to greatly inhibit ABTS radical (IC50 equal to 0.019 mg mL^{-1}), displaying, in the aqueous medium, an increase in its scavenger properties by almost one order of magnitude compared to the organic one. Lyophilized olive mill wastewaters reactive species and tara gum chains were involved in an eco-friendly grafting reaction to synthesize a polymeric conjugate that was characterized by spectroscopic, calorimetric, and toxicity studies. In vitro acute oral toxicity was tested against 3T3 fibroblasts and Caco-2 cells, confirming that the polymers do not have any effect on cell viability at the dietary use concentrations. Antioxidant properties of the polymeric conjugate were also evaluated, suggesting its employment as a thickening agent, in the preparation of pear puree-based pudding. High performance of consistency and relevant antioxidant features over time (28 days) were detected in the milk-based foodstuff, in comparison with its non-functional counterparts, confirming lyophilized olive mill wastewaters as an attractive source to achieve high-performing functional foods.

Vincenzo Russo et al. [7] studied the role of territoriality and safety perception on the intention to buy dairy products with certification marks. Over the years, the territorial origins of agri-food products have become a consolidated marketing model which stands as an alternative to mass production. References to territory, whether on the packaging or in advertising, have become an increasingly popular way for marketers to differentiate products, by attributing specific characteristics to them, derived from specific cultural identities and traditions. The aim of their study was to capture the possible differences between two groups, Italian and French, in the perception and intention to buy products with certification marks. The authors tested a multi-group structural equations model, assessing the mediation of the Perceived Product Safety between Packaging with reference to Territoriality and Intention to Buy. The authors' findings show that in both groups Packaging with reference to Territoriality has a positive association with Intention to Buy and Perceived Product Safety and that Perceived Product Safety has a positive association with Intention to Buy. The difference is the mediation of Perceived Product Safety, present only in the Italian group. This opens important considerations on the role of the perception of safety, particularly in the pandemic period, in the presentation of products, particularly in products with certification marks linked to sustainability and territoriality.

Tao et al. [8] wrote an article on Big Data in Food Industry. The authors underlined that a huge amount of data is being produced in the food industry, but the application of big data—regulatory, food enterprise, and food-related media data—is still in its infancy. Each data source has the potential to develop the food industry, and big data has broad application prospects in areas such as social co-governance, exploit of consumption markets, quantitative production, new dishes, take-out services, precise nutrition, and health management. However, there are urgent problems in technology, health, and sustainable development that need to be solved to enable the application of big data to the food industry. The results showed the great potential for big data in the food industry. Big data has particularly broad application prospects in social co-governance of the food industry, quantitative production, exploitation of consumption markets, new dishes, take-out services, and precise nutrition and health management.

Simone Mancini with the research team [9] discussed the future of edible insects in Europe. They started from the evidence that the effects of population increase and food production on the environment have prompted various international organizations to focus on the future potential for more environmentally friendly and alternative protein products. One of those alternatives might be edible insects. Entomophagy, the practice of eating insects by humans, is common in some places but has traditionally been shunned in others, such as European countries. The last decade has seen a growing interest from the public and private sectors in the research in the sphere of edible insects, as well as significant steps forward from the legislative perspective. In the EU, edible insects are considered novel foods, therefore a specific request and procedure must be followed to place them on the market; in fact, until now, four requests regarding insects as a novel food have been approved. Insects could also be used as feed for livestock, helping to increase food production without burdening the environment (indirect entomophagy). Market perspectives for the middle of this decade indicate that most of the demand will be from the feed sector (as pet food or livestock feed production). Undoubtedly, this sector is gaining momentum and its potential relies not only on food, but also on feed in the context of a circular economy.

Clodoveo and her collaborators [10] presented an overview of innovative packaging methods aimed to increase the shelf-life of cook-chill foods. Analyzing the changing citizen habits, it is clear that the consumption of meals prepared, packaged, and consumed inside and outside the home is increasing globally. This is a result of rapid changes in lifestyles as well as innovations in advanced food technologies that have enabled the food industry to produce more sustainable and healthy fresh packaged convenience foods. This paper presents an overview of the technologies and compatible packaging systems that are designed to increase the shelf-life of foods prepared by cook–chill technologies. The concept of shelf-life is discussed and techniques to increase the shelf life of products are presented including active packaging strategies.

The last paper of this collection has been written by Clodoveo et al. [11]. and is entitled "The Tower of Babel of Pharma-Food Study on Extra Virgin Olive Oil Polyphenols". Much research has been conducted to reveal the functional properties of extra virgin olive oil polyphenols on human health once extra virgin olive oil is consumed regularly as part of a balanced diet, as in the Mediterranean lifestyle. Despite the huge variety of research conducted, only one effect of extra virgin olive oil polyphenols has been formally approved by EFSA as a health claim. This is probably because EFSA's scientific opinion is entrusted to scientific expertise about food and medical sciences, which adopt very different investigative methods and experimental languages, generating a gap in the scientific communication that is essential for the enhancement of the potentially useful effects of extra virgin olive oil polyphenols on health. Through the model of the Tower of Babel, we propose a challenge for science communication, capable of disrupting the barriers between different scientific areas and building bridges through transparent data analysis from the different investigative methodologies at each stage of health benefits assessment. The goal of this work is the strategic, distinctive, and cost-effective integration of interdisciplinary experiences and technologies into a highly harmonious workflow, organized to build a factual understanding that translates, because of trade, into health benefits for buyers, promoting extra virgin olive oils as having certified health benefits, not just as condiments.

As summarized above, the collection of 11 articles that make up this Special Issue devoted to "Innovations in the Food System: Exploring the Future of Food" underscores the progress that has been made toward a better understanding of the trends in the global food industry, oriented to a sustainable production assisted by economic and technological tools. surely these first evidences will be developed and enriched in the near future and additional studies are expected to cover a broad range of approaches to the design of innovation in the Food systems.

Funding: This research received no external funding.

Conflicts of Interest: The authors declare no conflict of interest.

References

1. Xu, J.; Cai, J.; Yao, G.; Dai, P. Strategy Optimization of Quality Improvement and Price Subsidy of Agri-Foods Supply Chain. *Foods* **2022**, *11*, 1761. [CrossRef] [PubMed]
2. Stella, G.; Torquati, B.; Paffarini, C.; Giordani, G.; Cecchini, L.; Poletti, R. "Food Village": An Innovative Alternative Food Network Based on Human Scale Development Economic Model. *Foods* **2022**, *11*, 1447. [CrossRef] [PubMed]
3. Maiorano, P.; Capezzuto, F.; Carluccio, A.; Calculli, C.; Cipriano, G.; Carlucci, R.; Ricci, P.; Sion, L.; Tursi, A.; D'Onghia, G. Food from the Depths of the Mediterranean: The Role of Habitats, Changes in the Sea-Bottom Temperature and Fishing Pressure. *Foods* **2022**, *11*, 1420. [CrossRef] [PubMed]
4. Ngew, E.; Phue, W.H.; Liu, Z.; George, S. Composite of Layered Double Hydroxide with Casein and Carboxymethylcellulose as a White Pigment for Food Application. *Foods* **2022**, *11*, 1120. [CrossRef] [PubMed]
5. Palocci, C.; Presser, K.; Kabza, A.; Pucci, E.; Zoani, C. A Search Engine Concept to Improve Food Traceability and Transparency: Preliminary Results. *Foods* **2022**, *11*, 989. [CrossRef] [PubMed]
6. Spizzirri, U.G.; Caputo, P.; Oliviero Rossi, C.; Crupi, P.; Muraglia, M.; Rago, V.; Malivindi, R.; Clodoveo, M.L.; Restuccia, D.; Aiello, F. A Tara Gum/Olive Mill Wastewaters Phytochemicals Conjugate as a New Ingredient for the Formulation of an Antioxidant-Enriched Pudding. *Foods* **2022**, *11*, 158. [CrossRef] [PubMed]
7. Russo, V.; Zito, M.; Bilucaglia, M.; Circi, R.; Bellati, M.; Marin, L.E.M.; Catania, E.; Licitra, G. Dairy Products with Certification Marks: The Role of Territoriality and Safety Perception on Intention to Buy. *Foods* **2021**, *10*, 2352. [CrossRef] [PubMed]
8. Tao, Q.; Ding, H.; Wang, H.; Cui, X. Application research: Big data in food industry. *Foods* **2021**, *10*, 2203. [CrossRef] [PubMed]
9. Mancini, S.; Sogari, G.; Espinosa Diaz, S.; Menozzi, D.; Paci, G.; Moruzzo, R. Exploring the Future of Edible Insects in Europe. *Foods* **2022**, *11*, 455. [CrossRef] [PubMed]
10. Clodoveo, M.L.; Muraglia, M.; Fino, V.; Curci, F.; Fracchiolla, G.; Corbo, F.F.R. Overview on innovative packaging methods aimed to increase the shelf-life of cook-chill foods. *Foods* **2021**, *10*, 2086. [CrossRef] [PubMed]
11. Clodoveo, M.L.; Muraglia, M.; Crupi, P.; Hbaieb, R.H.; De Santis, S.; Desantis, A.; Corbo, F. The Tower of Babel of Pharma-Food Study on Extra Virgin Olive Oil Polyphenols. *Foods* **2022**, *11*, 1915. [CrossRef] [PubMed]

MDPI

Article

Optimization of a Calcium-Based Treatment Method for Jellyfish to Design Food for the Future

Francesca Anna Ramires [1,†], Stefania De Domenico [1,2,†], Danilo Migoni [2], Francesco Paolo Fanizzi [2], Dror L. Angel [3], Rasa Slizyte [4], Katja Klun [5], Gianluca Bleve [1,*] and Antonella Leone [1,6]

1 Unit of Lecce, National Research Council–Institute of Sciences of Food Production, 73100 Lecce, Italy
2 Department of Biological and Environmental Sciences and Technologies (DiSTeBA), University of Salento, 73100 Lecce, Italy
3 Department of Maritime Civilizations, School of Archaeology and Maritime Cultures, University of Haifa, Haifa 31905, Israel
4 Department of Fisheries and New Biomarine Industry, SINTEF Ocean, Brattørkaia 17C, 7010 Trondheim, Norway
5 National Institute of Biology, Marine Biology Station, 6330 Piran, Slovenia
6 Local Unit of Lecce, Consorzio Nazionale Interuniversitario per le Scienze del Mare (CoNISMa), 73100 Lecce, Italy
* Correspondence: gianluca.bleve@ispa.cnr.it
† These authors contributed equally to this work.

Abstract: Edible jellyfish are a traditional Southeast Asian food, usually prepared as a rehydrated product using a salt and alum mixture, whereas they are uncommon in Western Countries and considered as a novel food in Europe. Here, a recently developed, new approach for jellyfish processing and stabilization with calcium salt brining was upgraded by modifying the pre-treatment step of freshly caught jellyfish and successfully applied to several edible species. Treated jellyfish obtained by the application of the optimized version of this method respected both quality and safety parameters set by EU law, including no pathogenic microorganisms, absence or negligible levels of histamine and of total volatile basic nitrogen, no heavy metals; and the total bacterial, yeast, and mold counts were either negligible or undetectable. Jellyfish treated by the presented method exhibited unique protein content, amino acid and fatty acid profiles, antioxidant activity, and texture. The optimized method, initially set up on *Rhiszostoma pulmo*, was also successfully applied to other edible jellyfish species (such as *Cotylorhiza tuberculata*, *Phyllorhiza punctata*, and *Rhopilema nomadica*) present in the Mediterranean Sea. This study discloses an innovative process for the preparation of jellyfish-based food products for potential future distribution in Europe.

Keywords: jellyfish; novel food; safety; quality; nutritional traits; organic calcium salts

Citation: Ramires, F.A.; De Domenico, S.; Migoni, D.; Fanizzi, F.P.; Angel, D.L.; Slizyte, R.; Klun, K.; Bleve, G.; Leone, A. Optimization of a Calcium-Based Treatment Method for Jellyfish to Design Food for the Future. *Foods* **2022**, *11*, 2697. https://doi.org/10.3390/foods11172697

Academic Editor: Maria Lisa Clodoveo

Received: 19 July 2022
Accepted: 1 September 2022
Published: 4 September 2022

1. Introduction

Jellyfish (JF) are mainly available and consumed as food in Asian countries. However, their use in food preparation has recently spread widely worldwide in the form of ready-to-use products, also attributed to the availability on internet market channels [1,2]. In recent years, JF food products have also become more popular in Western Countries [3], possibly due to an increase in JF populations related to environmental factors, such as rising temperatures, marine pollution, oxygen depletion, and a reduction of marine predator populations [4].

Jellyfish blooms and the invasive behavior of some species make them a good candidate for potential resources for food and other applications [5–7].

JF for human consumption are generally prepared by separating the umbrella from the oral arms and washing them extensively in order to eliminate mucus, gonads, sand, and superficial microorganisms. Then, these highly perishable JF tissues are treated with mixtures of NaCl and aluminum salts (alum) [1,8] in order to stabilize them, thus extending their shelf life, reducing any microbial issues, and promoting the expected organoleptic characteristics and texture so highly appreciated by Eastern people. In Asia, although ancient recipes and empirical procedures are still followed [9], new methods still based on the use of alum have been developed [1], since Eastern cuisines pay high attention to product texture and taste. However, the traditionally preserved JF products available on the market contain high levels of aluminum, which is strongly bound to the tissue [10] and cannot be eliminated through the usual washes applied before consumption.

The research on new JF stabilization procedures and treatments for food uses is recently moving to limit the use of alum, due to its toxicity [11,12], and to obtain semi-finished and finished products closer to Western Countries' style and expectations. Pedersen et al. [13] reported the possible substitution of alum with other tanning salts, such as iron salts, with a mechanism similar to a tanning process. In addition, the same authors produced alum-free crisps by soaking the jellyfish in ethanol and drying it afterward.

At present, JF is a novel food in Europe [14] and is limited by several issues, such as the (*i*) very high aluminum content of Asian traditionally preserved JF products and (*ii*) lack of safe stabilization methods for treating and processing JF tissues according to EU safety standards. Consequently, the development of a new, safe, and validated technology for processing JF could encourage regulatory authorities to approve the use and commercialization of edible JF species.

In our previous work, we proposed new parameters for the risk assessment of JF as food in Europe [15]. They were newly identified and applied to JF and JF-derived products not already included in the European regulation on seafood safety [16,17]. More recently, we proposed a new procedure [18] to process JF raw materials using calcium salts, which were selected from the food additives allowed in several Western Countries (the EU, Australia, USA, and New Zealand). It was observed that calcium salts were able to work as firming and stabilizing agents for JF biomasses, thus opening the opportunity to prepare safe semi-finished products suitable for subsequent food applications. Bleve et al. [18] set up the procedure under controlled conditions, thus demonstrating the microbiological safety of the method.

In this paper, the above-mentioned method was further optimized by modifying the JF pre-treatment step and including washes with sterile seawater; furthermore, the whole procedure was validated by using several JF species with different characteristics. Two different strategies are described here in order to substitute the use of sterile seawater for JF washing, with this step being quite challenging and not applicable on an industrial scale. Additionally, JF treated with calcium salts were tested for safety and quality aspects—the treated tissues were analyzed for protein, fatty acids, amino acids, element content, and antioxidant activity. The process efficacy was initially tested on the model species *Rhizostoma pulmo* and successively verified on three other potentially edible JF species (*Cothyloriza tuberculata*, *Phylloriza punctata*, and *Rophilema nomadica*). A comparison of the JF treated with the optimized processing method proposed here with JF prepared with the traditional Asian methods was also carried out.

2. Materials and Methods

2.1. Sample Collection and Pre-Treatment

Rhizostoma pulmo, Macrì 1778 (Cnidaria, Scyphozoa, Rhizostomatidiae) specimens were hand-collected from a motorboat in the Ionian Sea (Ginosa Marina, Italy 40°24′37.5″ N 16°53′04.2″ E) using a nylon landing net (3.5 cm mesh size) during samplings in the 2019–2020 summer period and processed by either of the two procedures described below.

In the absence of specific slaughter guidelines for cnidarians [19], the traditional method used in Asian Countries to kill the jellyfish [1] was applied to *R. pulmo* specimens by cutting the oral arms from the umbrella and removing the gastric content. For *Cotylorhiza tuberculata*, *Phyllorhiza punctata*, and *Rhopilema nomadica* samples, also sampled during the 2019–2020 summer period, whole JF were immediately frozen at −40 °C. At least 5 different specimens from each species were used; for *Rhizostoma pulmo*, randomized sampling was conducted as in Leone et al. [20].

All the procedures followed in this study are summarized in the scheme reported in Figure 1.

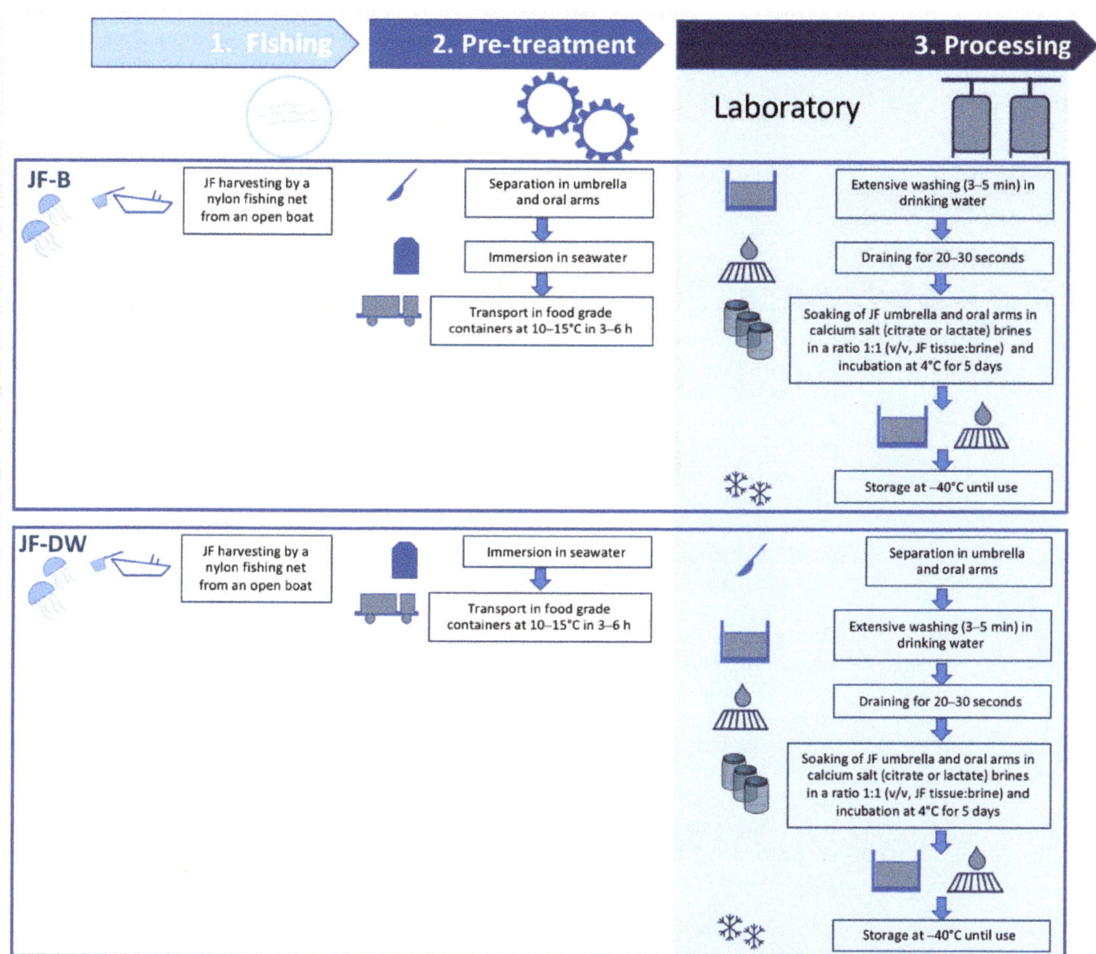

Figure 1. Diagram illustrating the procedure of the JF treatment procedures.

Procedure 1 (JF-B). JF oral arms were separated from the umbrellas, immersed in refrigerated seawater, and transported to the laboratory (max 3–6 h). At the laboratory, jellyfish parts were extensively washed for 3–5 min in drinking water.

Procedure 2 (JF-DW). Whole JF were immersed in refrigerated seawater immediately after capture and transported to the laboratory (max 3–6 h). At the laboratory, the jellyfish were immersed in drinking water, the umbrellas were separated from the oral arms, and these parts were extensively washed for 3–5 min.

Following both procedures, the umbrellas and oral arms were placed in sterile food-grade plastic bags and stored at −80 °C or immediately treated in the newly formulated brines with a firming agent.

Procedure 2 (JF-DW) was also applied to *Cotylorhiza tuberculata* samples harvested from the Gulf of Trieste (northern Adriatic Sea, Slovenia) and the Ionian Sea (Italy), and to *Rhopilema nomadica* and *Phyllorhiza punctata* harvested from the Eastern Mediterranean (Israeli coastal waters), frozen at −40 °C, and shipped on dry ice to Italy. Frozen material was thawed overnight on ice and then extensively washed with drinking water.

Aliquots of all untreated and treated JF samples (see below) were freeze-dried in order to analyze the elemental, lipid, and amino acid compositions or immediately frozen to evaluate the protein content and antioxidant activity. A commercial product, jellyfish stored in brine, from Japan (Salt-Alum Jp) was also analyzed and used for comparison.

2.2. Microbiological Analyses of Pre-Treated Jellyfish

Ten grams from each JF sample were added to 90 mL of buffered peptone water (Biolife Italiana, Milano, Italy) as a diluent (1:10). For total bacterial counts (TBCs), samples were diluted and plated by the pour plate technique on plate count agar (PCA) (Biolife Italiana, Milano, Italy) at pH 7.0 and incubated at 30 °C for 72 h; the enumeration of yeast and molds was performed by incubation at 25 °C for 5 days on dichloran Rose–Bengal chloramphenicol agar (DRBC, Thermo Fisher Scientific, Monza, Italy). The presence of Enterobacteriaceae, *Escherichia coli*, *Salmonella enteritidis*, Coagulase-positive staphylococci, *Staphylococcus aureus*, *Vibrio* spp. (*V. cholerae*, *fluvialis*, *parahemolyticus*, and *vulnificus*), *Bacillus* spp. (*B. cereus*, *turigensis*, *megaterium*, and *subtilis*), *Shewanella putrefacens*, *Aereomonas hydrophila*, and *Pseudomonas fluorescens* was assessed following the procedure described by Bleve et al. [18].

For the determination of halophilic microorganisms, JF samples were homogenized with a sterilized blender, and 25 g of each sample was added to peptone seawater 0.1% (w/v peptone) and artificial seawater. All samples and their respective serial dilutions were plated in different media dissolved in artificial seawater as described by Bleve et al. [15,18]. For each plate, the number of colony-forming units (CFU) per gram of JF was determined.

The JF samples were also submitted to an accredited external laboratory for independent analyses (Laboratori Artas Società Cooperativa, Poggiardo, Lecce, Italy). Ten grams of each JF sample were added to 90 mL of buffered peptone water (Biolife Italiana, Milano, Italy) as a diluent (1:10) and homogenized for 2 min in a Stomacher in accordance with specific standard methods for the total bacterial count (UNI EN ISO 4833-1:2013), coliforms (ISO 4832:2006), β-glucuronidase-positive *Escherichia coli* (ISO 16649-2:2001), coagulase-positive staphylococci (UNI EN ISO 6888-2:1999), and yeast and molds (ISO 21527-1:2008, ISO 21527-2:2008). For the detection of the pathogenic bacteria *Salmonella* spp. (UNI EN ISO 6579-1:2017) and *Listeria monocytogenes* (ISO 11290-1:2017), 25 g of jellyfish samples were suspended in 225 mL of buffered peptone water (Biolife Italiana, Milano, Italy) and Fraser broth at half concentration (Biomerieux, Marcy l'Etoile, France), respectively, as diluents.

2.3. Jellyfish Treatment in Brine

Solutions of calcium citrate and calcium lactate were prepared using 0.1 M calcium-citrate solution or 0.1 M calcium–lactate hydrate solution and adjusted to pH 5.0 by using the corresponding alpha organic acids, 1 M citric acid or 1 M lactic acid 85% (v/v), all from Sigma-Aldrich (Darmstadt, Germany). These concentrations were arbitrarily chosen and tested for JF stabilization treatment, as already previously described by Bleve et al. [18].

In order to determine these treatment conditions, JF tissue samples were soaked in calcium salt brines at different pH values ranging from 3 to 6 for 10 days. A starting brine pH value of 5 was determined as the best compromise, obtained after the evaluation of the effects produced by calcium salts on JF tissue's traits, such as texture and appearance, and safety aspects. More acidic pH values (<5) exerted undesirable effects on JF consistency by damaging (attacking and corroding) the tissue (data not shown).

Pre-treated jellyfish (umbrella and oral arms washed for 3–5 min in drinking water, JF-B, or JF-DW) were immersed in brines at a 1:1 ratio (v/v, JF tissue:brine) in food-grade glass or plastic containers. These technological phases were arbitrarily transferred from the conditions generally used for vegetable stabilization treatment and directly tested for the first time for their possible adaptation to JF tissues. The containers were stored at 4 °C for 5 days and then JF tissues (umbrella and oral arms) were removed, washed with drinking water to eliminate excess salts, sealed in food-grade plastic bags, and stored at −80 °C for further tests. Aliquots of each sample were also freeze-dried and stored.

2.4. JF Treatment with NaCl and Aluminum Salt

R. pulmo specimens were also processed by the traditional Asian method [21,22] using salt and alum (Salt–Alum JF-DW). Briefly, *R. pulmo* umbrellas were separated from the oral arms and extensively washed for 3–5 min with drinking water. The washed umbrellas were covered with a salt mix containing 90% (*w*/*w*) NaCl and 10% $KAl(SO_4)_2 \cdot 12H_2O$ (alum) (*w*/*w*) (Cruciani Prodotti Crual, Roma, Italy) using about 100 g of salt–alum mix per 1 kg of JF biomass and incubated at 4 °C in a food-grade glass container. After 4 days, brines released from the JF tissues were removed and the umbrellas were covered with a salt mixture containing 92.5% (*w*/*w*) NaCl and 7.5% alum (*w*/*w*). After 4 days, the same procedure was repeated, but the percentage of alum in the salt mix was reduced to 5% (*w*/*w*) and finally to 2.5% (*w*/*w*). At the end of the process, the salted jellyfish samples were left to dry on a draining rack at room temperature for 4 days, inverting them several times to drain and remove excess water. The entire process took 20 days. Aliquots of each sample were also lyophilized.

2.5. Physical–Chemical Analyses

The histamine concentration in the JF was determined according to the AOAC N° 021402 2014 method (HistaSure ELISA, LDN, Germany). The total volatile base nitrogen (TVBN) was determined by treating each jellyfish sample (100 g) with 0.6 M perchloric acid (Merck KGaA, Darmstadt, Germany). After alkalinization, the extract was exposed to steam distillation and an acid receiver absorbed the volatile base components. The TVBN concentration was determined by the titration of the absorbed bases [23].

For salinity and pH determination, 15 g of JF tissue was collected and stored at −80 °C for further analysis. The texture was measured as described in Bleve et al. [18], using a digital penetrometer (model 53205, TR Turoni, Srl Forlì, Italy). The penetration test was performed using a three-bar probe (3 × 22 mm) for a total plunger area of 1.98 cm^2 by operating on samples consisting of radial triangular slices of the JF tissues. Firmness values were reported as the means of three different measures, expressed in Newtons (N). All analyses were carried out in triplicate.

2.6. Elemental Analyses

The elemental composition (Al, As, B, Ba, Ca, Cd, Cr, Cu, Fe, Hg, K, Mg, Mn, Na, Ni, Pb, Sr, V, and Zn) of the JF samples was measured using inductively coupled plasma–atomic emission spectroscopy (ICP-AES). Lyophilized JF samples were weighed and mixed with 4 mL of H_2O_2 and 6 mL of super-pure HNO_3 69%, digested at 180 °C for 10 min using a microwave digestion system (START D, Milestone Srl, Sorisole (BG), Italy), cooled, diluted with super pure water, and filtered through 0.45 µm syringe filters. A spectrometer (Thermo Fisher Scientific, iCap 6000 Series, Monza, Italy) was previously calibrated for quantitative analysis with five standard solutions containing known concentrations of the elements (0.001, 0.01, 0.1, 0.5, and 1.0 mg/L). The calibration lines showed correlation coefficients (r) greater than 0.99 for all the measured elements. The analysis results were expressed as the average (+/− standard deviation of three different measurements) element concentrations, expressed in ppm (mg/kg of sample weight). JF supernatants obtained after the centrifugation of the JF samples were also analyzed, corresponding to the lyophilized samples.

2.7. Protein Content Determination

Jellyfish samples were homogenized in a blender and diluted in distilled water until a homogeneous solution was obtained. The total protein content in each sample was evaluated using the Bradford assay [24], set for an Infinite 200 PRO microplate reader (TECAN, Männedorf, Switzerland) and using bovine serum albumin (BSA) as a standard.

2.8. Antioxidant Activity Determination

In the diluted JF samples (protein assays), the antioxidant activity (AA) was evaluated using the Trolox Equivalent Antioxidant Capacity (TEAC) method based on the radical cation ABTS•+ and Trolox as a standard. The assay was adapted for an Infinite 200 PRO microplate reader (TECAN, Männedorf, Switzerland), and both samples and the standard were assayed as described in De Domenico et al. [25]. The results were expressed in nmol of Trolox equivalents per gram of JF fresh weight (nmol TE/g FW).

2.9. Lipid Extraction

Total lipids were extracted using a modified Bligh and Dyer method [26]: lyophilized samples (200 mg) were mixed with 12 mL of a solution of chloroform:methanol (2:1) and 3 mL KCl (0.88%), shaken, and centrifuged at 5140× g for 5 min. The lower phase was set aside, and the upper phase was subjected to further extraction with one volume of a solution of chloroform:methanol (2:1, v/v). After phase separation, the lower phase was isolated and added to the first one, and mixed with one-quarter volume of a solution of methanol:water (1:1, v/v). The lower phase was dried using nitrogen and analyzed for lipid composition.

Fatty Acids Analysis

Fatty acid methyl esters (FAME) were obtained according to Leone et al. [27] using boron trifluoride (BF_3), as follows. The total lipid extract in hexane (200 μL) was saponified at 90 °C for 20 min with 0.5 M KOH in methanol (3 mL) with a known quantity of internal standard (methyl-tricosanoate). Fatty acids were methylated with 2 mL of BF_3 in methanol (14%), and the samples were evaporated under a stream of nitrogen and dissolved in 50 μL of hexane, and 1 μL was analyzed by gas chromatography–mass spectrometry (GC-MS). GC–MS analyses were performed using an AGILENT 5977E gas chromatograph (Agilent Technologies, Santa Clara, CA, USA) on a VF-WAXms (60 m, 0.25 mm i.d., 0.25 mm film thickness, Agilent) with the following parameters: the column temperature was maintained at 160 °C for 1 min, programmed at 4 °C/min to 240 °C for 30 min. Helium was used as a carrier gas at a constant flow rate of 1 mL/min. The mass spectrometer was operated in the electron impact mode with a scan range of 50–700 m/z. The temperature of the MS source and quadrupole were set at 230 °C and 150 °C, respectively. Analyses were performed in the full-scan mode. Compounds were identified by comparing the retention times of the chromatographic peaks with those of authentic standards (F.A.M.E. Mix C8-C24, Sigma-Aldrich Corporation, St. Louis, MO, USA) analyzed under the same conditions. The MS fragmentation patterns were compared with those of pure compounds, and a mass spectrum database search was performed using the National Institute of Standards and Technology (NIST) MS 98 spectral database.

2.10. Amino Acid Analysis

The amino acid profile in lyophilized samples was analyzed with an HPLC system (Agilent Infinity 1260, Agilent Technologies) coupled with an online post-column derivatization module (Pinnacle PCX, Pickering Laboratories, Mountain View, CA, USA), using ninhydrin (Trione) as a derivatizing reagent and a Na^+ ion-exchange column (4.6 × 110 mm, 5 μm). Eighteen standard amino acids, ammonia, and taurine were quantified from standard curves measured with the amino acid standards. Prior to the analysis, the samples were hydrolyzed in 6 M HCl containing 0.4% mercaptoethanol for 24 h at 110 °C (HCl hydrolysis). Glutamine and asparagine were converted to glutamic and aspartic acid, respectively.

Cysteine (Cys) was quantified as cystin (Cys-Cys). The samples were filtered via a micro filter, the pH was adjusted to 2.2, and the samples were further diluted with a citrate buffer (pH 2.2) for HPLC analysis.

2.11. Statistical Analysis

All data presented are the means of three independent replicates (n = 3). Statistical analysis was based on one-way analysis of variance. Tukey's post hoc method was applied to establish significant differences among the means ($p < 0.05$, $p < 0.01$ and $p < 0.001$). All statistical comparisons were performed using Sigma-Stat, version 3.11 (Systat Software Inc., Chicago, IL, USA).

Statistical analyses on the protein content and antioxidant activity were performed in Graphpad Prism 6.0 using an analysis of variance (ANOVA) followed by Dunnett's multiple comparison post hoc test to compare each treatment with the control, and Bonferroni's multiple comparison post hoc test to compare the samples with each other. Differences were considered statistically significant for p values of < 0.05. All assays were replicated (n = 3) and data were represented as the mean ± standard deviation (SD). Principal component analysis (PCA) to compare important physical parameters and chemical compounds associated with the samples was carried out using XLSTAT software (Addinsoft Inc., Long Island City, NY, USA).

3. Results and Discussion

3.1. Safety Traits of Rhizostoma Pulmo JF Treated Samples

Two pre-treatment strategies are described here in order to set up an optimized stabilization method for jellyfish intended for food use. In particular, in the first approach (JF-B, procedure 1), jellyfish were immediately washed with refrigerated seawater just after being harvested, and their umbrellas were promptly separated from their oral arms, an operation that can be carried out on the freshly caught jellyfish directly on board. This approach can greatly reduce the ashore disposal of possibly large quantities of JF by-products, thus returning unused JF material directly to the sea. In the second approach (JF-DW, procedure 2), whole JF were transported in chilled seawater to the laboratory, where the use of drinking water was tested for washing JF after transport.

This study proposes the combination of washing with drinking water, a procedure commonly used in the fishing industry, and subsequent treatment with calcium salts. It was observed that this method helped to stabilize the JF tissues, to improve their texture and nutraceutical traits, and to reduce undesired microorganisms in the processed JF products. Moreover, this approach represents an optimization of the recently proposed method for stabilizing and processing JF as food products for human consumption [18].

The main phases of the two procedures for JF preparation proposed in this study are presented in Figure 1 from the starting material to the final products.

In both pre-treatments proposed in this article, JF samples were washed with drinking water, although at different times. In the JF-B procedure, this washing step occurred after cleaning, cutting, and storing the animals in seawater, and before placing them in calcium salt solutions in the laboratory; in the JF-DW approach, instead, whole JF were firstly transported to the laboratory in chilled seawater and were then immersed in drinking water for the time necessary to wash and prepare the JF tissues before soaking them in calcium salt solutions.

At the starting point, both JF-B and JF-DW pre-treatments ensured negligible levels of JF-associated microorganisms, also in terms of halophilic microbes (Tables S1 and S2). This evidence demonstrates a better ability of both JF-B and JF-DW pre-treatments in reducing the initial JF microbial load compared with the seawater treatment (JF-SW) used in Bleve et al. [18]. Moreover, the use of calcium lactate and calcium citrate brines prevented the growth of potential pathogens (*Vibrio* spp., *Salmonella* spp., *Listeria monocytogenes*, and staphylococci) and spoiling microbial contaminants in both JF-B and JF-DW (Tables S1 and S2). The latter evidence was verified by applying the accredited standard parameters established by the law in force for food safety and process hygiene criteria to the Ca-Lactate and Ca-Citrate *R. pulmo* samples treated with both JF-B and JF-DW methods [15]. The approach with calcium salt brines had already been explored by Bleve et al. [18], where calcium lactate E327 and calcium citrate E333 were successfully tested on the edible JF species *R. pulmo*, being both included among the list of food additives and firming agents permitted in the European Union, U.S.A., Australia, and New Zealand.

The total counts of staphylococci were acceptable and showed a similar trend in all JF samples. *Escherichia coli*, coliforms, yeast, molds, and the pathogens *Salmonella* spp. and *L. monocytogenes* were not detected in any of the tested samples treated with either the JF-B or JF-DW methods (Table 1).

Table 1. Safety and quality parameters applied to JF-B (JF directly pre-treated on the boat) and JF-DW (JF washed with drinking water). Samples were treated for 5 days with calcium citrate (Ca-Citrate) or calcium lactate (Ca-Lactate) brines, following accredited conventional assays used for seafood and fish-derived products (as already described by Bleve et al. [15]). The different letters in line indicate significant differences between samples ($p < 0.05$).

Accredited Analysis	Ca-Citrate		Ca-Lactate	
	JF-B	**JF-DW**	**JF-B**	**JF-DW**
	CFU/g	**CFU/g**	**CFU/g**	**CFU/g**
Total bacteria	<10 (a)	$1.30 \times 10^3 \pm 1.23 \times 10^1$ (b)	<10 (a)	$3.70 \times 10^2 \pm 2.31$ (c)
Coliforms	<10 (a)	<10 (a)	<10 (a)	<10 (a)
Escherichia coli	<10 (a)	<10 (a)	<10 (a)	<10 (a)
Staphylococci	$1.60 \times 10^2 \pm 5.21$ (a)	$1.00 \times 10^2 \pm 7.25$ (a)	$1.00 \times 10^2 \pm 8.32$ (a)	$7.30 \times 10^1 \pm 6.15$ (a)
Yeast and Molds	<10 (a)	<10(a)	<10 (a)	<10 (a)
	presence/25 g	**presence/25 g**	**presence/25 g**	**presence/25 g**
***Salmonella* spp.**	0 (a)	0 (a)	0 (a)	0 (a)
Listeria monocytogenes	0 (a)	0 (a)	0 (a)	0 (a)
	mg/Kg	**mg/Kg**	**mg/Kg**	**mg/Kg**
Histamine	<3 (a)	<3 (a)	<3 (a)	<3 (a)
	mg/100 g	**mg/100 g**	**mg/100 g**	**mg/100 g**
TBVN	<0.1 (a)	<0.1 (a)	2.5 ± 0.1 (b)	<0.1 (a)

Several other studies have reported the presence of both bacteria [28] and fungi associated with JF tissues (body and mucus), which may present a risk to humans [15].

Histamine and TVNB were not detected in the tested JF samples (<3 mg/Kg and <0.1 mg/100 g, respectively) (Table 1), thus indicating that there was no tissue degradation and also confirming that the used procedure maintained the freshness of the JF raw material.

In order to compare this optimized process with the traditional Asian procedure, a batch of *R. pulmo* was treated in parallel using mixtures of NaCl and alum for tissue stabilization (Salt-Alum JF-DW) as described by Hsieh et al. [8] and Pedersen et al. [13]. The Salt-Alum JF-DW samples exhibited low counts of *Bacillus* spp. (4×10^1 CFU/g) and discrete levels of halophilic bacteria (4.4–9.9 $\times 10^2$ CFU/g) and yeasts (2×10^2–10^3 CFU/g), although no microbial pathogens were detected (Table S3).

Additional tested parameters [15,18], including the total bacterial count, yeasts, Enterobacteriaceae, *Vibrio* spp., coagulase-positive staphylococci, and *Bacillus* spp., indicated that the Salt-Alum JF-DW samples were safe for consumption.

3.2. Chemical–Physical Characteristics of Treated R. pulmo Samples

The two proposed pre-treatments (JF-B and JF-DW) exerted different effects on the texture, pH, and salinity of the obtained samples (Table 2).

Table 2. Texture, pH, and salinity values of *R. pulmo* JF-B (JF directly pre-treated on the boat) and JF-DW (JF washed with drinking water). JF samples were untreated and treated with brines containing different calcium salts at 5 days of treatment (Ca-Citrate: calcium citrate; Ca-Lactate: calcium lactate), and JF-DW was treated with salt–alum (obtained after 20 days at 4 °C and 2 days of air drying, as described in the Material and Methods section).

Pre-Treatments	Treatments			
	Untreated	Ca-Citrate	Ca-Lactate	Salt-Alum
		Texture (N)		
JF-B	−74 ± 15 (a)	−34 ± 11 (b)	−53 ± 8 (b)	n.d.
JF-DW	−75 ± 10 (a)	−137 ± 15 (b)	−117 ± 5 (b)	−43 ± 11(c)
		Salinity (%)		
JF-B	3.5 ± 0.1 (a)	2 ± 0.2 (b)	2.4 ± 0.1 (b)	n.d.
JF-DW	3.5 ± 0.2 (a)	1.5 ± 0.3 (b)	1.8 ± 0.2 (b)	2.5 ± 0.1 (c)
		pH		
JF-B	7.1 ± 0.2 (a)	4.72 ± 0.2 (b)	7.06 ± 0.3 (a)	n.d.
JF-DW	6.9 ± 0.3 (a)	5.2 ± 0.1 (b)	5.56 ± 0.4 (b)	3.65 ± 0.4 (c)

The different letters in line indicate significant differences between the samples ($p < 0.05$). N: Newton; n.d., not determined.

The optimized calcium citrate and calcium lactate brine treatments exerted different effects on the chemical–physical features of the JF tissues. The preliminary results reported here showed that salt treatment in the JF-B samples reduced the tissue texture (in terms of penetration force); moreover, salinity values of around 2% were measured and the pH values were very different between the two calcium treatments (Table 2). In Ca-Citrate and Ca-Lactate JF-DW samples instead, both brine preparations increased the tissue texture, achieving values of 1.8 and 1.6-fold, respectively, higher than the untreated samples. Additionally, these samples showed reduced salinity values and pH values equal to the initial ones (Table 2). As a result, the JF-DW procedure was selected as the preferred pre-treatment method for further experiments.

Lee et al. [29] already demonstrated the ability of calcium-based food additives (including calcium acetate, calcium carbonate, calcium–casein, calcium chloride, calcium citrate, calcium lactate, calcium sulfate, and calcium phosphate) to improve gelation and polymerization, as occurs during the preparation of surimi from codfish. The increased texture in both Ca-Citrate and Ca-Lactate JF-DW samples, in terms of higher penetration force, can be considered a good index of quality, since those products became denser and more manageable for the subsequent steps. However, Ca-Citrate and Ca-Lactate JF-DW samples showed a gel-like consistency very different from the rubbery and elastic texture of Salt-Alum JF-DW produced following the traditional Asian method. The pH was maintained at around 5 in the Ca-Citrate-treated JF-DW, whereas higher pH levels were obtained in the Ca-Lactate samples. Although being higher than those of the Salt-Alum JF-DW samples, these pH values ensured the expected safety level requested for the semi-finished product (as already shown in Table 1 and Table S1). The appearance of both semi-finished products obtained either by the methods proposed here (Ca-Citrate and Ca-Lactate JF-DW) and by Salt-Alum JF-DW are shown in Figure S1. The two JF-B and JF-DW pre-treatments exerted opposite effects on the texture of the Ca-Citrate- and Ca-Lactate-treated samples in comparison with the corresponding JF-SW samples [18]. In fact, the texture (in terms of penetration force) increased in the JF-DW samples, whilst it decreased in the JF-B samples. The latter evidence seems to indicate that prolonged exposure of JF tissues to drinking water during the JF-B procedure could affect their structure. Regarding the pH values, the JF-B samples showed values very close to those of JF-SW samples, thus evidencing a significant difference with respect to the samples treated with either of the two calcium salts [18]. Additionally, the JF-DW samples exhibited similar pH values, independent of the calcium salts used for the treatment.

3.3. Nutritional Analyses of Treated *R. pulmo* Samples

In order to characterize their nutritional values, the JF samples treated by different procedures were analyzed to evaluate their protein content, amino acid composition, antioxidant activity, lipid content, and fatty acids composition. The moisture contents of the different JF samples were: 96.88 ± 1.12 g/100 g FW for the untreated *R. pulmo* JF-DW, 97.85 ± 0.35 g/100 g FW for Ca-Citrate *R. pulmo* JF-DW, 97.6 ± 0.2 g/100 g FW for Ca-Lactate *R. pulmo* JF-DW, 81.94 ± 0.56 g/100 g FW for Salt-Alum *R. pulmo* JF-DW, and 73 ± 0.61 g/100 g FW for Salt-Alum Jp. These data reveal that there were no statistically significant differences existing between the untreated and Ca-Citrate and Ca-Lactate JF-DW *R. pulmo* samples, whereas a substantial reduction in moisture was obtained in both alum-treated JF samples, thus revealing a further difference between the final products obtained by the two types of procedures. The different moisture contents were considered during the analyses and comparisons of nutrient compounds.

3.3.1. Protein Content, Amino Acid Composition, and Antioxidant Activity

The *R. pulmo* tissues washed with drinking water only (JF-DW) contained 253.2 mg protein per 100 g of fresh weight (FW) (Figure 2a). This value is comparable with the protein content detected by Bleve et al. [18] in JF samples pre-treated with seawater, thus demonstrating that the step of washing with drinking water did not affect the initial protein content. Calcium salt treatment (Ca-Lactate JF-DW and Ca-Citrate JF-DW) significantly reduced the protein content to 60% of the initial value in both samples (88 mg/100 g FW), whereas the traditional salt–alum treatment (Salt-Alum JF-DW) did not affect the protein content (272.9 mg/100 g FW). The commercial ready-to-eat jellyfish sample from Japan (Salt-Alum Jp) contained 178.2 mg protein/100 g FW (Figure 2a), slightly lower than that of the JF-DW and Salt-Alum JF-DW samples, but higher than those of the Ca-Lactate JF-DW and Ca-Citrate JF-DW samples. This evidence suggests that washing with drinking water followed by treatment with calcium salts treatment could lead to a loss of proteins in the processed JF. On the contrary, JF washed with sea water (JF-SW) followed by a 5-day soaking step with Ca-Citrate and Ca-Lactate did not show a significant loss in protein level [18]. This result can be probably explained by a combination of two simultaneous events occurring during the treatments: on one hand, the leakage of solubilized proteinaceous compounds into the brines, and on the other hand, the release of small peptides [30] due to the local denaturation of collagen, being highly susceptible to enzymatic proteolysis [31] under these conditions.

The amino acid composition and content (calculated as the dry weight percentage of lyophilized *R. pulmo* samples) were assayed in untreated JF, JF-DW, and in calcium salts-treated JF (Ca-Citrate JF-DW and Ca-Lactate JF-DW). The total content of amino acids increased from the untreated JF (6%) to the JF washed with drinking water (JF-DW) (9.2%), Ca-Citrate JF-DW (15.4 ± 0.7%), and Ca-Lactate JF-DW (15.3 ± 0.8%) (Table S4). Washing in drinking water and soaking in calcium salt solutions could also cause a leakage of soluble non-proteinaceous components and increase the proteinaceous/amino acid percentage on a dry-weight basis.

The percentages of the taurine, leucine, tyrosine, phenylalanine, and lysine amino acids were higher in the fresh and untreated JF than in the treated JF samples. Increases in the percentages of proline, hydroxyproline, and glycine were also observed in JF-DW and the Ca-treated samples (Ca-Citrate JF-DW and Ca-Lactate JF-DW), with proline and hydroxyproline being abundant in collagen [32], a protein that JF are rich in.

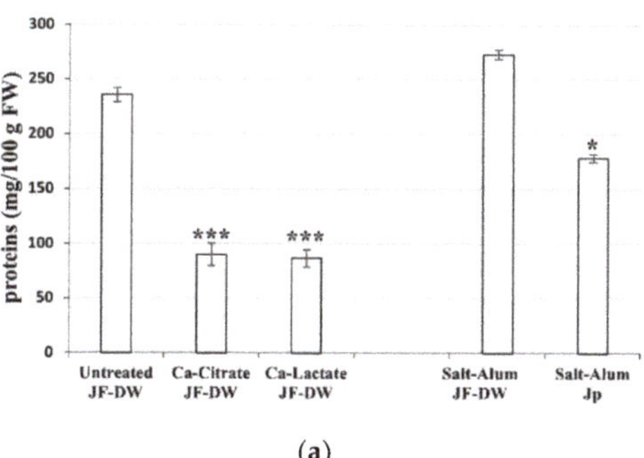

(a)

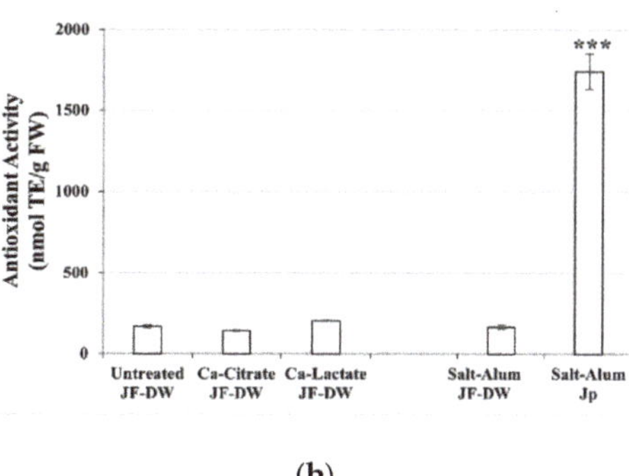

(b)

Figure 2. (**a**) Protein content expressed in mg per 100 g of fresh weight (mg/100 g FW) and (**b**) antioxidant activity expressed in nmol TE per gram of fresh weight (nmol TE/g FW) in differently treated samples of *R. pulmo* jellyfish. JF-DW, *R. pulmo* washed with drinking water; Ca-Lactate JF-DW and Ca-Citrate JF-DW, JF samples treated with calcium lactate or calcium citrate brines for 5 days, respectively; Salt-Alum JF-DW, samples of *R. pulmo* JF-DW treated with salt and alum following the traditional method; Salt-alum Jp, commercial ready-to-eat sample of JF produced in Japan by a traditional alum-based method. Values are the means of three independent measurements ± standard deviation. ANOVA statistical test followed by Dunnett's multiple comparison post hoc test were used to compare each treatment with the control (* $p < 0.05$ and *** $p < 0.001$).

Antioxidant activity (AA) was also evaluated in the same JF samples and expressed in nanomoles of Trolox equivalent per gram of fresh weight (nmol TE/g FW, Figure 2b). Both treatments with either calcium salts or the salt–alum treatment showed similar antioxidant values to those found in JF-DW of approximately 200 nmol TE/g FW. This result indicates that washing the JF with drinking water and successively soaking them in calcium salt-based brines did not affect their antioxidant activity. Furthermore, JF-DW-pre-treated samples showed antioxidant activity levels comparable to those obtained previously by applying the JF-SW procedure [18]. These results could confirm that proteolytic events, possibly due to the treatment, together with the release of small peptides [30], did not affect the final antioxidant activity of the samples. Moreover, the pattern of those released small peptides may be different from those of the low-molecular-weight JF peptides obtained by the controlled enzymatic hydrolysis of jellyfish collagen, as described by De Domenico et al. [25].

It was also observed that the antioxidant activity in the commercial ready-to-eat jellyfish (Salt-Alum Jp) was much higher than that in the treated *R. pulmo* JF-DW (Figure 2b). This evidence could be related to several factors, such as the different JF species used in the commercial product and its high dehydration level, or artificial antioxidants possibly added as preservatives by the manufacturer.

3.3.2. Fatty Acids Composition

In JF-DW, saturated fatty acids (SFAs) accounted for about 50% of total fatty acids (FA), followed by polyunsaturated fatty acids (PUFAs, about 45%) and a small amount of mono-unsaturated fatty acids (MUFAs, 4.3% of the total FA) (Table 3). In *R. pulmo* samples, there was an increase in the SFA percentage, from 50.4% (JF-DW) to 79.3 and 64.4% in Ca-Citrate JF-DW and Ca-Lactate JF-DW, respectively. The SFA content in Salt-Alum JF-DW *R. pulmo* was 81.3% and that in the commercial ready-to-eat jellyfish was 87%. Moreover, Salt-Alum JF-DW and Salt-Alum Jp samples exhibited a more complex lipid profile, since they contained several SFAs that were absent from JF-DW, such as nonadecanoic acid (C19:0), arachidic acid (C20:0), behenic acid (C22:0), and lignoceric acid (C24:0). These differences should be mainly due to the alum treatment of *R. pulmo*. The total MUFA content generally increased in all treated JF samples compared with untreated JF-DW, thus indicating that the initial content of oleic acid (C18:1) was preserved, and also that iso-oleic acid (C18:1 trans-10), palmitoleic acid (C16:1), and vaccenic acid (C18:1 cis-11) appeared.

Table 3. Fatty acid composition of *R. pulmo* JF samples. They were washed with drinking water (JF-DW), treated with brines containing calcium salts (Ca-Citrate JF-DW and Ca-Lactate JF-DW), or treated with the salt–alum method (Salt-Alum JF-DW); a commercial JF sample from Japan treated by the salt–alum method (Salt-Alum Jp) was also tested. Fatty acid composition data are expressed as the percentage of the total fatty acids ± SD.

	Fatty Acids Composition (%)				
	Rhizostoma Pulmo Samples				Commercial JF
	JF-DW	Ca-Citrate JF-DW	Ca-Lactate JF-DW	Salt-Alum JF-DW	Salt-Alum Jp
Saturated FA (SFA)					
Myristic acid C14:0	4.0 ± 0.4	7.5 ± 0.8	7.2 ± 0.7	4.1 ± 0.4	2.6 ± 0.3
Pentadecanoic acid C15:0	—	—	—	1.7 ± 0.2	—
Palmitic acid C16:0	23.5 ± 2.5	34.9 ± 4.3	28.4 ± 0.3	35.4 ± 3.5	31.5 ± 3.1
Margaric acid C17:0	1.1 ± 0.2	1.3 ± 0.3	3.6 ± 0.1	4.8 ± 0.5	2.5 ± 0.3
Stearic acid C18:0	21.8 ± 2.1	35.6 ± 4.2	22.4 ± 2.3	32.9 ± 3.3	47.2 ± 4.8
Nonadecanoic acid C19:0	—	—	2.8 ± 0.3	0.7 ± 0.1	0.6 ± 0.1
Arachidic acid C20:0	—	—	—	1.1 ± 0.1	1.5 ± 0.1
Behenic acid C22:0	—	—	—	0.3 ± 0.1	0.7 ± 0.1
Lignoceric acid C24:0	—	—	—	0.3 ± 0.1	0.6 ± 0.1
Total SFA	50.4 ± 5.1	79.3 ± 0.8	64.4 ± 6.5	81.3 ± 8.2	87.1 ± 8.7
Monounsaturated FA (MUFA)					
Palmitoleic acid C16:1 (ω7)	—	2.7± 0.3	4.5± 0.5	3.5 ± 0.4	2.9 ± 0.3
Oleic acid C18:1 (ω9)	2.5 ± 2.1	2.0 ± 0.2	3.1 ± 0.3	1.4 ± 0.1	—
Iso-oleic acid C18:1 trans-10	1.8 ± 0.2	—	2.0 ± 0.2	—	1.3 ± 0.1
Vaccenic acid C18:1 cis-11 (ω7)	—	1.1 ± 0.1	—	5.3 ± 0.5	3.8 ± 0.4
Trans-vaccenic acid C18:1 trans-13	—	—	—	2.8 ± 0.3	—
Paullinic acid C20:1 (ω7)	—	—	—	0.5 ± 0.1	—
Total MUFA	4.3 ± 0.5	5.8 ± 0.6	9.6 ± 0.1	13.9 ± 1.4	8.0 ± 0.8
Polyunsaturated FA (PUFA)					
Linoleic acid C18:2 (ω6)	—	—	3.1 ± 0.5	2.0 ± 0.2	3.9 ± 0.4
Isolinoleic acid C18:2 trans 8.11	—	3.9 ± 0.4	—	—	—
Linolenic acid C18:3 (ω3)	—	—	2.0 ± 0.2	—	—
Stearidonic acid C18:4 (ω3)	—	—	2.8 ± 0.3	—	—
Eicosadienoic acid C20:2 (ω6)	—	—	—	—	—
Arachidonic acid C20:4 (ω6)	33.8 ± 3.4	3.6 ± 0.4	4.5 ± 0.1	2.0 ± 0.2	1.0 ± 0.1
Eicosapentaenoic acid C20:5 (ω3)	5.4 ± 0.4	4.0 ± 0.4	6.3 ± 0.6	0.4 ± 0.1	—
Docosapentaenoic acid C22:5 (ω3)	2.1 ± 0.2	—	1.0 ± 0.1	—	—
Docosahexaenoic acid C22:6 (ω3)	4.1 ± 0.4	3.4 ± 0.3	6.3 ± 0.6	0.4 ± 0.1	—
Total PUFA	45.4 ± 4.5	14.9 ± 1.5	26.0 ± 1.9	4.8 ± 0.5	4.9 ± 0.5
Σω6	33.8 ± 3.4	3.6 ± 0.4	7.6 ± 0.6	4.0 ± 0.4	4.9 ± 0.5
Σω3	11.6 ± 1.0	7.4 ± 0.7	18.4 ± 1.8	0.9 ± 0.2	—
ω6/ω3	2.9	0.5	0.4	4.6	4.9
Total Lipids (%DW)	8.3 ± 0.9	13.2 ± 1.2	12.5 ± 1.3	6.3 ± 0.5	3.6 ± 0.4

The increase in the saturated fatty acids content of all treated samples could be due to lipid oxidation, which may lead to isomerization events and the production of new SFA and PUFA species when lipid carbon chains break up and unsaturated FAs are converted to SFAs [33].

The total PUFA content decreased after all the salt treatments. In the Ca-Lactate JF-DW and Ca-Citrate JF-DW samples, the PUFA content decreased to 14.9% and 26%, respectively, from the initial value of 45% in untreated JF-DW. Moreover, PUFAs were heavily reduced in both Salt-Alum JF-DW and Salt-Alum Jp to 4.8 and 4.9%, respectively. However, despite the decrease in quantity, the PUFA composition was still preserved in the calcium-salt-treated JF. Linoleic (C18:2), linolenic (C18:3, ALA), and stearidonic (C18:4) acids were detected in the calcium-treated samples, but not in the JF-DW, while the contents of other nutritionally relevant FAs, such as arachidonic (C20:4), eicosapentaenoic (C20:5, EPA), docosapentaenoic (C22:5, DPA), and docosahexaenoic (C22:6, DHA) acids, were maintained or increased. Interestingly, the novel ω3-PUFA stearidonic acid (C18:4) was detected in the Ca-Lactate JF-DW sample. This FA species is the substrate for the conversion of alpha-linolenic acid (ALA) into longer ω3-PUFAs (EPA, DPA, and DHA) in humans, and it has attracted great interest in recent years because it is obtained only from plants [34].

Overall, in comparison with the corresponding samples obtained by the same authors following the JF-SW method [18], JF-DW pre-treatment in both calcium salt samples exhibited increased values in terms of the total MUFA percentage. This effect was more pronounced for the total PUFA percentages, where increases of 2.3- and 3.3-fold were obtained for JF-DW Ca-citrate and Ca-lactate samples in comparison with the corresponding JF-SW samples [18].

In addition, only the treatments with calcium salts yielded ω6/ω3 ratios less than 1 (0.4 and 0.5 for Ca-Citrate and Ca-Lactate JF-DW, respectively), which represents a healthy composition, as suggested by the nutritional recommendations. The calcium salt-based treatments increased the total lipids concentration in the samples and improved the ratio of essential fatty acids (EFA) naturally present in the untreated material (ω6/ω3 = 2.9). Conversely, the previous not-optimized method proposed by Bleve et al. [18] reported ω6/ω3 ratios of 3.5 and 1.4 for JF-SW Ca-citrate and Ca-lactate treatments, respectively, and 4.6 for JF-DW-Salt-Alum and 4.9 Salt-Alum Jp (traditional salt-alum-based treatment), which are definitely well above the recommended ratio of ω6/ω3 < 1. Since dietary ω3 PUFAs and a balanced ω6/ω3 ratio are needed for the maintenance of human health, the combination of JF-DW and calcium salt treatment proposed here preserved these compounds better than JF-SW and the traditional salt–alum methods. In addition, the total lipid content increased in the samples treated with calcium salts in comparison with the untreated JF-DW (Table 3) and even decreased in the salt–alum-treated samples, thus indicating the protective effects of the calcium salt process on JF lipids, as compared with the traditional method.

3.3.3. Element Content

The profiles of some elements associated with JF samples treated with calcium citrate and calcium lactate brines, as well as with JF treated with salt–alum, revealed the absence or very low levels of cadmium (Cd), lead (Pb) and mercury (Hg) in JF tissues (Table 4). In *R. pulmo* treated with the traditional salt–alum traditional method (Salt-Alum JF-DW), a value of Pb corresponding to 9.748 ppm (or mg/Kg) of dry weight was observed. Notably, the Ca-Citrate JF-DW and Ca-Lactate JF-DW showed lower contents of metals compared with the samples produced with the traditional alum-based process (Salt-Alum JF-DW and Salt-Alum Jp). This evidence was important to demonstrate the unique features of the products obtained by the optimized method described here, since the element composition of the food matrix can directly impact human health and is therefore closely related to food safety.

Table 4. Elements evaluated in the samples of *R. pulmo* subjected to different treatments. JF-DW, JF samples washed with drinking water; Ca-Citrate JF-DW and Ca-Lactate JF-DW, *R. pulmo* JF treated with brines containing calcium citrate or calcium lactate, respectively; Salt-Alum JF-DW, JF samples treated with an alum-based procedure; Salt-Alum Jp, commercial JF treated with alum from Japan. Data are expressed in ppm ± SD; the different letters in line indicate significant differences between samples ($p < 0.05$).

	Rhizostoma Pulmo			Commercial
Elements	**Ca-Citrate JF-DW**	**Ca-Lactate JF-DW**	**Salt-Alum JF-DW**	**Salt-Alum Jp**
		Ppm		
Ag	0 (a)	0 (a)	8.554 ± 0.432 (b)	9.844 ± 0.432 (c)
Al	0.05722 ± 0.00411 (a)	0.20558 ± 0.04742 (a)	5213.552 ± 157.573 (b)	5979.045 ± 104.524 (c)
As	0.02667 ± 0.00103 (a)	0.03526 ± 0.00016 (a)	0.202 ± 0.002 (b)	0.236 ± 0.019 (c)
B	0.71483 ± 0.13397 (a)	0.71211 ± 0.16684 (a)	33.981 ± 0.255 (b)	84.628 ± 0.838 (c)
Ba	0.02983 ± 0.00486 (a)	0.02834 ± 0.00413 (a)	44.510 ± 0.289 (b)	8.693± 0.350 (c)
Bi	0	0	0	0
Ca	1480.38 ± 51.1962 (b)	736.053 ± 16.5789 (a)	632.589± 0.178 (c)	141.498 ± 0.090 (d)
Cd	0.0016 ± 0.00036 (a)	0.00163 ± 0.00021 (a)	0 (b)	0 (b)
Co	0.00055 ± 0.00008 (b)	0.00071 ± 0.00008 (a)	0 (c)	0 (c)
Cr	0.00072 ± 0.00005 (a)	0.00226 ±0.00009 (a)	87.544 ± 0.187 (b)	11.906 ± 0.131 (c)
Cu	0.02392 ± 0.00483 (a)	0.02484 ± 0.00563 (a)	8.180 ± 0.037 (b)	8.578 ± 0.124 (c)
Fe	0.09768 ± 0.04256 (a)	0.18979 ± 0.01947 (a)	284.792 ± 8.383 (b)	10.189 ± 1.345 (c)
Hg	0 (b)	0.01079 ± 0.00789 (a)	0 (b)	0 (b)
In	0	0	0	0
K	116.364 ± 29.4737 (a)	121.895 ± 40.5789 (a)	12021.737± 135.355 (b)	1241.373 ± 18.463(c)
Li	0.02682 ± 0.00457 (a)	0.02668 ± 0.00526 (a)	1.145 ± 0.008 (b)	0.536 ± 0.018 (c)
Mg	642.584 ± 91.866 (a)	545.632 ± 156.474 (a)	1664.770 ± 62.521 (b)	352.398 ± 9.879 (a,d)
Mn	0.00653 ± 0.00022 (a)	0.02137 ± 0.00153 (a)	140.681 ± 0.071 (b)	3.532 ± 0.203 (c)
Mo	0.0028 ± 0.00008 (a)	0.00339 ± 0.00008 (a)	2.084 ± 0.100 (b)	5.286 ± 0.551 (c)
Na	3858.61 ± 715.55 (a)	3165.26 ± 1127.89 (a)	>10000 (b)	>10000 (b)
Ni	0.00069 ± 0.00017 (b)	0.00147 ± 0.00011 (a)	0 (c)	0 (c)
Pb	0 (a)	0 (a)	9.748 ± 0.372 (b)	0 (a)
Sr	2.88278 ± 0.47129 (a)	2.77105 ± 0.48684 (a)	25.813 ± 9.1158 (b)	5.960 ± 0.157 (a,b)
Te	0.00012 ± 0.0002 (b)	0.00103 ± 0.0005 (a)	0 (b,c)	0 (b,c)
Tl	0	0	0	0
V	0.47493 ± 0.08632 (a)	0.46234 ± 0.10503 (a)	12.180 ± 0.034 (b)	81.073 ± 0.002 (b)
Zn	0.34144 ± 0.03876 (a)	0.47313 ± 0.02434 (a)	36.019 ± 0.221 (b)	2.606 ± 0.033 (b)

Chromium (Cr) is considered an essential element, playing a role in the maintenance of carbohydrates, fats, and protein metabolism. However, the levels of this element in Asian-style produced JF should be supervised, since the European suggested daily intake range for humans is 25–200 mg/day [35]. The same considerations can be applied to Vanadium (V), which has a mean dietary intake of about 10–20 µg/person/day or 0.2–0.3 µg/kg body weight/day. Studies in humans revealed gastrointestinal disturbances deriving from the oral intake of vanadium compounds, as well as adverse effects on kidneys and other organs in rats, at relatively low doses. These compounds are not considered essential for humans.

All the tested JF samples did not contain significant levels of Pb, Cd, and Hg, which are considered critical contaminants in foodstuffs [36,37].

Notably, very low levels of aluminum were detected in the Ca-Citrate and Ca-Lactate JF-DW samples. In accordance with other studies [10,38], the data reported here showed very high levels of aluminum in both salt–alum-treated JF, the Salt-Alum JF-DW, and the commercial Salt-Alum Jp, as expected. Regarding the use of alum as a structuring agent for human food, allowed as aluminum sulfates (E 520–523) and sodium aluminum phosphate (E 541), the European Union is very restrictive due to the possible neurotoxic effects of aluminum salts [11], whereas the Joint FAO/WHO Expert Committee on Food Additives (JECFA) set a provisional tolerable weekly intake (PTWI) of 2 mg/kg of body weight [39].

3.4. Principal Component Analysis Applied to JF Treated Samples

A principal component analysis (PCA) was carried out in order to thoroughly compare the JF products obtained by the different treatment methods. The PCA was applied to many relevant parameters, such as the texture, pH, salinity, protein content, fatty acid content, antioxidant activity, and metal content, that can describe the JF samples: JF-DW, Ca-Lactate JF-DW, Ca-Citrate JF-DW, JF treated by the traditional salt–alum method (Salt-Alum JF-DW), and the commercial JF (Salt-Alum Jp) (Figure 3).

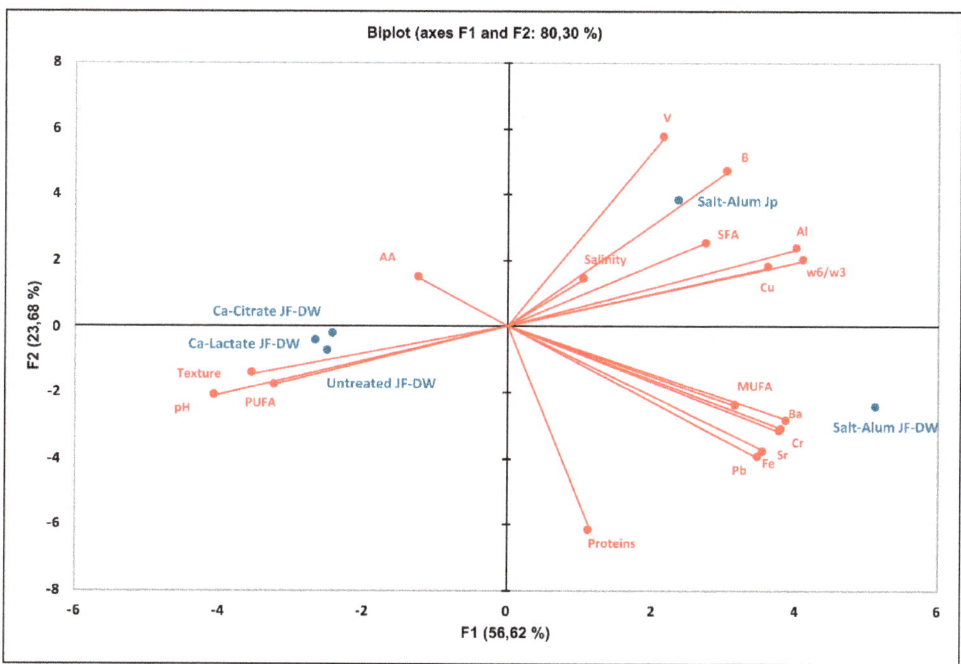

Figure 3. PCA of parameters associated with all treated JF samples. The PCA variables were the data obtained from the analysis of the values of physical traits and the concentrations of chemical compounds at the end of the process. The figure displays the sample scores and variable loadings in the planes formed by PC1–PC2.

A bi-plot is used to show the projection of the variables on the plane defined by the first and second principal components. The total variance of the two main components was 80.3 % (Figure 3). PC1 clustered samples treated with both calcium salts (Ca-Lactate and Ca-Citrate JF-DW) with the untreated JF-DW on the negative semi-axis of the first component, discriminating them from the two salt–alum-treated JF. The clustered group of untreated JF-DW, Ca-Lactate-JF-DW, and Ca-Citrate JF-DW samples were evidently located in the portion of the plane characterized by the pH, texture, antioxidant activity (AA), and PUFA content. The second and the third groups containing Salt-Alum JF-DW and the commercial Salt-Alum Jp, respectively, were located in the opposite portion of the plane, mainly associated with metal ions, particularly aluminum, salinity, SFAs, and MUFAs, and an unfavorable $\omega6/\omega3$ ratio.

The application of this unsupervised technique disclosed the considerable difference between the products obtained by the newly proposed method and the traditional methods used to prepare JF for human food uses. The Ca-Lactate-JF-DW and Ca-Citrate JF-DW final products were very close to the fresh untreated JF and were characterized by peculiar food safety and food quality traits.

3.5. Application of the New Treatment to Different JF Species

The described JF treatments were also applied to three other putatively edible JF species: *Cotylorhiza tuberculate*, *Rhopilema nomadica*, and *Phyllorhiza punctata*. These three scyphozoa are consumed by humans as food in different areas of the world [1] and were chosen in this study as their presence has also been detected in the Mediterranean Sea.

Food safety and several product features (such as microbiology, texture, pH, and salinity), together with basic nutritional characteristics (such as the fatty acids profile, protein content, and antioxidant activity) were assessed for these samples to evaluate them as potential sea-derived food.

Safety and quality traits. The loads of potential pathogenic species, spoilage microbes, and halophilic microbes were not significant in the untreated samples for any of the analyzed JF (Tables S5–S7). After calcium citrate and lactate treatments, the final microbial counts in terms of halophilic species were very low in all three tested JF species, thus suggesting that brine treatment exerts efficient microbial control in all of the different analyzed JF species, despite their distinctive features.

Calcium citrate and calcium lactate treatments improved the texture value of *Cotylorhiza tuberculata* samples by 2.1- and 1.8-fold, respectively. On the other hand, the same treatments led to a decrease in the texture value in the *Rhopilema nomadica* and *Phyllorhiza punctata* samples (Table S8). The latter results are probably related to the non-optimal conditions for storing the samples, having been shipped as frozen material from Israel to Italy and subject to an extended transit time. Therefore, additional tests are planned to confirm the applicability of the proposed procedures to *R. nomadica* and *P. punctata* species.

Nutritional traits. *R. nomadica* and *C. tuberculata* washed with drinking water (JF-DW) showed protein contents of 176.8 mg/100 g FW and 170.3 mg/100 g FW, respectively, (Figure S2a), values that were slightly lower than that measured for *R. pulmo* JF-DW (Figure 2a. On the other hand, *P. punctata* showed a protein content about 2.2-fold higher than that of the other two analyzed species (393.6 mg/100 g FW, Figure S2a). The calcium salt brine treatments decreased the protein content by 60–70% in almost all three JF species, as was previously observed also for *R. pulmo* samples (Figure 2a), with the only exception of *R. nomadica* treated with calcium citrate, where the original value was preserved (Figure S2a).

Additionally, *R. nomadica* treated with calcium citrate kept the same antioxidant activity (AA) value (about 130 nmol TE/g FW, Figure S2b) as that before treatment, whilst the same treatment on *P. punctata* and *C. tuberculata* caused a reduction in AA of about 50% (Ca-Citrate JF-DW, Figure S2b). Furthermore, both *R. nomadica* and *P. punctata* showed a reduction of 40% in the initial AA after calcium lactate treatment, whereas in *C. tuberculate*, the reduction was only about 20%. These data, together with the values obtained for *R. pulmo* (Figure 2), suggested that the calcium brines significantly affected the protein concentration and the choice of calcium salt for the treatment should be adapted to the different JF species.

The FA profiles, reported as the percentage of total FA, were very different between the three JF species (Table S9). *R. nomadica* and *C. tuberculata* JF-DW showed a higher content of SFA than *P. punctata*. PUFAs detected in the JF were probably correlated with the presence of symbiotic species of zooxanthellae microalgae in *C. tuberculata* and *P. punctata*. Calcium salt treatments, mainly calcium citrate, led to a slight increase in the SFA percentage and a reduction in the PUFA content in *R. nomadica* and *C. tuberculata*. In *R. nomadica*, these treatments also led to a reduction in MUFA. On the contrary, *C. tuberculata*, which did not contain any detectable MUFAs in untreated material, contained iso-oleic acid (C18:1 trans-10) and the PUFA isolinolenic acid (C18:2 trans 8,11) when treated with calcium lactate. *P. punctata* treated with calcium salts exhibited a decrease in the SFA content and an increase in the PUFA contents, whereas the levels of MUFAs did not vary compared with the untreated sample. Remarkably, the MUFA iso-oleic acid (C18:1) and PUFA eicosadienoic acid (C20:2) appeared in *P. punctata* Ca-Lactate JF-DW, although they were not initially detectable in the untreated JF sample. However, as shown in Table S9, the essential fatty acids' (EFAs') ω-6 to ω-3 ratio in untreated *P. punctata* was less than 1 (ω6/ω3 < 1) and was maintained after the samples were treated with calcium salts (0.4 and 0.3 for Ca-Citrate and Ca-Lactate JFDW, respectively). Among all other treatments, *C. tuberculata* JF-DW treated with calcium citrate was the only species having yielded a favorable ratio of ω6/ω3 < 1 (0.7, Table S9). On the contrary, both *R. nomadica* and *C. tuberculata* samples treated with calcium lactate always achieved a ω6/ω3 ratio higher than the recommended score of 1 (Table S9).

4. Conclusions

An optimized method that combines pre-treatment with drinking water followed by a soaking step in calcium salt brine was proposed to stabilize and treat JF for possible food uses in Europe and Western countries. The described procedure for JF-DW pre-treatment improved the fundamental aspects of JF tissue stabilization. The significant reductions in any microbiological growth and undesired enzymatic risks, increased texture values, and desirable antioxidant and fatty acid profiles are some relevant improvements with respect to the very recently proposed JF-SW method [18]. Moreover, the presented approach allowed the content of toxic heavy metals, and especially aluminum, to be strongly reduced. This new, safe approach was initially set up on *R. pulmo* JF species, and later successfully applied to other JF species present in the Mediterranean Sea, thus leading to a preliminary validation of the proposed method. The products obtained by the method described here were used for the formulation of new food prototypes. The characterization of the safety, quality, nutritional, and sensory aspects is ongoing. This study can offer a contribution to fill the knowledge gap in the assessment of JF use as human food in Western countries, even though further important evidence needs to be gathered in terms of toxicological and allergen testing. In addition, the technological simplicity of this process will allow its application in poor coastal environments. As a potential future application, a new commercial kit based on the procedure proposed here could be easily developed and retailed by the same fish shops selling freshly harvested edible JF as a helpful tool for consumers interested in preparing homemade and safe JF-based dishes.

5. Patents

The optimized method for JF treatment and stabilization described in this manuscript is part of the European Patent EP 3763224, deposited on 2020 [40].

Supplementary Materials: The following supporting information can be downloaded at: https://www.mdpi.com/article/10.3390/foods11172697/s1, Figure S1. *Rhizostoma pulmo* jellyfish products obtained by the new proposed method. Ca-Citrate JF-DW: JF after treatment with calcium citrate brine; Ca-Lactate JF-DW: JF after treatment with calcium lactate brine; Salt-Alum JF-DW: JF obtained by the salt–alum-based traditional method. Figure S2. Proteins content (a) and Antioxidant activity (b) in *Cotylorhiza tuberculata*, *Rhopilema nomadica* and *Phyllorhiza punctata* JF-DW samples and the corresponding calcium salt treated samples (Ca-Lactate JF-DW and Ca-Citrate JF-DW). Protein contents were expressed as mg per 100 g of fresh weight (mg/100 g FW) and antioxidant activity was expressed as nmol TE per gram of FW (nmol TE/g FW). Values are the means of three independent measurements, ±standard deviation. ANOVA statistic test followed by Bonferroni's multiple comparison post-hoc test was used to compare each treatment with the others for each JF species. Table S1. Microbiological analyses of JF-DW (JF transported on lab and pre-treated with drinking water) untreated and treated with brines containing different calcium salts at 5 days treatment at 4 °C (Ca-Citrate: calcium citrate; Ca-Lactate: calcium lactate). The different letters in line indicate significant differences between samples ($p < 0.05$). TBC: total bacterial count at 30 °C; sCMA: saline Corn Meal Agar; sSDA: saline Sabouraud Dextrose Agar. Table S2. Microbiological analyses of JF-B (JF directly pre-treated on boat) untreated and treated with brines containing different calcium salts at 5 days treatment at 4 °C (Ca-Citrate: calcium citrate; Ca-Lactate: calcium lactate). The different letters in line indicate significant differences between samples ($p < 0.05$). TBC: total bacterial count at 30 °C; sCMA: saline Corn Meal Agar; sSDA: saline Sabouraud Dextrose Agar. Table S3. Microbiological analyses of *Rhizostoma pulmo* JF sample washed in drinking water and then treated with NaCl-alum (Salt-Alum

JF-DW). This sample was obtained after 20 days at 4 °C and 2 days air drying (as described in Material and Methods section). TBC: total bacterial count at 30 °C; sCMA: saline Corn Meal Agar; sSDA: saline Sabouraud Dextrose Agar. Table S4. Amino acid composition of *Rhizostoma pulmo* JF fresh and untreated (Untreated JF), washed in drinking water (JF-DW), and treated with calcium citrate (Ca-Citrate JF-DW) and calcium lactate (Ca-Lactate JF-DW). Data are mean ± Standard deviation (±SD) of four independent analyses. AA. amino acids; Nd. Not detected. Table S5. Microbiological analyses of *Cotylorhiza tuberculata* JF sample washed with drinking water (JF-DW) and treated with brines containing calcium salts (Ca-Citrate: calcium citrate; Ca-Lactate: calcium lactate). The different letters in line indicate significant differences between samples ($p < 0.05$). TBC: total bacterial count at 30 °C; sCMA: saline Corn Meal Agar; sSDA: saline Sabouraud Dextrose Agar. Table S6. Microbiological analyses of *Rhopilema nomadica* JF sample washed with drinking water (JF-DW) and treated with brines containing calcium salts (Ca-Citrate: calcium citrate; Ca-Lactate: calcium lactate). The different letters in line indicate significant differences between samples ($p < 0.05$). TBC: total bacterial count at 30 °C; sCMA: saline Corn Meal Agar; sSDA: saline Sabouraud Dextrose Agar. Table S7. Microbiological analyses of *Phyllorhiza punctata* JF sample washed with drinking water (JF-DW) and treated with brines containing calcium salts (Ca-Citrate: calcium citrate; Ca-Lactate: calcium lactate). The different letters in line indicate significant differences between samples ($p < 0.05$). TBC: total bacterial count at 30 °C; sCMA: saline Corn Meal Agar; sSDA: saline Sabouraud Dextrose Agar. Table S8. Texture, salinity and pH values of *Cotylorhiza tuberculata. Rhopilema nomadica* and *Phyllorhiza punctata* JF samples washed with drinking water (JF-DW) and treated with brines containing calcium salts at 5 days treatment (Ca-Citrate: calcium citrate; Ca-Lactate: calcium lactate). The different letters in line indicate significant differences between samples ($p < 0.05$). Table S9. Comparison of the fatty acid composition of and *Rhopilema nomadica. Phylloriza punctata* and *Cothyloriza tuberculata* JF samples washed with drinking water (JF-DW) and treated with brines containing calcium salts at 5 days treatment (Ca-Citrate: calcium citrate; Ca-Lactate: calcium lactate). Fatty acid composition data are expressed as percentage of the total fatty acids ± SD.

Author Contributions: Conceptualization, G.B. and A.L.; methodology, F.A.R., G.B., S.D.D., A.L., D.M., R.S. and K.K.; validation, F.A.R. and S.D.D.; investigation, F.A.R., G.B., S.D.D. and A.L.; data curation, F.A.R. and S.D.D.; writing—original draft preparation, G.B., S.D.D. and A.L.; writing—review and editing, G.B., A.L., F.P.F. and D.L.A.; funding acquisition, A.L. All authors have read and agreed to the published version of the manuscript.

Funding: This research was funded by the European Commission, H2020 program, Research and Innovation Action Project "GoJelly—A gelatinous solution to plastic pollution", grant number 774499. S.D.D. was supported in her research partially by the GoJelly project and partially by the Apulian Regional program "Research for Innovation—REFIN" (practice code number AFD9B120, project idea UNISAL119).

Institutional Review Board Statement: The jellyfish *Rhizostoma pulmo* is an invertebrate (Cnidaria) which is not an endangered or protected species; therefore, no permit was needed for sampling, which was conducted during jellyfish blooms. In the absence of specific guidelines for the slaughter of cnidarians [19], a method was employed that minimizes the potential for distress or pain in all animal taxa, based on the latest new taxa-specific knowledge of their physiology and anatomy.

Informed Consent Statement: Not applicable.

Data Availability Statement: All related data and methods are presented in this paper. Additional inquiries should be addressed to the corresponding author.

Acknowledgments: The authors thank Leone D'Amico for his valuable technical assistance in JF treatment, and Mariella Quarto and Salvatore Lisi for administrative assistance. The authors also deeply thank Mauro Bleve for the manuscript's English revision. The elemental analyses were supported by the Consorzio Interuniversitario di Ricerca in Chimica dei Metalli nei Sistemi Biologici (C.I.R.C.M.S.B.).

Conflicts of Interest: The authors declare no conflict of interest.

References

1. Brotz, L.; Schiariti, A.; López-Martínez, J.; Álvarez-Tello, J.; Peggy Hsieh, J.H.; Jones, R.P.; Quiñones, J.; Dong, Z.; Morandini, A.C.; Preciado, M.; et al. Jellyfish fisheries in the Americas: Origin, state of the art, and perspectives on new fishing grounds. *Rev. Fish Biol. Fish.* **2017**, *27*, 1–29. [CrossRef]
2. Armani, A.; D'Amico, P.; Castigliego, L.; Sheng, G.; Gianfaldoni, D.; Guidi, A. Mislabeling of an "unlabelable" seafood sold on the European market: The jellyfish. *Food Control* **2012**, *26*, 247–251. [CrossRef]
3. Torri, L.; Tuccillo, F.; Bonelli, S.; Piraino, S.; Leone, A. The attitudes of Italian consumers towards jellyfish as novel food. *Food Qual. Prefer.* **2019**, *79*, 103782. [CrossRef]
4. Mills, C.E. Jellyfish blooms: Are populations increasing globally in response to changing ocean conditions? *Hydrobiology* **2001**, *451*, 55–68. [CrossRef]
5. Bosch-Belmar, M.; Milisenda, G.; Basso, L.; Doyle, T.K.; Leone, A.; Piraino, S. Jellyfish Impacts on Marine Aquaculture and Fisheries. *Rev. Fish. Sci. Aquac.* **2020**, *29*, 242–259. [CrossRef]
6. Youssef, J.; Keller, S.; Spence, C. Making sustainable foods (such as jellyfish) delicious. *Int. J. Gastron. Food Sci.* **2019**, *16*, 100141. [CrossRef]
7. Raposo, A.; Coimbra, A.; Amaral, L.; Gonçalves, A.; Morais, Z. Eating jellyfish; safety, chemical and sensory properties. *J. Sci. Food Agric.* **2018**, *98*, 3973–3978. [CrossRef]
8. Hsieh, Y.H.P.; Leong, F.M.; Rudloe, J. Jellyfish as food. *Hydrobiologia* **2001**, *451*, 11–17. [CrossRef]
9. Rudloe, J. Jellyfish: A new fishery for the Florida panhandle. In *A Report to the US Department of Commerce Economic Development Administration*; EDA Project No. 04-06-03801; EDA: Washington, DC, USA, 1992; p. 35.
10. Zhang, H.; Tang, J.; Huang, L.; Shen, X.; Zhang, R.; Chen, J. Aluminium in food and daily dietary intake assessment from 15 food groups in Zhejiang Province, China. *Food Addit. Contam. B Surveill.* **2016**, *9*, 73–78. [CrossRef]
11. Regulation (EU) No 231/2012 of the European Parliament and of the Council of 9 March 2012 laying down specifications for food additives listed in Annexes II and III to Regulation (EC) No 1333/2008. *Off. J. Eur. Union L 83* **2012**, *55*, 1.
12. Stahl, T.; Taschan, H.; Brunn, H. Aluminium content of selected foods and food products. *Environ. Sci. Eur.* **2011**, *23*, 37. [CrossRef]
13. Pedersen, M.T.; Brewer, J.R.; Duelund, L.; Hansen, P.L. On the gastrophysics of jellyfish preparation. *Int. J. Gastron. Food Sci.* **2017**, *9*, 34–38. [CrossRef]
14. Regulation (EU) No 2283/2015 of the European Parliament and of the Council of 25 November 2015. On Novel Foods, Amending Regulation (EU) N. 1169/2011 of the European Parliament and of the Council and Repealing Regulation (EC) N. 258/97 of the European Parliament and of the Council and Commission Regulation (EC) N. 1852/2001. *Off. J. Eur. Union L 327* **2015**, *58*, 1–22.
15. Bleve, G.; Ramires, F.A.; Gallo, A.; Leone, A. Identification of Safety and Quality Parameters for Preparation of Jellyfish Based Novel Food Products. *Foods* **2019**, *8*, 263. [CrossRef]
16. Regulation (EU) No 1441/2007 of the European Parliament and of the Council of 5 December 2007 amending Regulation (EC) N. 2073/2005 on microbiological criteria for foodstuffs. *Off. J. Eur. Union L 322* **2007**, *50*, 12.
17. Regulation (EU) No 2073/2005 of the European Parliament and of the Council of 15 November 2005 on microbiological criteria for foodstuffs. *Off. J. Eur. Union L 338* **2005**, *48*, 1.
18. Bleve, G.; Ramires, F.A.; De Domenico, S.; Leone, A. An alum-free jellyfish treatment for food applications. *Front. Nutr.* **2021**, *8*, 718798. [CrossRef]
19. Murray, M.J. Euthanasia. In *Invertebrate Medicine*, 3rd ed.; Lewbart, G., Ed.; Wiley-Blackwell: Hoboken, NJ, USA, 2022.
20. Leone, A.; Lecci, R.M.; Milisenda, G.; Piraino, S. Mediterranean jellyfish as novel food: Effects of thermal processing on antioxidant, phenolic, and protein contents. *Eur. Food Res. Technol.* **2019**, *245*, 1611–1627. [CrossRef]
21. Schiariti, A. Life History and Population Dynamics of *Lychnorhiza lucerna*—An Alternative Fishery Resource? Ph.D. Thesis, Universidad de Buenos Aires, Buenos Aires, Argentina, 2008. (In Spanish).
22. Subasinghe, S. Jellyfish processing. *Infofish Int.* **1992**, *4*, 63–65.
23. Regulation (EC) No 2074/2005 of the European Parliament and of the Council of 5 December 2005 Laying Down Implementing Measures for Certain Products Under Regulation (EC) No 853/2004 of the European Parliament and of the Council and for the Organization of Official Controls Under Regulation (EC) No 854/2004 of the European Parliament and of the Council and Regulation (EC) No 882/2004 of the European Parliament and of the Council, Derogating from Regulation (EC) No 852/2004 of the European Parliament and of the Council and Amending Regulations (EC) No 853/2004 and (EC) No 854/2004. *Off. J. Eur. Union L 338* **2005**, *48*, 27–59.
24. Bradford, M.M. A rapid and sensitive method for the quantitation of microgram quantities of protein utilizing the principle of protein-dye binding. *Anal. Biochem.* **1976**, *72*, 248–254. [CrossRef]
25. De Domenico, S.; De Rinaldis, G.; Paulmery, M.; Piraino, S.; Leone, A. Barrel Jellyfish (*Rhizostoma pulmo*) as source of antioxidant peptides. *Mar. Drugs* **2019**, *17*, 134. [CrossRef] [PubMed]
26. Bligh, E.G.; Dyer, W.J. A rapid method of total lipid extraction and purification. *Can. J. Biochem. Physiol.* **1959**, *37*, 911–917. [CrossRef] [PubMed]
27. Leone, A.; Lecci, R.M.; Durante, M.; Meli, F.; Piraino, S. The bright side of gelatinous blooms: Nutraceutical value and antioxidant properties of three Mediterranean jellyfish (Scyphozoa). *Mar. Drugs* **2015**, *13*, 4654–4681. [CrossRef] [PubMed]
28. Kos Kramar, M.; Tinta, T.; Lučić, D.; Malej, A.; Turk, V. Bacteria associated with moon jellyfish during bloom and post-bloom periods in the Gulf of Trieste (northern Adriatic). *PLoS ONE* **2019**, *14*, e0198056. [CrossRef]

29. Lee, N.; Park, J.W. Calcium compounds to improve gel functionality of Pacific whiting and Alaska pollock surimi. *J. Food Sci.* **1998**, *63*, 969–974. [CrossRef]
30. Toldrà, F. The role of muscle enzymes in dry-cured meat products with different drying conditions. *Trends Food Sci. Technol.* **2006**, *17*, 164–168. [CrossRef]
31. Kirkness, M.W.H.; Forde, N.R. Single-Molecule Assay for Proteolytic Susceptibility: Force-Induced Collagen Destabilization. *Biophys. J.* **2018**, *114*, 570–576. [CrossRef]
32. Holmgren, S.K.; Taylor, K.M.; Bretscher, L.E.; Raines, R.T. Code for collagen's stability deciphered. *Nature* **1998**, *392*, 666–667. [CrossRef]
33. Ahmed, M.; Pickova, J.; Ahmad, T.; Liaquat, M.; Farid, A.; Jahangir, M. Oxidation of Lipids in Foods. *SJA* **2016**, *32*, 230–238. [CrossRef]
34. Whelan, J. Dietary stearidonic acid is a long chain (n-3) polyunsaturated fatty acid with potential health benefits. *J. Nutr.* **2009**, *139*, 5–10. [CrossRef] [PubMed]
35. EFSA CONTAM Panel (EFSA Panel on Contaminants in the Food Chain). Scientific Opinion on the risks to public health related to the presence of chromium in food and drinking water. *EFSA J.* **2014**, *12*, 3595.
36. Regulation (EU) No 420/2011 of the European Parliament and of the Council of 29 April 2011 amending Regulation (EC) No 1881/2006 setting maximum levels for certain contaminants in foodstuffs. *Off. J. Eur. Union L 111* **2011**, *54*, 3–16.
37. Regulation (EU) No 488/2014 of the European Parliament and of the Council of 12 May 2014 amending Regulation (EC) No 1881/2006 as regards maximum levels of cadmium in foodstuffs. *Off. J. Eur. Union L 138* **2014**, *57*, 75–79.
38. Yang, M.; Jiang, L.; Huang, H.; Zeng, S.; Qiu, F.; Yu, M.; Li, X.; Wei, S. Dietary exposure to aluminium and health risk assessment in the residents of Shenzhen, China. *PLoS ONE* **2014**, *9*, e89715. [CrossRef]
39. Joint FAO/WHO Expert Committee on Food Additives. *Evaluation of Certain Food Additives and Contaminants: Seventy-Fourth Report of the Joint FAO/WHO Expert Committee on Food Additives*; WHO: Rome, Italy, 2011; p. 16.
40. Leone, A.; Bleve, G.; Gallo, A.; Ramires, F.A.; De Domenico, S.; Perbellini, E. Method for the Treatment of Jellyfish Intended for Human Consumption without the Use of Aluminium Salts and Products/Ingredients Obtained by this Process. European Patent 3763224; European Application Number EP 3763224 A1, 13 January 2021. Available online: https://data.epo.org/publication-server/rest/v1.0/publication-dates/20210113/patents/EP3763224NWA1/document.pdf (accessed on 1 July 2022).

foods

MDPI

Article

The Role of the Emotional Sequence in the Communication of the Territorial Cheeses: A Neuromarketing Approach

Vincenzo Russo [1,2], Marco Bilucaglia [1,2,†], Riccardo Circi [2,†], Mara Bellati [3,*], Riccardo Valesi [4], Rita Laureanti [5], Giuseppe Licitra [6] and Margherita Zito [1,2]

1 Department of Business, Law, Economics and Consumer Behaviour "Carlo A. Ricciardi", Università IULM, 20143 Milan, Italy
2 Behavior and Brain Lab IULM—Neuromarketing Research Center, Università IULM, 20143 Milan, Italy
3 Institute of Agricultural Biology and Biotechnology (IBBA), National Research Council of Italy (CNR), 20133 Milan, Italy
4 Department of Management, Università degli Studi di Bergamo, 24129 Bergamo, Italy
5 Departments of Electronics, Information and Bioengineering (DEIB), Politecnico di Milano, 20133 Milano, Italy
6 Departmentf of Agricolture, Food and Enviroment (Di3A), Università di Catania, 95123 Catania, Italy
* Correspondence: mara.bellati@ibba.cnr.it
† These authors contributed equally to this work.

Abstract: Over the past few years, many studies have shown how territoriality can be considered a driver for purchasing agri-food products. Products with certification of origin are perceived as more sustainable, safer and of better quality. At the same time, producers of traditional products often belong to small entities that struggle to compete with large multinational food corporations, having less budget to allocate to product promotion. In this study, we propose a neuromarketing approach, showing how the use of these techniques can help in choosing the most effective commercial in terms of likeability and ability to activate mnemonic processes. Two commercials were filmed for the purpose of this study. They differed from each other in terms of emotional sequence. The first aimed primarily at eliciting positive emotions derived from the product description. The second aimed to generate negative emotions during the early stages, highlighting the negative consequences of humans' loss of contact with nature and tradition and then eliciting positive emotions by presenting cheese production using traditional techniques as a solution to the problem. Based on the literature on the emotional sequences in social advertising, we hypothesised that the second commercial would generate an overall better emotional reaction and activate mnemonic processes to a greater extent. Our results partially support the research hypotheses, providing useful insights both to marketers and for future research on the topic.

Keywords: emotions; neuromarketing; traditional cheese; territoriality; agri-food products

Citation: Russo, V.; Bilucaglia, M.; Circi, R.; Bellati, M.; Valesi, R.; Laureanti, R.; Licitra, G.; Zito, M. The Role of the Emotional Sequence in the Communication of the Territorial Cheeses: A Neuromarketing Approach. *Foods* **2022**, *11*, 2349. https://doi.org/10.3390/foods11152349

Academic Editor: Maria Lisa Clodoveo

Received: 16 June 2022
Accepted: 3 August 2022
Published: 5 August 2022

1. Introduction

Over the past few years, rootedness in the territory has become a very important motivational driver for the purchase of agri-food products [1–4]. In order to protect the rights of both consumers and producers, the European Union (EC Regulations 2081/92 and 2082/92) has identified three different designations that certify the products with a strong territorial identity: the Protected Designations of Origin, the Protected Geographical Indications, and the Traditional Specialities Guaranteed [5]. Those certified products are often associated with appealing concepts such as quality [6], tradition [7], sustainability [8], safety [2] and cultural identification [9]. Nevertheless, they struggle to compete with their industrial competitors in terms of budget allocation for communication campaigns since they are often produced by small and rural entities [10].

The communication campaigns make extensive use of emotions since they mediate and moderate consumer decision-making processes [11]. The effectiveness of commercials in generating emotions has been shown to be a good sales predictor [12]. In fact, emotions have a strong impact on message perception [13], increasing the likelihood that the advertised product or brand will attract attention and be remembered [14–16].

In recent years, much research has focused on how the sequence of opposite emotions (negative emotions followed by positive emotions or positive emotions followed by negative emotions) can be effective in persuading consumers [17]. Some evidence suggests that negative emotions followed by positive emotions are more effective because people perceive emotions based on an initial reference point [18,19]. For example, charity advertising most often tends to elicit negative emotions from the description of a problematic situation and then generate positive emotions from the description of the possibility of making actions to help people in need [20,21]. Although the issues and psychological mechanisms underlying the charity advertising are very different from those of agri-food products, similar concepts can be, in principle, applied to both. Just as the act of giving can help solve a problematic situation, the consumption of agri-food products rooted in the local area can be presented as a possible solution to problems such as pollution, communication asymmetry between consumer and producer, and the low perception of safety associated with industrial products [22–25].

In the past, the effectiveness of the communication campaigns in eliciting specific emotions had been studied mainly by assessing consumers' conscious responses. However, the limitations of "classic" survey instruments used by marketers have been discussed in the literature for a long time now. Questionnaires [26,27], interviews [28,29] and focus groups [30] have been shown to be reliable only within certain limits, due to both the impossibility of obtaining detailed and/or truthful opinions from people [31] and the need to rely on subjective interpretations of interviewers that may not reflect the real internal dynamics of the consumer [30]. The lack of reliable methods for predicting consumer behaviour can have serious consequences: of the new products launched in the market, between 40% and 80% are doomed to fail, causing economic damage to companies quantifiable in the order of billions of dollars [32–34]. For these reasons, in recent years, there has been a growing interest in neuromarketing techniques [34,35].

We refer to neuromarketing as the use of neuroscience tools and insights to provide answers to challenges in business practices, especially in advertising and marketing research [36]. This discipline studies the latent mental processes underlying consumer behaviour [37,38]. The emergence of this strand of literature is due, on the one hand, to the need of marketers to identify methods that can better predict the success of marketing campaigns and, on the other hand, the need of neuroscientists to develop methods and techniques that can increase our knowledge of the brain [34,39–41]. Neuromarketing aims to overcome the limitations of traditional marketing methodologies by directly investigating emotional reactions using tools capable of detecting electrophysiological variables [40–43].

Neuromarketing applications are widespread. They include the evaluation of static advertising in both digital and printed format [44], radio and video commercials [45], as well as product packaging [46]. In addition to the profit-making sector, no-profit organisations operating in the charity [47] and social utility [48] have taken advantage of neuromarketing. Within the food and beverage sector, neuromarketing investigated the effectiveness of the food packages in communicating key factors such as the health content of the labels and the presence of additives [49], as well as the consumption sustainability [50] and the territoriality [1].

To the best of our knowledge, no study has ever assessed the effectiveness of certified agri-food product communication using neuromarketing techniques. In addition, no study has ever investigated the role of the emotional sequence in the communication of certified agri-food products.

This study fills these gaps in the literature, evaluating the emotional impact of two different video commercials created to promote certified cheeses from Southern Italy. Both the commercials focused on themes such as references to territory, production techniques and natural landscapes as key communication drivers. References to territoriality are often used because the specificity of the area of origin and the limited production area help endow the product with special characteristics in the eyes of the consumer [51]. Emphasis on production techniques was placed to emphasise sustainability in terms of respect for the environment and support for local people [22,23]. Finally, since certified products are also characterised in terms of eco-sustainability and environment preservation, references to nature are often present in their promotion [52].

Although the two videos focused on the same themes, they differed in terms of emotional sequence. The first, named "Rewind", was designed to elicit mostly positive emotions, focusing mainly on aspects related to the goodness of the product and production techniques. The second, named "The Myth", was designed to elicit an emotional sequence from initial negativity to positivity in the end: the theme of the territoriality is proposed as a solution to the problems represented by the loss of contact of humans with territories and traditions. A detailed description of the videos can be found in Appendix A.

As common practice in neuromarketing studies [53], we collected three electrophysiological signals: the electroencephalogram (EEG), the skin conductance (SC) and the photoplethysmogram (PPG). Four different indices were calculated from the above-mentioned signals:

- Approach-Withdrawal Index (AWI): an EEG-based index associated with the instinctive reaction of approaching towards or moving away from a stimulus [54,55];
- Memorisation Index (MI): an EEG-based index associated with successful memory encoding [56,57];
- Hearthrate (HR): a PPG-based index associated with the emotional valence [58]
- Emotional Index (EI): a compound SC- and PPG-based index that summarises the emotional degree (either positive or negative) [59].

We compared both the overall videos and the sequences corresponding to the four narrative themes (i.e., territory, product, production techniques, natural landscapes) to explore the impact of the emotional sequence on the affect (AWI, HR and EI indices) and the memorisation (MI index).

This study was intended to help producers identify the most effective communication strategy for territorial agri-food products. We believe it represents an added value, especially for small local realities, helping them to optimise investments and reduce the gap with the big food corporations.

2. Materials and Methods

2.1. Research Hypotheses

In the text below, we referred to the videos "Rewind" and "The Myth", as, respectively, "R" and "M". The themes of "territory", "product", "production techniques" and "natural landscapes" were shortened to, respectively, "Tr", "Pr" "Prd" and "Nt".

The general aim of the study was detailed in terms of the following six research Hypotheses (H1)–(H6):

Hypothesis 1 (H1). *The video characterised by a negative–positive emotional sequence (M) generates, overall, a more positive emotional reaction than the commercial predominantly focused on positive emotions (R). We expect greater AWI, EI and HR values in M than in R;*

Hypothesis 2 (H2). *The themes Tr, Prd, and Nt will generate a more positive reaction in video M than R. For each theme, we expect greater AWI, EI and HR values in M than R;*

Hypothesis 3 (H3). *Pr sequences generate a more positive emotional reaction in M than in R. We expect greater AWI EI and HR values in M than in R;*

We also expect an impact on the salience, and thus on their memorisation, of the elements of the emotional sequence. Therefore, we hypothesised that:

Hypothesis 4 (H4). *The video M activates greater memorisation processes than R. We expect greater MI values in M than R.*

Hypothesis 5 (H5). *The themes Tr, Prd, and Nt will activate greater memorisation processes in M than R. For each theme, we expect greater MI values in M than R;*

Hypothesis 6 (H6). *Pr activates greater memorisation processes in M than R. We expect greater MI values in M than in R.*

2.2. Instrumentation

We recorded the EEG using an NVX-52 device (Medical Computer Systems, Ltd., Moscow, Russia) at a sample frequency of 2 kHz and a resolution of 24 bits. We placed 38 active Ag/AgCl electrodes on the scalp according to the 10–20 system [60] by means of an elastic cap, in addition to two Ag/AgCl earlobes electrodes and one Ag/AgCl adhesive patch that served, respectively, as reference and ground.

We recorded the SC and the PPG signals using, respectively, the GSRSens (Medical Computer Systems, Ltd.) and FpSens (Medical Computer Systems, Ltd.) sensors, both connected to the auxiliary inputs of the NVX-52. We placed the two Ag/AgCl electrodes of the GSRSens on the index and ring finger from the non-dominant hand and the FpSens on the middle finger from the same hand. Both the GSR and PPG signals were acquired synchronously to the EEG at the same sample frequency and resolution. The recordings were controlled by the NeoRec software (Medical Computer Systems, Ltd.).

We used the iMotions software (iMotions, A/V) to deliver the stimuli. At the beginning of the experiment, iMotions generated a TTL pulse that was fed into the digital inputs of the NVX-52 using the ESB (EEG Synchronisation Box) [61]. This served to perform an off-line synchronisation between the recorded data and the stimuli timestamps.

2.3. Study Population and Experimental Protocol

Forty healthy Italian subjects (20 males) with ages ranging from 33 to 56 years (M = 45.67, SD = 7.36) were enrolled in the experiment. The subjects were randomly divided into two sub-groups of 20 subjects each. The groups did not differ in terms of mean age and gender proportions, as verified by the Mann–Whitney (W = 200.500, p = 1.000) and chi-squared ($\chi^2(1) = 0.000, p = 1.000$) tests, respectively.

The sample size was selected after a sensitivity analysis that was performed using the G*Power software [62] with the following input parameters:

- $\alpha = 0.05$;
- $(1 - \beta) = 0.95$;
- Total sample size = 40;
- Number of groups = 2;
- Correlation among the repeated measures = 0.5;
- Nonsphericity correction = 1.

The computed effect size was $f = 0.235$, corresponding to a medium value [63].

The study protocol followed the Helsinki declaration and informed written consent was obtained from each participant.

Each subject sat on a chair placed in front of a 23.8-inch monitor (FlexScan EV2451, Eizo KK) located in a 7 $\times$ 3 m experimental room, artificially lit by florescence lights and in the absence of any natural light. Two experimenters positioned the SC, PPG and EEG sensors and checked the quality of the signals before starting the recording. The contact impedance of the EEG electrodes was measured and ensured to be less than 10 kΩ [64].

At the beginning of experiment, the subject performed a 60-s-long eye-closed baseline (EYC), followed by a 2-min-long neutral baseline (BSL). Then, according to the group splitting, either the M or R video was proposed.

2.4. Video Segmentation

For each video, the sequences corresponding to the 4 narrative themes (Nt, Tr, Pr, Prd) were identified and manually marked by 2 independent judges using the Boris software [65]. In order to compute the inter-rater reliability, the Cohen's κ was evaluated within a 2s-long sliding window. We obtained values of $\kappa = 0.83$, and $\kappa = 0.86$ for video M and R, respectively, corresponding to a strong agreement [66]. Onsets and durations of the themes were built as the intersection between the chunks identified by the two raters. Finally, the onsets and the durations of both the EYC and BSL epochs, as well as the 4 themes, were exported for the subsequent analyses.

2.5. EEG Processing

The EEG was processed using Matlab (The Matwhorks, Inc., Natick, MA, USA) and the EEGLab toolbox [67], following a previously proposed standard pipeline [54,68].

First, the data were re-referenced to the linked earlobes and down-sampled to 512 Hz. Then, a band-pass filter (0.1–30 Hz) and a notch filter (50 and 100 Hz) were applied in order to remove the physiological and external noise. The Artefact Subspace Reconstruction (ASR) with a default cut-off parameter (k = 20) was applied in order to remove non-stationary artefacts [69]. The data were then decomposed into Independent Components (ICs) using the SOBI algorithm [70]. By using the neural-net based classifier ICLabel [71], artefactual ICs were identified as those with brain probability $\texttt{Pr\{brain\}} \leq 0.7$ and set to zero, while non-artefactual ICs were back-projected to the original sensor space. A Current Source Density (CSD) reference was then applied in order to increase the spatial resolution of the EEG at the sensor level [72].

Finally, the cleaned EEG was epoched according to the onset and the duration of the EYC and BSL stimuli, as well as the narrative sequences. For each subject, we computed the Individual Alpha Frequency (IAF), which is defined as the centre of gravity of the PSD within the extended alpha range (7.5–12.5 Hz) [73]. In the IAF calculation, we considered, as PSD, the mean PSD aver-aged across all the occipital channels. The occipital PSDs were computed using the EYC data. Finally, we computed 2 canonical EEG bands as: $\delta = [0;\ \mathrm{IAF} - 6]$ Hz and $\alpha = [\mathrm{IAF} - 2;\ \mathrm{IAF} + 2]$ Hz [74].

In order to have the highest temporal resolution, all indices were computed following the filtering approach, which is based on filtering and averaging an appropriate set of EEG channels to produce a cluster [54]. The Hilbert Transform was applied to the filtered channels before the averaging to compute the smoothed instant power [75]. The AWI was obtained by subtracting the α-filtered right-frontal (FP2, F4, F8, FT8, FC4) and left-frontal (FP1, F3, F7, FT7, FC3) clusters [54], while MI was obtained as the θ-filtered left-frontal (FP1, F3, F7, FT7, FC3) cluster [57].

2.6. SC and PPG Processing

The SC and PPG signals were processed using Matlab (Mathworks, Inc.), following a previously proposed standard pipeline [68,76].

The SC signal was first band-pass filtered (0.001–0.35 Hz); then, a threshold for SC extreme values (0.05–60 μS) and extreme rate of changes (±8 μS/s) was used in order to detect artefacts [77]. The artefactual points were replaced by a linear interpolation using adjacent points. From artefact-corrected SC, the tonic Skin-Conductance Level (SCL) was extracted by means of the cvxEDA algorithm [78].

The BVP signal was first low pass filtered (5 Hz); then, all peaks were identified using the Pan–Tompkins algorithm [79], and the instant HR was computed from the inverse of the peak-to-peak distance. Finally, the HR signal was linearly interpolated and filtered with a 2s-long moving average filter in order to obtain a smoother signal.

By means of a trigonometric transformation, SCL and HR were converted into the uni-dimensional EI [57].

2.7. Baseline Normalisation

AWI, MI, HR and EI signals were epoched according to the narrative sequences and z-score transformed with respect to the BSL as [76]:

$$x'(t) = [x(t) - m_{BSL}]/s_{BSL} \quad (1)$$

where $x'(t)$ is the z-score transformed signal, $x(t)$ is the original signal, m_{BSL} is the temporal mean of $x(t)$ in the BSL epoch and s_{BSL} is the temporal standard deviation of $x(t)$ in the BSL epoch.

Then, the signals were temporally averaged across each narrative sequence in order to obtain a condensed stimulus-related index [80]. For each Video × Theme combination, outliers were identified by means of the inter-quantile range (IQR) criterion as points outside the interval [Q1 − 1.5 × IQR; Q3 + 1.5 × IQR], where Q1 is the first quartile, Q3 is the third quartile and IQR = Q3 − Q1 [81].

2.8. Statistical Analyses

The statistical analyses were performed using JASP v.0.14 [82]. Each index was analysed by a two-way mixed ANOVA, considering the Video as a between-subject factor (two levels: M, R) and the narrative theme (hereinafter, Theme) as the within factor (four levels: Nature, Territory, Product, and Production). Prior to the analyses, the sphericity of the Theme and the equality of variances of the Video were assessed by the Levene's and Mauchly's tests, respectively. In the case of sphericity violations, the Greenhouse-Geisser correction based on the sphericity estimator ω was applied [83]. All the post-hoc comparisons were Holm-corrected. In the following section, all the significant differences were provided either as mean (M) and standard deviation (SD) or marginal mean (MM) and standard error (SE).

After the processing phase, 3 subjects were excluded from further analysis due to the excessive noise in their physiological signals. The final sample consisted of 37 subjects (19 males) with ages ranging from 33 to 56 years (M = 45.24, SD = 7.48). The M and R subgroups groups still did not differ in terms of mean age and gender proportions, as verified by means of the Mann–Whithney (W = 198.500, p = 0.411) and chi-squared ($\chi^2(1) = 0.026$, p = 0.873) tests, respectively.

3. Results

3.1. EEG-Related Indices

The AWI did not show any significant main effect or interactions. The MI showed a significant main effect for the video ($F(33,1) = 5.493$, $p = 0.025$) and a significant interaction of the theme × video ($F(2.250, 74.246) = 5.711$, $p = 0.004$, $\omega = 0.750$). Post-hoc comparisons showed a significant ($p = 0.025$) difference between M (MM = 0.060, SE = 0.079) and R (MM = −0.202, SE = 0.079). Nature × R (M = 0.001, SD = 0.251) and Territory × R (M = −0.420, SD = 0.480) showed a significant ($p = 0.027$) difference, similarly to Product × R (M = −0.310, SD = 0.365, n = 17) and Territory × M (M = 0.230, SD = 0.481) - p = 0.015, as well as Territory × M (M = 0.230, SD = 0.481) and Territory × R (M = −0.420, SD = 0.480) - p = 0.001. The following Figure 1 and Table 1 show, respectively, the descriptive plot with standard error bars and the descriptive statistics of the MI, split for video and theme.

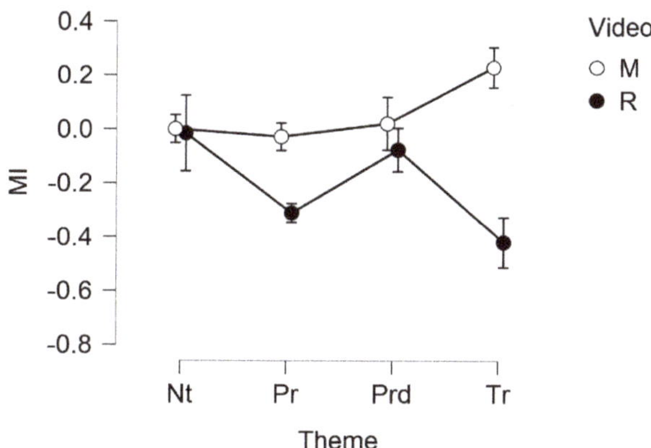

Figure 1. Descriptive plot with error bars of the MI, split for the video (M = The Myth, R = Rewind) and the theme (Nt = Nature, Pr = Product, Prd = Production, Tr = Territoriality). The vertical axis is expressed as unit-less z-scores.

Table 1. Descriptive statistics (M = mean, SD = standard deviation, n = number) of the MI, split for the video (M = The Myth, R = Rewind) and theme (Nt = Nature, Pr = Product, Prd = Production, Tr = Territoriality). All values are expressed as unitless z-scores.

Theme	Video	M	SD	n
Nt	M	0.001	0.251	18
	R	−0.015	0.687	17
Pr	M	−0.028	0.307	18
	R	−0.310	0.365	17
Prd	M	0.021	0.515	18
	R	−0.076	0.347	17
Tr	M	0.230	0.481	18
	R	−0.420	0.480	17

3.2. SC- and BVP-Related Indices

HR showed a significant main effect of the theme ($F(1.504, 45.135) = 3.669$, $p = 0.045$, $\omega = 0.501$) and the video ($F(1, 30) = 15.263$, $p < 0.001$). Post hoc comparisons found a significant ($p = 0.013$) difference between Product (MM = −0.640, SE = 0.167) and Territory (MM = −0.046, SE = 0.167), as well as between M (MM = 0.081, SE = 0.172) and R (MM = −0.867, SE = 0.172), $p < 0.001$. The following Figure 2 and Table 2 show, respectively, the descriptive plot with standard error bars and the descriptive statistics of the HR, split for video and theme.

Table 2. Descriptive statistics (M = mean, SD = standard deviation, n = number) of the HR, split for the video (M = The Myth, R = Rewind) and theme (Nt = Nature, Pr = Product, Prd = Production, Tr = Territoriality). All values are expressed as unitless z-scores.

Theme	Video	M	SD	n
Nt	M	0.087	0.735	16
	R	−1.090	0.906	16
Pr	M	−0.310	0.538	16
	R	−0.970	0.866	16
Prd	M	0.316	0.513	16
	R	−1.081	0.803	16
Tr	M	0.232	1.241	16
	R	−0.324	1.514	16

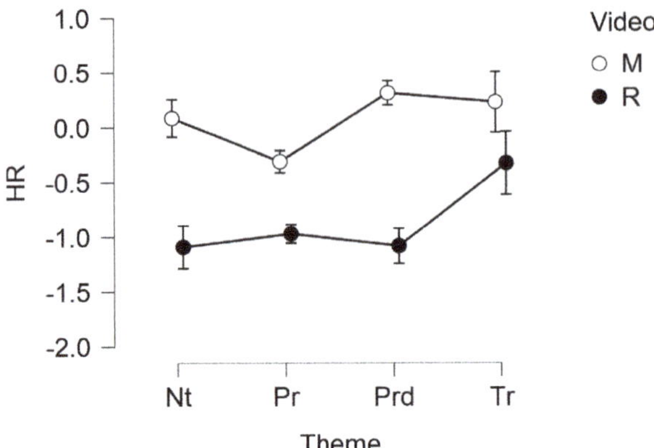

Figure 2. Descriptive plot with error bars of the HR, split for the video (M = The Myth, R = Rewind) and the theme (Nt = Nature, Pr = Product, Prd = Production, Tr = Territoriality). The vertical axis is expressed as unit-less z-scores.

EI showed a significant main effect of the video (F(1, 31) = 7.728, $p = 0.009$). Post hoc comparisons showed a significant difference between M (MM = −0.033, SE = 0.062) and R (MM = −0.279, SE = 0.062), $p = 0.009$. The following Figure 3 and Table 3 show, respectively, the descriptive plot with standard error bars and the descriptive statistics of the EI, split for video and theme.

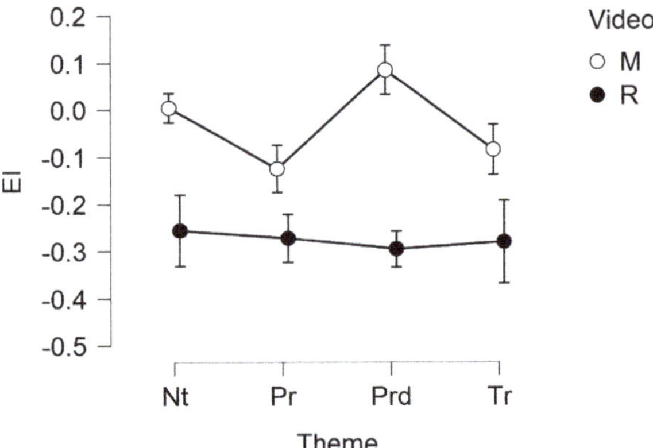

Figure 3. Descriptive plot with error bars of the EI, split for the video (M = The Myth, R = Rewind) and the theme (Nt = Nature, Pr = Product, Prd = Production, Tr = Territoriality). The vertical axis is expressed as unit-less z-scores.

Table 3. Descriptive statistics (M = mean, SD = standard deviation, n = number) of the EI, split for the video (M = The Myth, R = Rewind) and theme (Nt = Nature, Pr = Product, Prd = Production, Tr = Territoriality). All values are expressed as unitless z-scores.

Theme	Video	M	SD	n
Nt	M	0.005	0.178	16
	R	−0.255	0.345	17
Pr	M	−0.124	0.129	16
	R	−0.272	0.352	17

Table 3. *Cont.*

Theme	Video	M	SD	n
Prd	M	0.085	0.224	16
	R	−0.295	0.286	17
Tr	M	−0.084	0.273	16
	R	−0.280	0.581	17

4. Discussion

In this study, we investigated the role of emotional sequence in the communication of traditional cheeses from Southern Italy. For this purpose, we compared several physiological indices (AWI, MI, HR and EI) of two groups of participants during the vision of two video commercials. The first group watched a video (R) mainly characterised by a positive emotional tone, with sequences focused on the product quality and the traditionality of production processes. The second group watched a video (M) characterised by initial negative emotions, elicited by sequences showing the consequences of losing contact with the territory and traditions, followed by positive emotions, obtained by showing the positive consequences of regaining contact with the traditions and the territory. The videos were segmented by two individual raters into four narrative themes (Nt, Pr, Prd and Tr), and the physiological indices were averaged across the duration of each theme. We advanced six research hypotheses (H1–H6) that compared several metrics (AWI, MI, HR and EI) across both the Video and the Video × Theme dimensions, as summarised in the following Table 4.

Table 4. Summary of the six research hypotheses H1–H6 that compared AWI, MI, HR and EI metrics across both the Video (M = The Myth, R = Rewind) and the Video × Theme (Nt = Nature, Pr = Product, Prd = Production, Tr = Territoriality) dimensions. The direction of the expected differences is also provided, alongside the associated significant *p*-values (n.s. stands for not-significant). The fully or partially confirmed hypotheses are marked as, † and *, respectively.

Research Hypothesis	Metric	Direction	*p*-Value
H1 *	AWI	M > R	n.s.
	HR	M > R	0.045
	EI	M > R	0.009
H2	AWI	$Nt_M > Nt_R$	n.s.
		$Prd_M > Prd_R$	n.s.
		$Tr_M > Tr_R$	n.s.
	HR	$Nt_M > Nt_R$	n.s.
		$Prd_M > Prd_R$	n.s.
		$Tr_M > Tr_R$	n.s.
	EI	$Nt_M > Nt_R$	n.s.
		$Prd_M > Prd_R$	n.s.
		$Tr_M > Tr_R$	n.s.
H3	AWI	$Pr_M > Pr_R$	n.s.
	HR	$Pr_M > Pr_R$	n.s.
	EI	$Pr_M > Pr_R$	n.s.
H4 †	MI	M > R	0.025
H5 *	MI	$Nt_M > Nt_R$	n.s.
		$Prd_M > Prd_R$	n.s.
		$Tr_M > Tr_R$	0.001
H6	MI	$Pr_M > Pr_R$	n.s.

The Video M showed, overall, greater EI and HR than R, while AWI did not show any significant difference. This partially supports the research hypothesis H1, which assumed a greater emotional reaction in the emotional sequence. The AWI results must not be read as a contradiction to those of EI and HR for at least two reasons. First, despite the fact that EI, HR and AWI can be associated with the same psychological construct of the emotional valence, they belong to different divisions of the nervous system: the autonomous nervous system (ANS) for EI and HR, and the central nervous system (CNS) for the AWI [53]. It was shown that these sub-systems are non-linearly related, and the degree of their coupling linearly depends on other factors, such as the levels of arousal of the emotionally-relevant stimuli [84]. Since the storytelling and the framing of the videos were not designed to elicit high levels of arousal, a low coupling between the ANS and CNS measures is expected. Second, some studies have questioned the appropriateness of the AWI as a measure of emotional valence [85]. Despite the fact that people are generally attracted to what elicits positive emotions and tend to turn away from what elicits negative emotions, it is also true that not all negative emotions cause a turning-away reaction. Anger, for example, despite being a negative emotion, generates an instinctive approach response [86]. Within the negative emotions of Video M, it is likely to expect the presence of anger, especially in the sequences related to the men's loss of contact with territories and traditions, as well as to the Godhead's punishment. The insignificant main effect of the Video could be, thus, due to the comparison of two positive AWI values, one associated with positive emotions and the other with anger.

The differences between M and R on EI, HR and AWI values related to Tr, Prd, Pr and Nt themes did not reach statistical significance, not supporting H2 or H3. A possible explanation could be related to the difference in the storytelling between the two videos that, according to past studies [87,88], has a strong role in mediating and/or moderating the emotional content of the video commercials. In statistical terms, the storytelling may have played the role of a confounding factor in decreasing the effect size associated with the interactions, leading to non-significant differences across the themes. This should be verified with a future confirmatory study based on stimuli with fixed storytelling but variable emotional sequence.

Compared to R, M showed an overall significantly greater MI, fully supporting H4, which assumed a different impact of the videos on the salience and, thus, the memory encoding. This is in line with previous researches on charity advertising that underlined the role of the emotion sequences in enhancing the overall salience [20,21]. Salience, in turn, plays a key role in the memorisation process: it was shown that maximal-saliency stimuli are associated with a greater recollection probability, and they facilitate access to memory representation at retrieval [89].

The M video showed a significantly greater MI than R only for the Tr theme, only partially supporting H5, which assumed greater memorisation of Nt, Prd and Tr themes in M. For the Pr theme, MI did not show a significant difference between M and R: this did not support H6, which assumed a greater memorisation in Video M. Similarly to what was discussed with H2 and H3, the different storytelling could have played a confounding role since, according to past studies [87,90], it also has a strong impact on the memorisation processes.

There is a chance that some research hypotheses have been rejected due to the characteristics of the sample, rather than the feature of the videos. In fact, it has been shown that gender and age play a significant role in emotional evaluation [91] and episodic memory recall [92]. A confirmatory study based on a four-way mixed ANOVA design with gender and age as additional between-subject factors is suggested to verify this supposition.

It is worth mentioning the limitations of the present study. We evaluated two videos that had never previously aired since they were shot specifically for this study. Additionally, the two creative contents differed not only in the emotional sequence but also in the storytelling and the framing. At the same time, this allowed us to investigate a situation very similar to what happens outside of laboratory contexts: consortiums for the protection of territorial products (or, in general, companies) rarely have to choose between creative proposals that differ in single separable variables; more often, they receive different proposals from several advertising agencies, and they need to choose those that have the highest probability of being remembered and generating functional emotions for the enhancement of their products. Although there are many practical implications of our approach, further basic research is needed for stronger support of our findings.

5. Conclusions

In this study, we compared two video commercials of traditional cheeses from Southern Italy using a neuromarketing approach in order to highlight the most effective one in terms of emotion and memorisation. Despite the fact that both the videos were composed of the same four narrative themes (i.e., territory, product, production techniques and natural landscapes), they differed in the emotional sequence: the first one was mainly characterised by a positive emotional tone, while the second one was characterised by an initial negative tone that turns to positive in the end. We found that the second video generates a better emotional reaction and memorisation than the first one. This is in line with the literature on charity advertising that showed how the negative–positive emotional sequence can boost the overall emotional perception and memorisation. Significant differences, however, emerged only when considering the videos as a whole and not when we compared individual themes, probably due to the difference in storytelling or in the personal characteristics of the sample. A future confirmatory study should verify these assumptions by fixing the storytelling while varying the emotional sequence, as well as taking into account gender and age as additional grouping factors. However, our results provide useful insights for those stakeholders who are engaged in the promotion of traditional agri-food products, especially small local realities, helping them to optimise investments and reduce the gap with the big food corporations. Effective communication should place emphasis on how the purchase of these products provides solutions to specific issues, rather than simply exalting the goodness of the products and the benefits associated with their consumption.

Author Contributions: Conceptualisation, V.R., M.B. (Marco Bilucaglia), R.C. and G.L.; data curation, M.B. (Marco Bilucaglia); formal analysis, M.B. (Marco Bilucaglia), R.C., R.V. and R.L.; funding acquisition, V.R. and G.L.; investigation, M.B. (Mara Bellati) and R.V.; methodology, M.B. (Marco Bilucaglia), R.C., M.B. (Mara Bellati) and M.Z.; project administration, M.B. (Mara Bellati); resources, V.R. and G.L.; software, M.B. (Marco Bilucaglia) and R.L.; supervision, V.R., G.L. and M.Z.; validation, R.V.; visualisation, M.B. (Marco Bilucaglia) and R.C.; writing—original draft, V.R., M.B. (Marco Bilucaglia), R.C., M.B. (Mara Bellati), R.V., R.L., G.L. and M.Z.; writing—review and editing, V.R., M.B. (Marco Bilucaglia), R.C. and M.Z. All authors have read and agreed to the published version of the manuscript.

Funding: This study was developed within the project "Development of a synergy model aimed to qualify and valorize the Natural Historic Cheese of southern Italy in the Sicilian, Sardinia, Calabria, Basilicata, and Campania regions—Canestrum Casei", funded by "Progetto AGER—Agroalimentare e Ricerca 2" (RIF. 2017-1144).

Institutional Review Board Statement: The study was conducted in accordance with the Declaration of Helsinki, and approved by the Ethics Committee of Università IULM.

Informed Consent Statement: Informed consent was obtained from all subjects involved in the study.

Data Availability Statement: All data are available upon request.

Conflicts of Interest: The authors declare no conflict of interest. The funders had no role in the design of the study; in the collection, analyses, or interpretation of data; in the writing of the manuscript, or in the decision to publish the results.

Appendix A. Video Description

Appendix A.1. Rewind

The video starts with a girl receiving a postal package that contains traditional cheeses from Southern Italy. The girl is shown so excited while unwrapping it that she decides to eat one of the products. At the moment of tasting, a scene change takes place, and the entire production process is shown in reverse (hence, the name Rewind), that is, from the end product to the raw materials. The main video sequences include the landscapes of Southern Italy, the animals grazing freely, the cow milking and the detailed cheese production phases—from the milk curdling to the cheese seasoning and marking.

Appendix A.2. The Myth

The video starts with a voice-over narrating a legend (hence, the name The Myth). The Godhead gifted the human beings a harmonious and clean planet, the Hearth. Humans soon started harnessing nature, dealing with disastrous consequences. Several images of fires, natural landscapes polluted by garbage heaps and melting glaciers are shown. Displeased by their behaviour, the Godhead decides to punish them by erasing their memories. Five great sages decide to redeem humankind by starting the production of local cheeses using traditional techniques as a way to live in harmony with nature and preserve memories. Therefore, several images of natural landscapes, pictures of ancient villages in Southern Italy, animals grazing in pristine areas and cheeses produced with traditional techniques are shown.

References

1. Russo, V.; Milani Marin, L.E.; Fici, A.; Bilucaglia, M.; Circi, R.; Rivetti, F.; Bellati, M.; Zito, M. Strategic communication and neuromarketing in the fisheries sector: Generating ideas from the territory. *Front. Commun.* **2021**, *6*, 49. [CrossRef]
2. Russo, V.; Zito, M.; Bilucaglia, M.; Circi, R.; Bellati, M.; Milani Marin, L.E.; Catania, E.; Licitra, G. Dairy Products with Certification Marks: The Role of Territoriality and Safety Perception on Intention to Buy. *Foods* **2021**, *10*, 2352. [CrossRef] [PubMed]
3. Bell, D.; Valentine, G. *Consuming Geographies: We Are Where We Eat*; Routledge: London, UK, 2013.
4. Cacciolatti, L.A.; Garcia, C.C.; Kalantzakis, M. Traditional food products: The effect of consumers' characteristics, product knowledge, and perceived value on actual purchase. *J. Int. Food Agribus. Mark.* **2015**, *27*, 155–176. [CrossRef]
5. Duvaleix-Treguer, S.; Emlinger, C.; Gaigné, C.; Latouche, K. On the competitiveness effects of quality labels: Evidence from French cheese industry. In Proceedings of the 30th International Conférence of Agricultural Economists, Vancouver, BC, Canada, 28 July–2 August 2018; p. 21.
6. Wägeli, S.; Janssen, M.; Hamm, U. Organic consumers' preferences and willingness-to-pay for locally produced animal products. *Int. J. Consum. Stud.* **2016**, *40*, 357–367. [CrossRef]
7. Bryła, P. The role of appeals to tradition in origin food marketing. A survey among Polish consumers. *Appetite* **2015**, *91*, 302–310. [CrossRef]
8. Pilone, V.; De Lucia, C.; Del Nobile, M.A.; Contò, F. Policy developments of consumer's acceptance of traditional products innovation: The case of environmental sustainability and shelf life extension of a PGI Italian cheese. *Trends Food Sci. Technol.* **2015**, *41*, 83–94. [CrossRef]
9. Van Loo, E.J.; Grebitus, C.; Roosen, J. Explaining attention and choice for origin labeled cheese by means of consumer ethnocentrism. *Food Qual. Prefer.* **2019**, *78*, 103716. [CrossRef]
10. Ray, C. Culture, Intellectual Property and Territorial Rural Development. *Sociol. Rural.* **1998**, *38*, 3–20. [CrossRef]
11. Bagozzi, R.P.; Gopinath, M.; Nyer, P.U. The role of emotions in marketing. *J. Acad. Mark. Sci.* **1999**, *27*, 184–206. [CrossRef]
12. Poels, K.; Dewitte, S. How to capture the heart? Reviewing 20 years of emotion measurement in advertising. *J. Advert. Res.* **2006**, *46*, 18–37. [CrossRef]
13. Lewinski, P.; Fransen, M.L.; Tan, E.S. Predicting advertising effectiveness by facial expressions in response to amusing persuasive stimuli. *J. Neurosci. Psychol. Econ.* **2014**, *7*, 1. [CrossRef]
14. Estes, Z.; Verges, M. Freeze or flee? Negative stimuli elicit selective responding. *Cognition* **2008**, *108*, 557–565. [CrossRef] [PubMed]
15. LeBlanc, V.R.; McConnell, M.M.; Monteiro, S.D. Predictable chaos: A review of the effects of emotions on attention, memory and decision making. *Adv. Health Sci. Educ.* **2015**, *20*, 265–282. [CrossRef] [PubMed]
16. Otamendi, F.J.; Sutil Martín, D.L. The emotional effectiveness of advertisement. *Front. Psychol.* **2020**, *11*, 2088. [CrossRef] [PubMed]
17. Labroo, A.A.; Ramanathan, S. The influence of experience and sequence of conflicting emotions on ad attitudes. *J. Consum. Res.* **2007**, *33*, 523–528. [CrossRef]
18. Bae, M. The effect of sequential structure in charity advertising on message elaboration and donation intention: The mediating role of empathy. *J. Promot. Manag.* **2021**, *27*, 177–209. [CrossRef]
19. Carrera, P.; Oceja, L. Drawing mixed emotions: Sequential or simultaneous experiences? *Cogn. Emot.* **2007**, *21*, 422–441. [CrossRef]
20. Bennett, R. Individual characteristics and the arousal of mixed emotions: Consequences for the effectiveness of charity fundraising advertisements. *Int. J. Nonprofit Volunt. Sect. Mark.* **2015**, *20*, 188–209. [CrossRef]
21. Merchant, A.; Ford, J.B.; Sargeant, A. Charitable organizations' storytelling influence on donors' emotions and intentions. *J. Bus. Res.* **2010**, *63*, 754–762. [CrossRef]

22. Allen, P.; Hinrichs, C. Buying into 'buy local': Engagements of United States local food initiatives. In *Alternative Food Geographies: Representation and Practice*; Emerald: Bingley, UK, 2007; pp. 255–272.
23. Jia, S.S. Local food campaign in a globalization context: A systematic review. *Sustainability* **2021**, *13*, 7487. [CrossRef]
24. Van Ittersum, K.; Meulenberg, M.T.; Van Trijp, H.C.; Candel, M.J. Consumers' appreciation of regional certification labels: A Pan-European study. *J. Agric. Econ.* **2007**, *58*, 1–23. [CrossRef]
25. Bimbo, F.; Roselli, L.; Carlucci, D.; de Gennaro, B.C. Consumer Misuse of Country-of-Origin Label: Insights from the Italian Extra-Virgin Olive Oil Market. *Nutrients* **2020**, *12*, 2150. [CrossRef] [PubMed]
26. Cummings, R.G.; Harrison, G.W.; Rutström, E.E. Homegrown values and hypothetical surveys: Is the dichotomous choice approach incentive-compatible? *Am. Econ. Rev.* **1995**, *85*, 260–266.
27. Nisbett, R.E.; Wilson, T.D. Telling more than we can know: Verbal reports on mental processes. *Psychol. Rev.* **1977**, *84*, 231. [CrossRef]
28. Johansson, P.; Hall, L.; Sikström, S.; Tärning, B.; Lind, A. How something can be said about telling more than we can know: On choice blindness and introspection. *Conscious. Cogn.* **2006**, *15*, 673–692. [CrossRef]
29. Neeley, S.M.; Cronley, M.L. *When Research Participants Don't Tell It like It Is: Pinpointing the Effects of Social Desirability Bias Using Self vs. Indirect-Questioning*; ACR North American Advances: Duluth, MN, USA, 2004.
30. Smithson, J. Using and analysing focus groups: Limitations and possibilities. *Int. J. Soc. Res. Methodol.* **2000**, *3*, 103–119. [CrossRef]
31. List, J.A.; Gallet, C.A. What Experimental Protocol Influence Disparities Between Actual and Hypothetical Stated Values? *Environ. Resour. Econ.* **2001**, *20*, 241–254. [CrossRef]
32. Castellion, G.; Markham, S.K. Perspective: New Product Failure Rates: Influence of *Argumentum ad Populum* and Self-Interest. *J. Prod. Innov. Manag.* **2012**, *30*, 976–979. [CrossRef]
33. Crawford, C.M. Marketing Research and the New Product Failure Rate. *J. Mark.* **1977**, *41*, 51. [CrossRef]
34. Hakim, A.; Levy, D.J. A gateway to consumers' minds: Achievements, caveats, and prospects of electroencephalography-based prediction in neuromarketing. *Wiley Interdiscip. Rev. Cogn. Sci.* **2019**, *10*, e1485. [CrossRef]
35. Lee, N.; Broderick, A.J.; Chamberlain, L. What is 'neuromarketing'? A discussion and agenda for future research. *Int. J. Psychophysiol.* **2007**, *63*, 199–204. [CrossRef] [PubMed]
36. Ramsøy, T.Z. Building a foundation for neuromarketing and consumer neuroscience research: How researchers can apply academic rigor to the neuroscientific study of advertising effects. *J. Advert. Res.* **2019**, *59*, 281–294. [CrossRef]
37. Cao, C.C.; Reimann, M. Data Triangulation in Consumer Neuroscience: Integrating Functional Neuroimaging With Meta-Analyses, Psychometrics, and Behavioral Data. *Front. Psychol.* **2020**, *11*, 550204. [CrossRef] [PubMed]
38. Plassmann, H.; Weber, B. Individual differences in marketing placebo effects: Evidence from brain imaging and behavioral experiments. *J. Mark. Res.* **2015**, *52*, 493–510. [CrossRef]
39. Hsu, M.Y.T. RETRACTED: Cognitive systems research for neuromarketing assessment on evaluating consumer learning theory with fMRI: Comparing how two Word-Of-Mouth strategies affect the human brain differently after a product harm crisis. *Cogn. Syst. Res.* **2018**, *49*, 49–64. [CrossRef]
40. Karmarkar, U.R.; Plassmann, H. Consumer neuroscience: Past, present, and future. *Organ. Res. Methods* **2019**, *22*, 174–195. [CrossRef]
41. Karmarkar, U.R.; Yoon, C. Consumer neuroscience: Advances in understanding consumer psychology. *Curr. Opin. Psychol.* **2016**, *10*, 160–165. [CrossRef]
42. Plassmann, H.; Venkatraman, V.; Huettel, S.; Yoon, C. Consumer Neuroscience: Applications, Challenges, and Possible Solutions. *J. Mark. Res.* **2015**, *52*, 427–435. [CrossRef]
43. Smith, A.; Bernheim, B.D.; Camerer, C.F.; Rangel, A. Neural Activity Reveals Preferences without Choices. *Am. Econ. J. Microecon.* **2014**, *6*, 1–36. [CrossRef]
44. Ciceri, A.; Russo, V.; Songa, G.; Gabrielli, G.; Clement, J. A Neuroscientific Method for Assessing Effectiveness of Digital vs. Print Ads: Using Biometric Techniques to Measure Cross-Media Ad Experience and Recall. *J. Advert. Res.* **2020**, *60*, 71–86. [CrossRef]
45. Russo, V.; Valesi, R.; Gallo, A.; Laureanti, R.; Zito, M. "The Theater of the Mind": The Effect of Radio Exposure on TV Advertising. *Soc. Sci.* **2020**, *9*, 123. [CrossRef]
46. Juarez, D.; Tur-Viñes, V.; Mengual, A. Neuromarketing Applied to Educational Toy Packaging. *Front. Psychol.* **2020**, *11*, 2077. [CrossRef] [PubMed]
47. Missaglia, A.L.; Oppo, A.; Mauri, M.; Ghiringhelli, B.; Ciceri, A.; Russo, V. The impact of emotions on recall: An empirical study on social ads. *J. Consum. Behav.* **2017**, *16*, 424–433. [CrossRef]
48. Modica, E.; Rossi, D.; Cartocci, G.; Perrotta, D.; Feo, P.D.; Mancini, M.; Aricò, P.; Inguscio, B.M.S.; Babiloni, F. Neurophysiological Profile of Antismoking Campaigns. *Comput. Intell. Neurosci.* **2018**, *2018*, 1–11. [CrossRef]
49. Stasi, A.; Songa, G.; Mauri, M.; Ciceri, A.; Diotallevi, F.; Nardone, G.; Russo, V. Neuromarketing empirical approaches and food choice: A systematic review. *Food Res. Int.* **2018**, *108*, 650–664. [CrossRef]
50. Pagan, N.M.; Pagan, K.M.; Teixeira, A.A.; de Moura Engracia Giraldi, J.; Stefanelli, N.O.; de Oliveira, J.H.C. Application of Neuroscience in the Area of Sustainability: Mapping the Territory. *Glob. J. Flex. Syst. Manag.* **2020**, *21*, 61–77. [CrossRef]
51. Aurier, P.; Fort, F.; Sirieix, L. Exploring terroir product meanings for the consumer. *Anthropol. Food* **2005**, *4*. [CrossRef]

52. Biasi, R.; Brunori, E.; Smiraglia, D.; Salvati, L. Linking traditional tree-crop landscapes and agro-biodiversity in central Italy using a database of typical and traditional products: A multiple risk assessment through a data mining analysis. *Biodivers. Conserv.* **2015**, *24*, 3009–3031. [CrossRef]
53. Cherubino, P.; Martinez-Levy, A.C.; Caratù, M.; Cartocci, G.; Flumeri, G.D.; Modica, E.; Rossi, D.; Mancini, M.; Trettel, A. Consumer Behaviour through the Eyes of Neurophysiological Measures: State-of-the-Art and Future Trends. *Comput. Intell. Neurosci.* **2019**, *2019*, 1–41. [CrossRef]
54. Bilucaglia, M.; Laureanti, R.; Zito, M.; Circi, R.; Fici, A.; Russo, V.; Mainardi, L. It's a Question of Methods: Computational Factors Influencing the Frontal Asymmetry in Measuring the Emotional Valence. In Proceedings of the 2021 43rd Annual International Conference of the IEEE Engineering in Medicine & Biology Society (EMBC), Mexico City, Mexico, 1–5 November 2021; pp. 575–578.
55. Eddie Harmon-Jones, E.; Gable, P.A.; Peterson, C. The role of asymmetric frontal cortical activity in emotion-related phenomena: A review and update. *Biol. Psychol.* **2010**, *84*, 451–462. [CrossRef]
56. Summerfield, C.; Mangels, J.A. Coherent theta-band EEG activity predicts item-context binding during encoding. *NeuroImage* **2005**, *24*, 692–703. [CrossRef] [PubMed]
57. Vecchiato, G.; Babiloni, F.; Astolfi, L.; Toppi, J.; Vecchiato, G.; Astolfi, L.; Cherubino, P.; Dai, J.; Kong, W.; Wei, D. Enhance of theta EEG spectral activity related to the memorization of commercial advertisings in Chinese and Italian subjects. In Proceedings of the 2011 4th International Conference on Biomedical Engineering and Informatics (BMEI), Shanghai, China, 15–17 October 2011.
58. Mauss, I.B.; Robinson, M.D. Measures of emotion: A review. *Cogn. Emot.* **2009**, *23*, 209–237. [CrossRef] [PubMed]
59. Vecchiato, G.; Maglione, A.G.; Cherubino, P.; Wasikowska, B.; Wawrzyniak, A.; Latuszynska, A.; Latuszynska, M.; Nermend, K.; Graziani, I.; Leucci, M.R.; et al. Neurophysiological tools to investigate consumer's gender differences during the observation of TV commercials. *Comput. Math. Methods Med.* **2014**, *2014*, 912981. [CrossRef] [PubMed]
60. Nuwer, M.R. 10-10 electrode system for EEG recording. *Clin. Neurophysiol.* **2018**, *129*, 1103. [CrossRef]
61. Bilucaglia, M.; Masi, R.; Stanislao, G.D.; Laureanti, R.; Fici, A.; Circi, R.; Zito, M.; Russo, V. ESB: A low-cost EEG Synchronization Box. *HardwareX* **2020**, *8*, e00125. [CrossRef]
62. Faul, F.; Erdfelder, E.; Lang, A.G.; Buchner, A. G*Power 3: A flexible statistical power analysis program for the social, behavioral, and biomedical sciences. *Behav. Res. Methods* **2007**, *39*, 175–191. [CrossRef]
63. Cohen, J. *Statistical Power Analysis for the Behavioral Sciences*; Routledge: London, UK, 2013.
64. Sinha, S.R.; Sullivan, L.; Sabau, D.; San-Juan, D.; Dombrowski, K.E.; Halford, J.J.; Hani, A.J.; Drislane, F.W.; Stecker, M.M. American Clinical Neurophysiology Society Guideline 1. *J. Clin. Neurophysiol.* **2016**, *33*, 303–307. [CrossRef]
65. Friard, O.; Gamba, M. BORIS: A free, versatile open-source event-logging software for video/audio coding and live observations. *Methods Ecol. Evol.* **2016**, *7*, 1325–1330. [CrossRef]
66. McHugh, M.L. Interrater reliability: The kappa statistic. *Biochem. Medica* **2012**, *22*, 276–282. [CrossRef]
67. Delorme, A.; Makeig, S. EEGLAB: An open source toolbox for analysis of single-trial EEG dynamics including independent component analysis. *J. Neurosci. Methods* **2004**, *134*, 9–21. [CrossRef]
68. Laureanti, R.; Bilucaglia, M.; Zito, M.; Circi, R.; Fici, A.; Rivetti, F.; Valesi, R.; Wahl, S.; Mainardi, L.T.; Russo, V. Yellow (Lens) Better: Bioelectrical and Biometrical Measures to Assess Arousing and Focusing Effects. In Proceedings of the 2021 43rd Annual International Conference of the IEEE Engineering in Medicine & Biology Society (EMBC), Mexico City, Mexico, 1–5 November 2021.
69. Chang, C.Y.; Hsu, S.H.; Pion-Tonachini, L.; Jung, T.P. Evaluation of Artifact Subspace Reconstruction for Automatic Artifact Components Removal in Multi-Channel EEG Recordings. *IEEE Trans. Biomed. Eng.* **2020**, *67*, 1114–1121. [CrossRef] [PubMed]
70. Urigüen, J.A.; Garcia-Zapirain, B. EEG artifact removal—State-of-the-art and guidelines. *J. Neural Eng.* **2015**, *12*, 031001. [CrossRef] [PubMed]
71. Pion-Tonachini, L.; Kreutz-Delgado, K.; Makeig, S. ICLabel: An automated electroencephalographic independent component classifier, dataset, and website. *NeuroImage* **2019**, *198*, 181–197. [CrossRef] [PubMed]
72. Kayser, J.; Tenke, C.E. On the benefits of using surface Laplacian (current source density) methodology in electrophysiology. *Int. J. Psychophysiol.* **2015**, *97*, 171–173. [CrossRef]
73. Klimesch, W. EEG alpha and theta oscillations reflect cognitive and memory performance: A review and analysis. *Brain Res. Rev.* **1999**, *29*, 169–195. [CrossRef]
74. Borghini, G.; Aricò, P.; Flumeri, G.D.; Sciaraffa, N.; Babiloni, F. Correlation and Similarity between Cerebral and Non-Cerebral Electrical Activity for User's States Assessment. *Sensors* **2019**, *19*, 704. [CrossRef]
75. Allen, J.J.; Cohen, M.X. Deconstructing the "resting" state: Exploring the temporal dynamics of frontal alpha asymmetry as an endophenotype for depression. *Front. Hum. Neurosci.* **2010**, *4*, 232. [CrossRef]
76. Bilucaglia, M.; Laureanti, R.; Zito, M.; Circi, R.; Fici, A.; Rivetti, F.; Valesi, R.; Wahl, S.; Russo, V. Looking through blue glasses: Bioelectrical measures to assess the awakening after a calm situation. In Proceedings of the 2019 41st Annual International Conference of the IEEE Engineering in Medicine and Biology Society (EMBC), Berlin, Germany, 23–27 July 2019; pp. 526–529.
77. Kleckner, I.R.; Jones, R.M.; Wilder-Smith, O.; Wormwood, J.B.; Akcakaya, M.; Quigley, K.S.; Lord, C.; Goodwin, M.S. Simple, Transparent, and Flexible Automated Quality Assessment Procedures for Ambulatory Electrodermal Activity Data. *IEEE Trans. Biomed. Eng.* **2018**, *65*, 1460–1467. [CrossRef]

78. Greco, A.; Valenza, G.; Lanata, A.; Scilingo, E.P.; Citi, L. CvxEDA: A convex optimization approach to electrodermal activity processing. *IEEE Trans. Biomed. Eng.* **2016**, *63*, 797–804. [CrossRef]
79. Pan, J.; Tompkins, W.J. A Real-Time QRS Detection Algorithm. *IEEE Trans. Biomed. Eng.* **1985**, *BME-32*, 230–236.
80. Zito, M.; Fici, A.; Bilucaglia, M.; Ambrogetti, F.S.; Russo, V. Assessing the emotional response in social communication: The role of neuromarketing. *Front. Psychol.* **2021**, *12*, 625570. [CrossRef]
81. Zito, M.; Bilucaglia, M.; Fici, A.; Gabrielli, G.; Russo, V. Job Assessment Through Bioelectrical Measures: A Neuromanagement Perspective. *Front. Psychol.* **2021**, *12*, 673012. [CrossRef]
82. Love, J.; Selker, R.; Marsman, M.; Jamil, T.; Dropmann, D.; Verhagen, J.; Ly, A.; Gronau, Q.F.; Šmíra, M.; Epskamp, S.; et al. JASP: Graphical statistical software for common statistical designs. *J. Stat. Softw.* **2019**, *88*, 1–17. [CrossRef]
83. Verma, J.P. *Repeated Measures Design for Empirical Researchers*; John Wiley & Sons: Hoboken, NJ, USA, 2015.
84. Valenza, G.; Greco, A.; Gentili, C.; Lanata, A.; Sebastiani, L.; Menicucci, D.; Gemignani, A.; Scilingo, E. Combining electroencephalographic activity and instantaneous heart rate for assessing brain–heart dynamics during visual emotional elicitation in healthy subjects. *Philos. Trans. R. Soc. A Math. Phys. Eng. Sci.* **2016**, *374*, 20150176. [CrossRef]
85. Winkler, I.; Jager, M.; Mihajlović, V.; Tsoneva, T. Frontal EEG Asymmetry Based Classification of Emotional Valence using Common Spatial Patterns. *World Acad. Sci. Eng. Technol. Int. J. Med. Health Biomed. Bioeng. Pharm. Eng.* **2010**, *4*, 420–425.
86. Harmon-Jones, E.; Allen, J.J.B. Anger and frontal brain activity: EEG asymmetry consistent with approach motivation despite negative affective valence. *J. Personal. Soc. Psychol.* **1998**, *74*, 1310–1316. [CrossRef]
87. Gordon, R.; Ciorciari, J.; van Laer, T. Using EEG to examine the role of attention, working memory, emotion, and imagination in narrative transportation. *Eur. J. Mark.* **2018**, *52*, 92–117. [CrossRef]
88. Slater, M.D.; Rouner, D. Entertainment—Education and elaboration likelihood: Understanding the processing of narrative persuasion. *Commun. Theory* **2002**, *12*, 173–191.
89. Pedale, T.; Santangelo, V. Perceptual salience affects the contents of working memory during free-recollection of objects from natural scenes. *Front. Hum. Neurosci.* **2015**, *9*, 60. [CrossRef]
90. Petrican, R.; Moscovitch, M.; Schimmack, U. Cognitive resources, valence, and memory retrieval of emotional events in older adults. *Psychol. Aging* **2008**, *23*, 585. [CrossRef]
91. Abbruzzese, L.; Magnani, N.; Robertson, I.H.; Mancuso, M. Age and gender differences in emotion recognition. *Front. Psychol.* **2021**, *10*, 2371. [CrossRef]
92. Graves, L.V.; Moreno, C.C.; Seewald, M.; Holden, H.M.; Van Etten, E.J.; Uttarwar, V.; McDonald, C.R.; Delano-Wood, L.; Bondi, M.W.; Woods, S.P.; et al. Effects of age and gender on recall and recognition discriminability. *Arch. Clin. Neuropsychol.* **2017**, *32*, 972–973. [CrossRef] [PubMed]

 foods

MDPI

Article

Strategy Optimization of Quality Improvement and Price Subsidy of Agri-Foods Supply Chain

Jing Xu [1,*], Jiajia Cai [1], Guanxin Yao [2] and Panqian Dai [1]

[1] Jiangsu Modern Logistics Research Base, Business School, Yangzhou University, Yangzhou 225127, China; caijiajia2910@163.com (J.C.); daipanqian@126.com (P.D.)
[2] Research Institute of China Grand Canal, Yangzhou University, Yangzhou 225127, China; gxyao@yzu.edu.cn
* Correspondence: xujing_sxy@yzu.edu.cn

Abstract: Based on the realistic concerns about the improvement of the quality of agricultural foods (agri-foods), the optimal supply quality and price subsidy strategies of producers and sellers for the two-level agricultural supply chain, composed of a producer and a seller, are studied. The differences in the quality safety, price, and market demand of agri-foods in the supply chain are compared and analyzed. The study found that the maximum profit of supply chain participants decreases with the increase of price elasticity of demand. When the quality of agri-foods is upgraded in a producer-led manner, the quality of agri-foods in the supply chain does not undergo substantial improvement, and the maximum profit of agri-foods operators is insensitive to the price elasticity of demand at this time. When the seller-led quality upgrading is launched, the maximum profit of the producer decreases with the increase of the quality elasticity of demand, the maximum profit of the seller increases with the increase of the quality elasticity of demand, and the total profit of the supply chain also increases with the increase of the quality elasticity of demand under the centralized decision situation. The quality and safety of agri-foods as well as the overall profit of the supply chain can be improved most effectively under the centralized control decision with the goal of maximizing the supply chain benefits. In terms of quality and price, quality improvement actions of agri-foods driven by supply-side producers are less effective than those driven by demand-side consumption. In addition, cost-sharing contracts can significantly improve the quality of agri-foods in the supply chain and make them more "high-quality and low-price" than before the adoption of cost-sharing contracts.

Keywords: agri-food supply chain; quality safety; price compensation; market demand; cost-sharing contracts

Citation: Xu, J.; Cai, J.; Yao, G.; Dai, P. Strategy Optimization of Quality Improvement and Price Subsidy of Agri-Foods Supply Chain. *Foods* **2022**, *11*, 1761. https://doi.org/10.3390/foods11121761

Academic Editor: Maria Lisa Clodoveo

Received: 16 May 2022
Accepted: 10 June 2022
Published: 15 June 2022

1. Introduction

For a long time, the phenomenon that agricultural foods (agri-foods) with good quality do not sell at a good price is common, and the main reason for this is the credence property of agricultural products [1]. Farmers need to invest more in quality to get high-quality agri-foods compared with ordinary produce. However, the quality of the produce cannot be easily identified by consumers. Due to asymmetric information, consumers are not willing to pay high prices for high-quality produces [2]. As a result, agricultural producers often lack the motivation to improve the quality of agri-foods, leading to the prevalence of homogeneous "big-ticket" products in the field of agricultural supply [3].

With the development of economy and society, consumers' income level and consciousness of food safety are rising, and the consumption demand for agricultural produce is gradually changing from concern for supply quantity to supply quality and diversity [4]. Further, the consumers' demand for premium quality agri-foods is increasingly strong, which has driven more and more participants in the agricultural supply chain to increase investment for the improvement of agri-foods quality. Some measures such as establishing production standard systems and adopting clean production technologies are used. After taking the quality improvement actions, agricultural operators apply for pollution-free, green, or organic certification as a way to increase the premium price of their produce and build agricultural brands. This information is ultimately passed on to the consumer in the form of labels on the produce [5–7]. Nowadays, in the field of practice, this quality improvement measure has been launched either by agricultural producers on their own initiative or by sellers who induce agricultural producers to improve their quality inputs in the form of high-price subscriptions [8]. However, there is still no relevant theory to guide the practice of which model is better. Therefore, which way is more beneficial to achieve quality and price of agri-foods? How do the participants in the agricultural supply chain make quality improvement and price subsidy decisions? Additionally, can appropriate contracts be designed to encourage producers and sellers to increase their investment in agricultural quality improvement actions? With these questions, we conducted this study.

2. Literature Review

2.1. The Improvement Strategy of Product Quality

A review of the literature revealed that how to improve product quality to enhance market competitiveness has been a hot research topic in business operation management. For example, Li et al. studied the supplier's product upgrade strategies considering consumers' strategic behavior [9]. Chen et al. studied the joint pricing and quality decision problems in a dual-channel supply chain [10]. Li et al. studied pricing and quality competition issues under brand differentiation [11]. These studies focus on consumer reactions to product quality upgrades in a highly competitive context. Unlike general products, the problem of structural imbalance on the supply side of agricultural produce has existed for many years. In other words, the oversupply of general mass agri-foods and the undersupply of high-quality agri-foods coexist. In addition, agricultural producers are mostly in a disadvantaged position, operating at low profit. Under the urgent call of consumers for high-quality and safe agri-foods, even if they make quality improvement behaviors, they need the supply chain to bear the cost of quality improvement inputs and put produce into the market with high quality and competitive price. Otherwise, the agricultural supply chain lacks the incentive for quality improvement. Although scholars also pay attention to the quality improvement of agricultural products, they mainly focus on the quality grading [12,13], quality and safety regulation [14], and related behaviors affecting the quality and safety of agricultural produce [15]. Moreover, data surveys and empirical tests are mainly used [16,17], which cannot effectively predict the long-term impact of agri-foods quality upgrading.

2.2. Game Study on Product Price and Demand

In terms of supply chain games considering price and demand, Abad [18] constructed a demand–price-sensitive supply chain coordination model and finally obtained a method to achieve the optimal strategy when buyers and sellers cooperate. Sajadi et al. discussed the issue of supply chain coordination when market demand is influenced by factors such as the size of the retailer's effort to develop the market and the retail price [19]. Chan and Dai et al. discussed the problems of optimal order lot size and optimal order lead time for achieving maximum profit in the supply chain in a cooperative game, respectively [20,21]. Mir Mehdi et al. constructed non-cooperative game models, such as the Nash game and the Stackelberg game, as well as a cooperative game model to study a two-level supply chain game problem in which demand is influenced by advertising and price [22]. Zhao et al. constructed an option contract model to solve the conflict of interest in the supply chain [23]. Leng et al. used the Nash and Stackelberg game model to study the multi-supplier but single-retailer supply chain coordination problem for short life-cycle products [24]. Esmaeili et al. studied the supply chain game problem when demand is dependent on price and marketing costs [25]. In general, the means and game methods of supply chain coordination under different conditions have been studied extensively and thoroughly, but not much research has been conducted on quality and safety levels of agri-foods and corresponding price compensation mechanisms. For instance, Ren et al. studied the performance of food safety management systems of Chinese food business operators [26]. Jacques et al. [27] discussed the quality and safety standards in the food industry. Though the quality of agri-foods is noticed, there is no mention of subsidies for the implementors of the agricultural produce improvement initiative, making it difficult to provide targeted guidance for quality improvement and cost–price compensation in agricultural supply chains. Therefore, we conducted the present study based on our previous research on the quality [2] and price compensation [28] mechanisms of agricultural produce.

2.3. *Influences after the Improvement of the Agri-Food Quality*

Some other researchers discussed the influence of the foods label that reflects the quality of produce. Messer et al. proposed that labeling produce can have both good and ugly effects [5]. The good effects include that the labeling of quality can reduce information asymmetry, and the ugly effects includes that the labeling and certifying process may be costly, which leads to reduced demand for high-quality agricultural produces. Aftab et al. analyzed the impact of rising food prices on consumer welfare in the most populous countries of South Asia and found that consumer welfare declines in all countries mainly for cereals and milk, as these food items are relatively less elastic to price fluctuations [29]. The labeling process and price increases associated with improved quality of agricultural produce will undoubtedly have an impact on social welfare and other stakeholders. However, this is not the focus of this paper. Different from these articles, we only focused on the quality improvement action though it is a key part of the produce before it is labeled. In addition, the quality improvement studied in this article is voluntary and endogenous, proposed by the participants of the agricultural supply chain. As a result, in order to focus more on the research topic, we will not discuss at great length the social impact of quality improvement actions.

Above all, compared to existing research, the novelty of our work is as follows: The "quality safety factor" and "price compensation factor" are introduced into the linear inverse demand function of agricultural produce for the first time. By constructing three different game models under decentralized and centralized decision-making situations, the optimal strategy for upgrading the quality of agricultural produce is analyzed, and a cost-sharing contract is designed to coordinate the supply chain. In this way, some theoretical reference is provided for the participants in the agricultural supply chain when making decisions about the quality improvement input and price compensation.

3. Materials and Methods

3.1. *Problem Description*

A two-level agricultural supply chain consisting of a risk-neutral agri-foods producer and a risk-neutral agri-foods seller is studied in this paper. Prior to the marketing season, the seller orders q quantities of common produce from the produce producer at a price of p_0. However, considering that market demand is influenced by both the quality level and retail price of agricultural produce, and more and more consumers prefer green and organic agri-foods, the "Agricultural Produce Quality and Safety Enhancement Initiative" is proposed for the agricultural supply chain. This initiative aims at encouraging agricultural producers to increase quality inputs to improve the quality of agri-foods. Thus, the agricultural produce is as follows:

(1) In order to respond to the quality and safety enhancement initiative, the agricultural producer needs to increase the quality inputs, such as upgrading production technology level, improving production management methods, etc. Referring to the study of Jiang [30], the research and development inputs for producing green products have a quadratic relationship with the greenness of the products. Therefore, it is assumed that the relationship between quality inputs u and quality safety degree g in agricultural production is $u = \frac{1}{2}zg^2$, where z is the production influence factor.

(2) When the quality improvement action is promoted, the agricultural producer will raise the price to maintain his own interests according to the input costs of quality improvement. At this point, the selling price of the producer is $w = (1+\alpha)w_0$, where α is the price compensation coefficient after quality inputs. In order to encourage quality improvement and maintain profits, the agri-foods seller will also raise prices accordingly. To simplify the model, it can be assumed that the seller of agricultural produce raises prices by the same amount as the producer, and at this time, the selling price of the seller is $p = (1+\alpha)p_0$.

(3) Although the seller of agricultural produce determines the basic market demand q based on past experience and order from the producer, natural losses such as bumping and spoilage cannot be avoided in the actual transportation and sales process due to the perishable nature of agri-foods. Thus, it is assumed that the actual market demand is d ($d < q$); i.e., the single-order quantity of the seller is greater than the actual market demand. Since agri-foods are perishable, it is assumed that all agri-foods unsold are lost and no longer converted into value.

(4) Assume that the actual market demand d is positively related to the basic market demand q and the quality safety degree g and negatively related to the retail price p. Referring to the linear inverse demand function, let $d = q - b(1+\alpha)p_0 + kg$, where b is the sensitivity coefficient of consumers to the price of agri-foods, and k is the sensitivity coefficient of consumers to the quality of agri-foods. Obviously, the higher the product quality is, the lower the price is, and the more consumers prefer to buy such produce. This assumption is in line with the reality.

3.2. Notation

Based on the problem description and assumptions above, the notations used in this study are described in Table 1 below.

Table 1. The description of the notations.

Notations	Descriptions
π_n	Profit for the agricultural producer
π_s	Profit for the agricultural foods seller
π_{ns}	Overall profitability of the agricultural supply chain
q	Basic market demand, i.e., the order quantity determined by the seller based on previous years' sales
d	Actual market demand, i.e., the amount of produce ultimately sold by the seller
p_0	Selling price of produce per unit for the seller before the quality improvement initiative
w_0	Selling price of produce per unit for the producer before the quality improvement initiative
c_n	Production cost per unit of agricultural produce
c_s	Cost of sales per unit of agricultural foods
u	Quality improvement inputs, i.e., the unit cost paid by producers to improve the quality of agricultural produce
g	Quality safety degree
α	Price compensation factor after quality inputs
k	Sensitivity coefficient of consumers to the quality of agricultural produce, i.e., quality elasticity of demand
b	Sensitivity coefficient of consumers to the price of agricultural produce, i.e., price elasticity of demand
z	Production impact factor

4. Model Construction and Solving

4.1. Decentralized Decision Model

According to the above assumptions, the profit function of agricultural producer after implementing quality improvement actions is

$$\pi_n = q\left((1+\alpha)w_0 - c_n - \frac{1}{2}zg^2\right), \tag{1}$$

The profit function of the agricultural produce seller is

$$\pi_s = p_0(1+\alpha)(q - bp_0(1+\alpha) + \text{kg}) - q((1+\alpha)w_0 + c_s), \tag{2}$$

Proposition 1. *There exists a quality safety degree g that maximizes the profit of the agricultural producer and a level of price subsidy α that maximizes the profit of the agri-foods seller.*

Proof. The profit function π_n is as follows:

$$\pi_n = q\left[(1+\alpha)w_0 - c_n - \frac{1}{2}zg^2\right]$$

The partial derivative of π_n with respect to g yields $\frac{\partial \pi_n}{\partial g} = -qzg \neq 0$, $\frac{\partial^2 \pi_n}{\partial^2 g} = -qz < 0$, $(q > 0, z > 0)$. Therefore, π_n is a strictly concave function on g; i.e., there exists a maximum value of the profit function π_n. □

The profit function π_s is as follows:

$$\pi_s = p_0(1+\alpha)(\text{q} - bp_0(1+\alpha) + \text{kg}) - \text{q}((1+\alpha)w_0 + c_s)$$

The partial derivative of π_s with respect to α yields $\frac{\partial \pi_s}{\partial \alpha} = -2(1+\alpha)b{p_0}^2 + (kg+q)p_0 - qw_0 \neq 0$, $\frac{\partial^2 \pi_s}{\partial^2 \alpha} = -b{p_0}^2 < 0$. Therefore, π_s is a strictly concave function on α; i.e., there exists a maximum value of the profit function π_s.

(1) Producer-led Model (Model 1)

In this case, the producer is the leader of the supply chain, and the seller is the follower, such as the family farm-dominated or cooperative-dominated agricultural supply chain. Hence, this is a Stackelberg game model dominated by agricultural producer. In this model, the producer determines the quality input u (or quality safety degree g), and then, the seller chooses the optimal price compensation factor α^* based on the producer's input, and the backward induction method is applied to solve the model.

When $\frac{\partial \pi_n}{\partial g} = -qzg = 0$, the optimal quality safety degree can be obtained as follows:

$$g_1^* = 0, \tag{3}$$

Substitute g_1^* into Equation (1); when $\frac{\partial \pi_s}{\partial \alpha} = p_0 q - 2b{p_0}^2\alpha - 2b{p_0}^2 - qw_0 = 0$, the optimal price subsidy factor can be obtained as follows:

$$\alpha_1^* = \frac{q(p_0 - w_0)}{2bp_0^2} - 1, \tag{4}$$

Obviously, the agricultural producer as the leader lacks the incentive to improve the quality in this case, but the seller still provides price subsidies due to information asymmetry. The market demand is $d_1^* = \frac{q(p_0+w_0)}{2p_0}$. Substituting Equations (3) and (4) into

Equations (1) and (2), the optimal profit of the agricultural producer and the optimal profit of the agri-foods seller are obtained as follows:

$$\pi_n^{1*} = \frac{q^2(p_0 - w_0)w_0}{2bp_0^2} - qc_n$$

$$\pi_s^{1*} = \frac{q^2p_0^2 - 2q^2p_0w_0 + q^2w_0^2}{4bp_0^2} - qc_s$$

Proposition 2. *When decentralized decision making is dominated by the agricultural producer, the maximum profit of both producer and seller decreases with the increase of price elasticity of consumer demand, and the maximum profit of the agricultural seller increases as the retail price per unit of produce increases. When $p_0 < 2w_0$, the maximum profit of the agricultural producer increases with the increase of the retail price per unit of agricultural produce. Likewise, when $p_0 > 2w_0$, the maximum profit of the agricultural producer decreases with the increase of the retail price per unit of agri-foods.*

Proof. From the equations $\frac{\partial \pi_n^{1*}}{\partial b} = -\frac{(p_0-w_0)w_0q^2}{2p_0^2b^2} < 0$, $\frac{\partial \pi_s^{1*}}{\partial b} = -\frac{(p_0-w_0)^2q^2}{4p_0^2b^2} < 0$, and $\frac{\partial \pi_s^{1*}}{\partial p_0} = \frac{(p_0-w_0)w_0q^2}{2p_0^3b} > 0$, it can be obtained that π_n^{1*}, π_s^{1*} are negatively correlated with b, while π_s^{1*} is positively correlated with p_0. When $p_0 < 2w_0$, $\frac{\partial \pi_n^{1*}}{\partial p_0} > 0$ can be known, and otherwise, $\frac{\partial \pi_n^{1*}}{\partial p_0} < 0$. □

Proposition 2 shows that the maximum profit of agricultural supply chain participants is closely related to the level of economic and social development and the consumption environment. When the income level of consumers is low (high price elasticity of demand), the profit of agricultural operators is also slimmer. The higher the price of the produce before quality improvement, the higher the profit for the produce seller to participate in quality improvement. While the profit of agricultural producers is mainly influenced by the level of their price appreciation in the supply chain before quality improvement, if the retail price of agri-foods exceeds the wholesale price by more than two times, the lower the retail price of agri-foods before the implementation of quality improvement actions, and the greater the producer's profit.

(2) Seller-led Model (Model 2)

In this case, the seller of agri-foods is the leader of the supply chain, while the producer is the follower, such as the "company + farmer" type of agricultural supply chain. Therefore, this is a Stackelberg game model dominated by agricultural seller. The price subsidy coefficient α is determined by the seller of agri-foods firstly, and then, the ideal quality safety degree g^* and the optimal quality input level are determined by the producer according to α, and the backward induction method is also used to solve the model.

When $\frac{\partial \pi_s}{\partial \alpha} = p_0q + p_0kg - 2bp_0{}^2\alpha - 2bp_0{}^2 - qw_0 = 0$, the optimal price subsidy system can be obtained as follows: $\alpha_2^*(g) = \frac{p_0q + kgp_0 - w_0q}{2bp_0^2} - 1$. Substitute $\alpha_2^*(g)$ into Equation (2); when $\frac{\partial \pi_n}{\partial g} = \frac{kw_0}{2bp_0} - zg = 0$, the optimal quality safety degree can be obtained as follows:

$$g_2^* = \frac{kw_0}{2zbp_0}, \tag{5}$$

Substitute Equation (5) into the optimal response function $\alpha_2^*(g)$ of the produce seller, and then, obtain the optimal price subsidy level for the quality improvement of agri-foods as follows:

$$\alpha_2^* = \frac{qzb(p_0 - w_0) + 0.5k^2w_0}{2zb^2p_0^2} - 1, \tag{6}$$

Thus, the actual market demand is $d_2^* = \frac{qzb(p_0+w_0)+0.5k^2w_0}{2zbp_0}$. Based on this, substituting Equations (5) and (6) into Equations (1) and (2), the optimal profit of the agricultural producer and the optimal profit of the agricultural seller are obtained as follows:

$$\pi_n^{2*} = \frac{(bqzp_0w_0 - qzbw_0^2 - 0.25k^2w_0^2)q}{2zb^2p_0^2} - qc_n$$

$$\pi_s^{2*} = \frac{q^2z^2(p_0 - w_0)^2b^2 + k^2qzbw_0(p_0 - w_0) + 0.25k^4w_0^2}{4z^2b^3p_0^2} - qc_s$$

Proposition 3. *When decentralized decision making is dominated by sellers of agri-foods, the maximum profit of both agricultural producer and seller decreases with the increase of production impact factor z. The maximum profit of agricultural producer decreases as the quality elasticity of demand increases, and the price elasticity of demand decreases. The maximum profit of agricultural seller increases as the quality elasticity of demand increases, and the price elasticity of demand decreases.*

Proof. From the equations $\frac{\partial \pi_n^{2*}}{\partial k} = -\frac{qkw_0^2}{4zp_0^2b^2} < 0$, $\frac{\partial \pi_n^{2*}}{\partial z} = -\frac{qk^2w_0^2}{8z^2p_0^2b^2} < 0$, and $\frac{\partial \pi_n^{2*}}{\partial b} = -\frac{q^2zbw_0(w_0-p_0)-0.25qk^2w_0^2}{2zp_0^2b^3} > 0$, it can be obtained that π_n^{2*} is negatively correlated with k and z and positively correlated with b. Additionally, π_s^{2*} is positively correlated with k and negatively correlated with z, and b can be known due to $\frac{\partial \pi_s^{2*}}{\partial k} = \frac{kqzbw_0(p_0-w_0)+0.5w_0^2k^3}{2z^2p_0^2b^3} > 0$, $\frac{\partial \pi_s^{2*}}{\partial z} = -\frac{k^2w_0\left(qbz(p_0-w_0)+0.5k^2w_0\right)}{4z^3p_0^2b^3} < 0$, and $\frac{\partial \pi_s^{2*}}{\partial b} = \frac{-q^2z^2b^2(p_0-w_0)^2-2qzbk^2w_0(p_0-w_0)-0.75k^4w_0^2}{4z^2p_0^2b^4} < 0$. □

Proposition 3 shows that the greater the quality elasticity of consumer demand, the less price sensitive consumers are; and the smaller the quality inputs required to improve the quality and safety of agricultural products, the greater the maximum profit for agricultural sellers is. When consumers pay more attention to quality, the supply chain profit distribution will be more unfavorable to producers, which will then force the producer to improve quality input, and the producer's cost increases, and the profit decreases in a short period of time.

4.2. Centralized Decision Model (Model 3)

In this case, self-interest maximization is no longer the decision-making goal of the participants of agricultural supply chain. Instead, centralized decision making is made through win-win cooperation to maximize the overall interests of the supply chain. At this time, the total profit function of the agricultural supply chain is:

$$\pi_{ns} = p_0(1+\alpha)(q - bp_0(1+\alpha) + \mathrm{kg}) - q((1+\alpha)w_0 + c_s) + q((1+\alpha)w_0 - c_n - \frac{1}{2}zg^2)$$

Proposition 4. *When $2bqz - k^2 > 0$, π_{ns} is concave in α and g. In this case, the overall profit function of the supply chain has a maximum value.*

Proof. The Hessian matrix of π_{ns} is $\begin{vmatrix} \frac{\partial^2 \pi_{\mathrm{ns}}}{\partial^2 g} & \frac{\partial^2 \pi_{\mathrm{ns}}}{\partial g \partial \alpha} \\ \frac{\partial^2 \pi_{\mathrm{ns}}}{\partial \alpha \partial g} & \frac{\partial^2 \pi_{\mathrm{ns}}}{\partial^2 \alpha} \end{vmatrix} = \begin{vmatrix} -qz & kp_0 \\ kp_0 & -2bp_0^2 \end{vmatrix} = (2bqz - k^2)p_0^2$. When $2bqz - k^2 > 0$, conditions $(2bqz - k^2)p_0^2 > 0$ and $-qz < 0$ can be satisfied, so the Hessian matrix is negative definite. The profit function π_{ns} is a joint concave function with respect to the price subsidy coefficient α and the quality safety degree g. Hence, there exists optimal solutions α_3^* and g_3^* to maximize the profit function. □

By combining the equations $\frac{\partial \pi_{ns}}{\partial \alpha} = p_0 q + p_0 k g - 2bp_0^2 - 2bp_0^2\ \alpha = 0$ and $\frac{d\pi_{ns}}{dg} = p_0 k + p_0 k\alpha - qzg = 0$, the optimal price subsidy coefficient and quality safety degree are obtained as follows:

$$\alpha_3^* = \frac{zq^2}{p_0(2bqz - k^2)} - 1$$

$$g_3^* = \frac{qk}{2bqz - k^2}$$

At this point, the actual market demand is $d_3^* = \frac{bzq^2}{2bqz-k^2}$. The maximum profit of the agricultural supply chain under centralized decision making is

$$\pi_{ns}{}^* = \frac{0.5zq^3}{2bqz - k^2} - (c_s + c_n)q$$

Proposition 5. *In the centralized model, the overall profit of the agricultural supply chain increases with the increase of the quality elasticity of demand k and decreases with the increase of the price elasticity of demand b and the production impact factor z. That is, when consumers are more concerned about quality and less concerned about price, and the smaller the cost of quality inputs required by agricultural producers to implement quality improvement actions, the greater the overall profitability of the supply chain.*

Proof. Due to $\frac{\partial \pi_{ns}{}^*}{\partial b} = -\frac{q^4z^2}{(2qzbp_0-k^2)^2} < 0$, $\frac{\partial \pi_{ns}{}^*}{\partial k} = \frac{kzq^3}{(2qzb-k^2)^2} > 0$, and $\frac{\partial \pi_{ns}{}^*}{\partial z} = -\frac{0.5k^2q^3}{(2qzb-k^2)^2} < 0$, $\pi_{ns}{}^*$ is positively correlated with k and negatively correlated with b and z. □

4.3. Comparison and Analysis

According to the calculation results above, summarizing the quality improvement and price subsidy decisions of the agricultural supply chain participants as well as the corresponding changes in market demand, Table 2 can be obtained below.

Table 2. Comparison of the three game models.

Variables	Model 1	Model 2	Model 3
α^*	$\frac{q(p_0-w_0)}{2bp_0^2} - 1$	$\frac{qzb(p_0-w_0)+0.5k^2w_0}{2zb^2p_0^2} - 1$	$\frac{zq^2}{p_0(2bqz-k^2)} - 1$
g^*	0	$\frac{kw_0}{2zbp_0}$	$\frac{qk}{2bqz-k^2}$
d^*	$\frac{q(p_0+w_0)}{2p_0}$	$\frac{qzb(p_0+w_0)+0.5k^2w_0}{2zbp_0}$	$\frac{bzq^2}{2bqz-k^2}$
π_n^*	$\frac{q^2(p_0-w_0)w_0}{2bp_0^2} - qc_n$	$\frac{(bqzp_0w_0-qzbw_0^2-0.25k^2w_0^2)q}{2zb^2p_0^2} - qc_n$	
π_s^*	$\frac{q^2(p_0-w_0)^2}{4bp_0^2} - qc_s$	$\frac{q^2z^2(p_0-w_0)^2b^2+k^2qzbw_0(p_0-w_0)+0.25k}{4z^2b^3p_0^2} - qc_s$	
π_{ns}^*			$\frac{0.5zq^3}{2bqz-k^2} - (c_s + c_n)q$

The following prerequisites can be derived from the results in Table 2:

(1) From $\alpha_1{}^* = \frac{q(p_0-w_0)}{2bp_0^2} - 1 > 0$, $q\ (p_0 - w_0) > 2bp_0^2$ can be easily obtained.

(2) From $\alpha_3{}^* = \frac{zq^2}{p_0(2bqz-k^2)} - 1 > 0$ and $g_3{}^* = \frac{qk}{2bqz-k^2} > 0$, $2bqz > k^2 > 2bqz - \frac{zq^2}{p_0}$ can be obtained.

Based on the discussion above, the following propositions can be known:

Proposition 6. *The quality of agri-foods in the centralized decision model is higher than that in the decentralized model when the participants of the agricultural supply chain make decisions to maximize the profit of the supply chain. In addition, the quality of agri-foods is higher when the* *seller is the leader than in the producer-led case. That is, the best quality safety degree satisfies* $g_1{}^* < g_2{}^* < g_3{}^*$.

Proof. Due to $g_2{}^* = \frac{kw_0}{2zbp_0}$, and since k, w_0, z, b, p_0 are positive, $g_2{}^* > g_1{}^* = 0$ is satisfied. Again, due to $g_3{}^* = \frac{qk}{2bqz-k^2}$, for $g_3{}^* > g_2{}^*$ to be true, condition $2bqzk(p_0 - w_0) + k^3 w_0 > 0$ should be satisfied. Since the selling price is greater than the buying price, i.e., $p_0 - w_0 > 0$, and k, w_0 are positive, $g_3{}^* > g_2{}^*$ can be known. That is, $g_1{}^* < g_2{}^* < g_3{}^*$ is proven. □

Proposition 7. *In general, the optimal price compensation factor satisfies* $\alpha_1{}^* < \alpha_2{}^*$ *and* $\alpha_1{}^* < \alpha_3{}^*$.

Proof. Due to $\alpha_1{}^* = \frac{q(p_0-w_0)}{2bp_0^2} - 1$ and $\alpha_2{}^* = \frac{qzb(p_0-w_0)+0.5k^2w_0}{2zb^2p_0^2} - 1$, it can be obtained that $\alpha_2{}^* - \alpha_1{}^* = \frac{w_0k^2}{4zb^2p_0^2} > 0$. That is, $\alpha_1{}^* < \alpha_2{}^*$. Again, due to $\alpha_3{}^* = \frac{zq^2}{p_0(2bqz-k^2)} - 1$, for $\alpha_1{}^* < \alpha_3{}^*$ to be true, condition $k^2(w_0 - p_0) < 2bzqw_0$ should be satisfied. Because $w_0 - p_0 < 0$, i.e., $k^2(w_0 - p_0) < 0$, as well as $2bzqw_0 > 0$, $\alpha_1{}^* < \alpha_3{}^*$ can therefore be proven. □

Proposition 7 shows that in this agricultural supply chain, the price compensation factors both in the Stackelberg model dominated by sellers and the centralized decision model are greater than those in the Stackelberg game model dominated by agricultural producers. That is, in terms of quality and price, the quality improvement actions driven by producers on the supply side are not as effective as the quality improvement actions driven by consumption on the demand side.

Corollary 1. *When the initial price of agri-foods* $p_0 > \frac{k^2}{4b^2z}$, *the price appreciation of agri-foods satisfies* $\alpha_1{}^* < \alpha_2{}^* < \alpha_3{}^*$.

Proof. For $\alpha_3{}^* > \alpha_2{}^*$ to be true, condition $\frac{zq^2}{p_0(2bqz-k^2)} > \frac{qzb(p_0-w_0)+0.5k^2w_0}{2zb^2p_0^2}$ should be satisfied, which is equivalent to proving that

$$0.5k^2w_0\left(2bqz - k^2\right) - 2z^2b^2q^2w_0 + bqzk^2(w_0 - p_0) < 0 \tag{7}$$

Firstly, $bqzk^2(w_0 - p_0) < 0$ can be easily known, so to prove (7) is true, that is, the proof of $0.5k^2w_0(2bqz - k^2) - 2z^2b^2q^2w_0 < 0$ is satisfied. Secondly, it is clear from the precondition that $\frac{zq^2}{p_0} > 2bqz - k^2$, so to prove $0.5k^2w_0(2bqz - k^2) - 2z^2b^2q^2w_0 < 0.5k^2w_0\frac{zq^2}{p_0} - 2z^2b^2q^2w_0$ is as same as to prove $zq^2k^2w_0 - 4z^2b^2q^2p_0w_0 < 0$, that is, to prove that $zw_0q^2(k^2 - 4zb^2p_0) < 0$, which means $p_0 > \frac{k^2}{4b^2z}$. Hence, when $p_0 > \frac{k^2}{4b^2z}$ is satisfied, $\alpha_2{}^* < \alpha_3{}^*$ is true, and $\alpha_1{}^* < \alpha_2{}^* < \alpha_3{}^*$ can be obtained finally. □

Corollary 1 shows that the price compensation factor in the centralized decision model is greater than the price compensation factor under the decentralized decision model dominated by sellers only when the market price of agri-foods is high, and the condition $p_0 > \frac{k^2}{4b^2z}$ is met. Combined with Proposition 6, it can be seen that for high-priced agri-foods, supply chain cooperation to improve the quality of agri-foods is most beneficial for price appreciation, and the quality and price of agri-foods can be realized to the greatest extent.

Corollary 2. *The actual market demand satisfies* $d_2{}^* > d_1{}^*$. *When* $0 < w_0 < \frac{k^2p_0}{2bzq-k^2}$, $d_3{}^* > d_1{}^*$, *and when* $w_0 > \frac{k^2p_0}{2bzq-k^2}$, $d_3{}^* < d_1{}^*$. *When* $0 < w_0 < \frac{2bzqk^2p_0}{4b^2z^2q^2-k^4}$, $d_3{}^* > d_2{}^*$, *and when* $w_0 > \frac{k^2p_0}{2bzq-k^2}$, $d_3{}^* < d_2{}^*$.

Proof. Due to $d_1{}^* = \frac{q(p_0+w_0)}{2p_0}$ and $d_2{}^* = \frac{qzb(p_0+w_0)+0.5k^2w_0}{2zbp_0}$, $d_2{}^* - d_1{}^* = \frac{w_0k^2}{4zbp_0} > 0$, which means $d_2{}^* > d_1{}^*$. Again, due to $d_3{}^* - d_1{}^* = \frac{0.5(w_0+p_0)qk^2-bzq^2w_0}{p_0(2bzq-k^2)}$, $0.5(w_0 + p_0)qk^2 - bzq^2w_0 < 0$ can be known, and then, $d_3{}^* < d_1{}^*$ is obtained. Since the condition $0.5(w_0 + p_0)qk^2 - bzq^2w_0 < 0$ is equivalent to $w_0 > \frac{k^2p_0}{2bzq-k^2}$, when $0 < w_0 < \frac{k^2p_0}{2bzq-k^2}$, $d_3{}^* > d_1{}^*$ can be obtained, and when $w_0 > \frac{k^2p_0}{2bzq-k^2}$, $d_3{}^* < d_1{}^*$ can be obtained. □

Similarly, from the equation $d_3{}^* - d_2{}^* = \frac{0.5bzqp_0k^2-b^2z^2q^2w_0+0.25w_0k^4}{bzp_0(2bzq-k^2)}$, $0.5bzqp_0k^2 - b^2z^2q^2w_0 + 0.25w_0k^4 < 0$ can be known, which proves $d_3{}^* < d_2{}^*$. Since the condition $0.5bzqp_0k^2 - b^2z^2q^2w_0 + 0.25w_0k^4 < 0$ is equivalent to $w_0 > \frac{2bzqk^2p_0}{4b^2z^2q^2-k^4}$, $d_3{}^* > d_2{}^*$ can be obtained when $0 < w_0 < \frac{2bzqk^2p_0}{4b^2z^2q^2-k^4}$, and $d_3{}^* < d_2{}^*$ can also be obtained when $w_0 > \frac{k^2p_0}{2bzq-k^2}$.

Corollary 2 shows that in case of decentralized models, a seller-led initiative to improve the quality of agri-foods is more beneficial to increase actual market sales than a producer-led one. The centralized decision is more favorable to increase the sales volume of agri-foods when the agricultural producers sell to sellers with low unit price. Furthermore, the actual market demand in the centralized model is less than that in the decentralized decision case when the agricultural producers sell to sellers with high-value agri-foods.

5. Contract Coordination Strategy Based on Cost Sharing of Agricultural Quality Improvement

According to the analysis above, the quality improvement of agri-foods driven by the demand side is more effective. Based on this, referring to the cost-sharing contract model of Yang et al. [31], it is assumed that agri-foods sellers are willing to share the quality improvement cost at a ratio of $\beta \in (0,1)$, which incentivizes producers to increase inputs on agri-food quality improvement. In the decentralized scenario dominated by sellers, the profit functions of the agricultural producers and sellers are

$$\pi_n = q\left((1+\alpha)w_0 - c_n - (1-\beta)\frac{1}{2}zg^2\right)$$

$$\pi_s = p_0(1+\alpha)(q - bp_0(1+\alpha) + kg) - q\left((1+\alpha)w_0 + c_s + \frac{1}{2}\beta zg^2\right)$$

Extending the solution method in Model 2, the optimal decisions and profits of producers and sellers of agri-foods under cost-sharing contracts can be obtained as follows:

$$g_d^* = \frac{0.5kw_0}{(1-\beta)zbp_0}$$

$$\alpha_d^* = \frac{0.25k^2w_0}{zb^2p_0{}^2(1-\beta)} - 1$$

At this point, the profits of the participants in the agricultural supply chain are:

$$\pi_n^{d*} = \frac{0.5k^2w_0{}^2q}{4zb^2p_0{}^2(1-\beta)} - qc_n$$

$$\pi_s^{d*} = \frac{((1-\beta)p_0 + (0.5\beta-1)w_0)w_0k^2qzb + 0.25w_0{}^2k^4}{4b^3p_0{}^2(1-\beta)^2z^2} - qc_s$$

Proposition 8. *The effect of cost contract coordination strategy of this agricultural supply chain depends on the original parameter. When* $2qbzw_0 = qbzp_0 + 0.75w_0k^2$*, the optimal cost-sharing coefficient* β^* *can be obtained.*

Proof. The second-order derivative of $\pi_s{}^*$ with respect to β can be obtained as follows: $\frac{\partial^2 \pi_s{}^*}{\partial^2 \beta} = \frac{k^2 w_0\left(-qbz\beta p_0+qbzp_0+0.5qbz\beta w_0-2qbzw_0+0.75w_0k^2\right)}{2b^3 p_0{}^2(\beta-1)^4 z^2}$. Considering the numerator $-qbz\beta p_0 + qbzp_0 + 0.5qbz\beta w_0 - 2qbzw_0 + 0.75w_0k^2$ as $\mathrm{F}(\beta)$, $\frac{\partial' \mathrm{F}(\beta)}{\partial' \beta} = -0.5qbz(p_0 - 0.5w_0)w_0k^2 < 0$ can thus be obtained easily, which indicates $\mathrm{F}(\beta)$ is a decreasing function about $\beta \in (0,1)$. When $\beta = 0$, $\mathrm{MaxF}(0) = qbzp_0 - 2qbzw_0 + 0.75w_0k^2$ can be known. When $\mathrm{MaxF}(0) = 0$, $qbz(2w_0 - p_0) = 0.75w_0k^2$ can be obtained. Obviously, when $qbz(2w_0 - p_0) = 0.75w_0k^2$ holds, if $\beta > 0$, then $\mathrm{F}(\beta) < \mathrm{F}(0) = 0$, at which point $\frac{\partial^2 \pi_s{}^*}{\partial^2 \beta} < 0$; i.e., there exists the optimal cost-sharing coefficient that makes $\pi_s{}^*$ maximize. Moreover, $\beta^* = \frac{qbz(2p_0-3w_0)+k^2 w_0}{qzb(2p_0-w_0)}$ can be obtained when $\frac{\partial \pi_s{}^*}{\partial \beta} = 0$. □

Proposition 9. *Compared to the decentralized model led by sellers, agri-foods in the supply chain coordination model based on cost-sharing contracts are of higher quality and cheaper price under specific conditions.*

Proof. It is known that $g^* - g_2^* = \frac{0.5k\beta w_0}{zbp_0(1-\beta)} > 0$ and $\alpha^* - \alpha_2^* = \frac{2zb^2(1-\beta)p_0{}^2 - zbq(1-\beta)p_0 + zbqw_0(1-\beta) + 0.5k^2 w_0 \beta}{2zb^2 p_0{}^2(1-\beta)}$. Considering the numerator $2zb^2(1-\beta)p_0{}^2 - zbq(1-\beta)p_0 + zbqw_0(1-\beta) + 0.5k^2 w_0\beta$ as $\gamma(p_0)$, $\gamma(p_0)$ is thus a parabola with an opening upward about p_0. When $p_0 = \frac{q}{4b}$, $\gamma(p_0)$ obtains the minimum value $Min\gamma(p_0) = -0.0625(1-\beta)zq^2 + 0.5zbqw_0(1-\beta) + 0.25k^2 w_0\beta$. Then, if $Min\gamma(p_0) \geq 0$, i.e., $w_0 \geq \frac{0.25zq^2(1-\beta)}{2zbq(1-\beta)+k^2\beta}$, $\alpha^* \geq \alpha_2^*$ holds. Otherwise, when $0 < w_0 < \frac{0.25zq^2(1-\beta)}{2zbq(1-\beta)+k^2\beta}$, $\alpha^* < \alpha_2^*$ can be obtained. □

From the equations $\pi_n^{d*} - \pi_n^{2*} = \frac{\left(0.5k^2 w_0(1-0.5\beta)+\left(qbzw_0-qbzp_0-0.5k^2 w_0{}^2\right)(1-\beta)\right)qw_0}{2(1-\beta)zb^2 p_0{}^2}$ and $\pi_s^{d*} - \pi_s^{2*} = \frac{-\left(qzb(\beta-1)(w_0-p_0)-0.5\beta k^2 w_0\right)^2+0.5\beta k^4 w_0{}^2}{4z^2 b^3 p_0{}^2(\beta-1)^2}$, it can be seen that the positive or negative sign on the right side of the equation cannot be judged directly, which depends on the value of the relevant parameters. Therefore, the basic cost-sharing contract cannot significantly increase the profit of the participants.

Proposition 9 shows that the introduction of cost-sharing contracts based on the decentralized decision model can significantly improve the quality of products in the agricultural supply chain, and in most cases, the selling price of agricultural products will not be higher than that before the introduction of cost-sharing contracts. At this time, agri-foods in the supply chain based on cost-sharing contracts are of better quality and cheaper price. However, it cannot significantly improve the profit of agricultural supply chain participants.

6. Numerical Study

In this section, numerical examples are used to verify the above conclusions and carry out sensitivity analysis on the parameters in the model.

6.1. Numerical Example

According to the research on the agricultural market, the transaction price of several vegetables between cooperatives and supermarkets, the retail price of supermarkets, and other costs incurred in the agricultural–supermarket interface were considered comprehensively, and reasonable parameters $q = 30$, $p_0 = 5$, $w_0 = 3$, $c_n = 2$, $c_s = 0.7$, $k = 1.2$, $b = 0.8$, and $z = 0.2$ were set. Based on the results of the theoretical analysis, the optimal values of quality safety degree g^*, price subsidy level α^*, market demand d^*, and total profit $\pi_n{}^*$, $\pi_s{}^*$, and $\pi_{ns}{}^*$ of producers, sellers, and supply chain under different models are shown in Table 3.

Table 3. Comparison of the three game models.

Variables	Model 1	Model 2	Model 3
α^*	0.5	0.8	3.4
g^*	0	2.3	4.4
d^*	24	25.4	17.6
π_n^*	75	59.8	
π_s^*	24	46.5	
π_{ns}^*	99	105.3	249.9

Variables with * are the optimal variable values.

As can be seen from Table 3, under this set of parameters, compared with the producer-dominated decentralized decision model, the quality of agri-foods is higher in the seller-dominated decision model, and the price compensation factor is also higher. The market sales volume does not change much. Although the profit of agricultural producers decreases slightly, the seller can obtain a higher profit, and the total profit of the supply chain increases.

The results of the algorithm can verify the conclusion of the above proposition. Obviously, the quality of agri-foods in the centralized model is the highest, followed by the Stackelberg model dominated by sellers, and finally the model dominated by agricultural producers. Although the actual market sales of centrally controlled cooperative game are not necessarily higher than those of non-cooperative game, the agricultural supply chain under the centralized game can still achieve more profits, i.e., the centralized decision model with the goal of maximizing the overall profit of the supply chain is more conducive to realizing the "quality and price" mechanism of agricultural supply.

6.2. Sensitivity Analysis

According to the assumptions above, b is the sensitivity of consumers to the price of agri-foods, while k is the sensitivity of consumers to the quality of agri-foods. By analyzing the sensitivity of b and k, the trend of the decision variables in the game model can be explored when the key parameters change.

6.2.1. Sensitivity Analysis on Consumer Price Sensitivity Factors

Based on the parameters above, the variation of each variable in the three models regarding the consumer price sensitivity factor b is explored. From Figure 1, it can be seen that the price compensation factor α and the quality of agri-foods g decrease with the increase of consumer price sensitivity factor b. The trend indicates that when consumers become more and more sensitive to the price of agri-foods, the selling price of agri-foods will decrease with the increase of consumer price elasticity in order to increase sales, and at the same time, agricultural producers will choose to reduce quality improvement inputs to reduce costs. However, in general, the centralized decision with the goal of maximizing the overall benefit of the supply chain is still the optimal strategy to realize the quality and price of agri-foods.

From Figure 2, it can be seen that the profit of the participants in the agricultural supply chain decreases as the price elasticity of demand b increases, and the overall profit of the supply chain under the centralized decision is the largest. Since the profit of sellers is relatively small under this set of parameters, the overall profit distribution of sellers in the supply chain is slightly lower than that of producers. When the quality of agri-foods is promoted by the seller, the seller gains more profit compared with the producer; i.e., the profit is inclined to the dominant players in the supply chain.

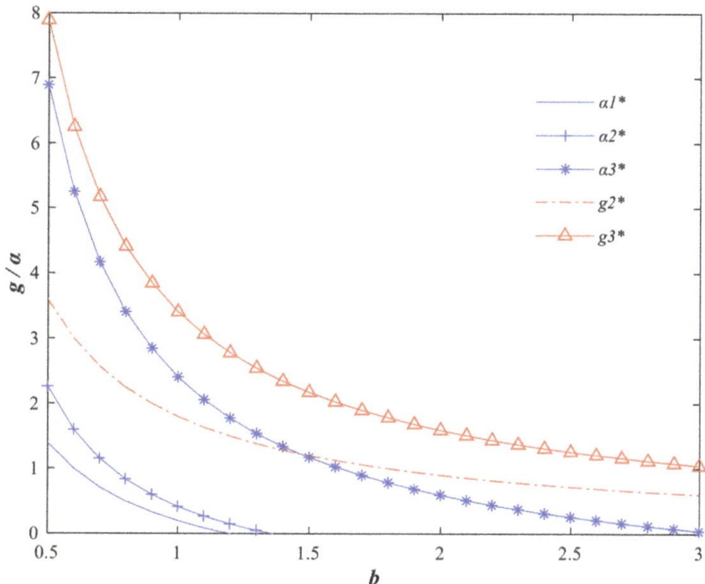

Figure 1. Variation of quality safety factor g and price compensation factor α with consumer price sensitivity factor b. (Variables with * are the optimal variable values.)

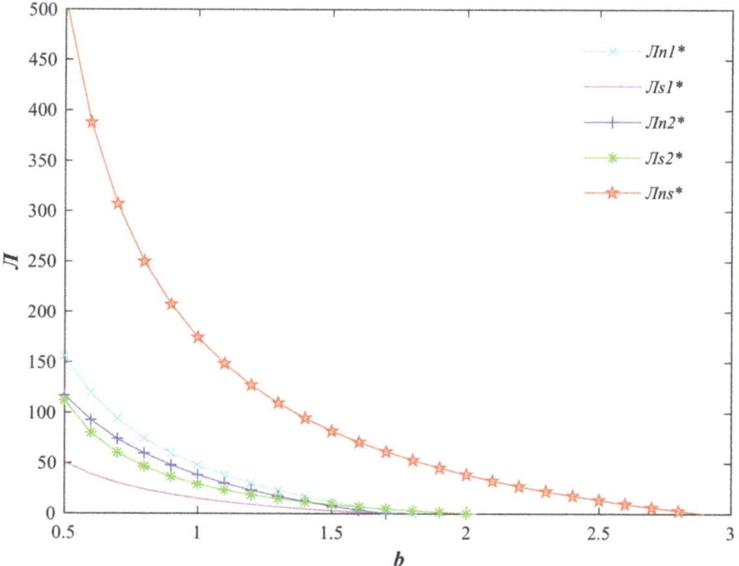

Figure 2. Variation of profits of agricultural supply chain participants with consumer price sensitivity factor b. (Variables with * are the optimal variable values).

6.2.2. Sensitivity Analysis on Consumer Quality Sensitivity Factors

Using the above parameters, the variation of each variable in the three models with respect to the consumer quality sensitivity factor k is analyzed. From Figure 3, it can be seen that with the increase of k, the price compensation factor α and quality safety

g in the producer-driven Stackelberg game model are at the lowest level and do not change. However, the price compensation factor α and quality safety degree g in the seller-dominated Stackelberg game model and the centralized decision model are both on the rise. In the producer-dominated Stackelberg game model, producers dominate the quality safety degree g, and sellers dominate the price compensation factor α. The trend shows that with consumers' increasing care about the quality of agri-foods, producers are concerned that sellers will not compensate them for the cost of their efforts to improve quality. Therefore, in order to ensure their own maximum gain, producers will be stingy to increase the quality improvement input, which ultimately indicates that the producer is insensitive to the consumer's quality demand elasticity. In Model 2 and 3, when consumers pay more and more attention on the quality of agri-foods, sellers will incentivize producers to improve the quality of agri-foods by increasing the price compensation factor.

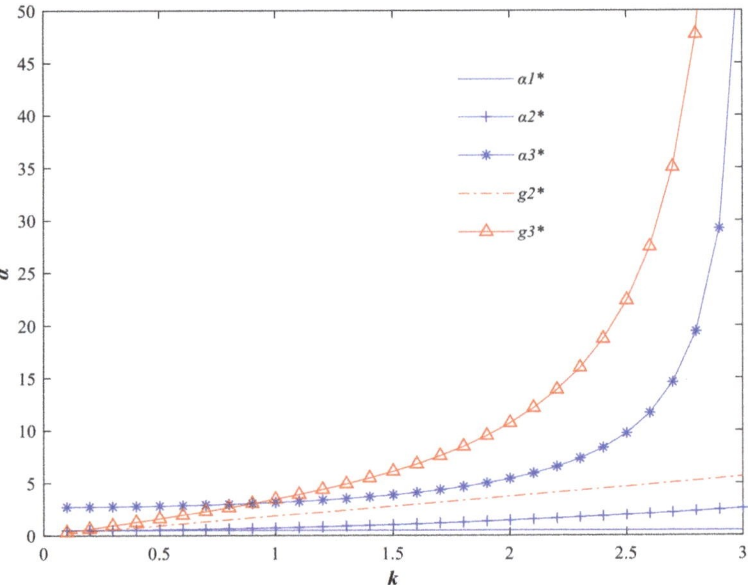

Figure 3. Variation of quality safety factor g and price compensation factor α with consumer quality sensitivity factor k. (Variables with * are the optimal variable values).

As can be seen in Figure 4, the producer is insensitive to the quality elasticity of demand because the producer does not actually make quality improvements in Model 1. The maximum profit of the seller in model 2 and the total profit of the supply chain in model 3 increase with the increase of the quality elasticity of demand, and the profit of the producer decreases with the increase of the quality elasticity of demand. When the quality elasticity of demand reaches a certain threshold, if the price increase is not large, the producers of agri-foods will be unprofitable due to the high quality input cost.

From Figure 5, it can be seen that the actual market demand for agri-foods in models 2 and 3 decreases with the increase of consumer price sensitivity and diverges into two stages with the increase of quality sensitivity.

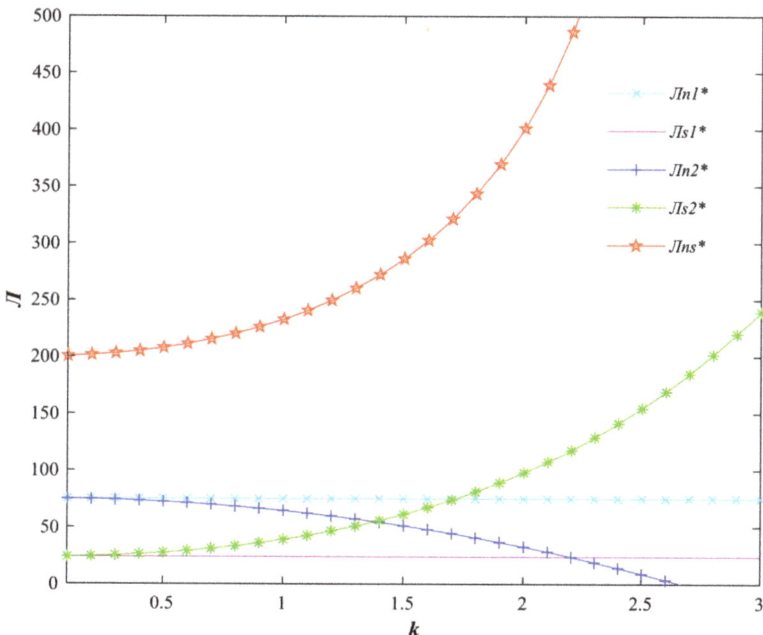

Figure 4. Variation of profit of agricultural supply chain participants with consumer quality sensitivity factor k. (Variables with * are the optimal variable values).

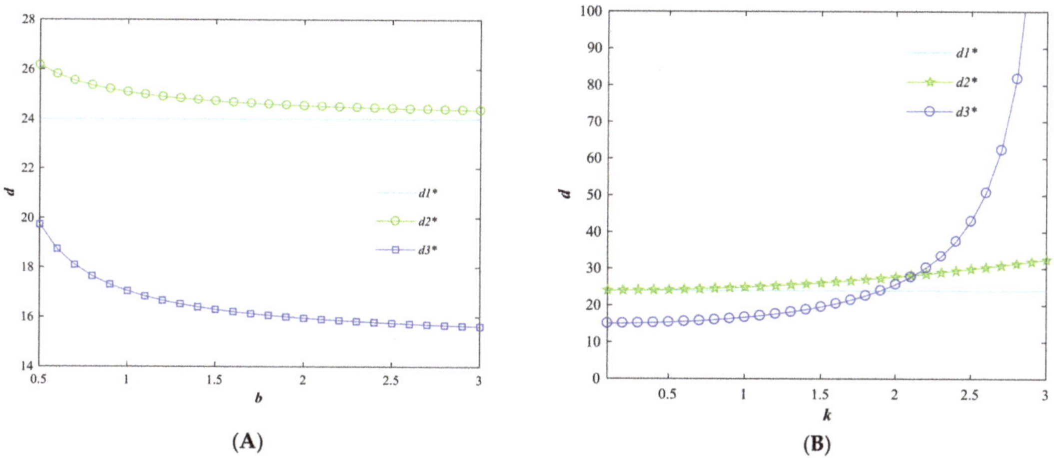

Figure 5. Variation of real market demand d with consumer price sensitivity factor b (**A**) and quality sensitivity factor k (**B**) (Variables with * are the optimal variable values).

In the first stage, when the price sensitivity of consumers is very low, the actual market demand in the centralized decision model is the lowest, followed by the game model dominated by producers, and finally the game model dominated by sellers. The second stage is when the quality sensitivity of consumers reaches a certain value, the actual market demand in the centralized decision model starts to exceed the actual market demand in the decentralized decision model with the increase of quality sensitivity.

This indicates that when consumers are not sensitive to the quality of agri-foods, low-quality agri-foods sell better. When consumers are more and more concerned about the quality of agri-foods, the "high-quality and high-priced" produce sells better.

6.3. Analysis of Other Parameters

6.3.1. The Impact of Agricultural Selling Prices on Profits

The above parameters were also used to analyze the profit changes of supply chain participants when the retail price gradually increases from 5 to 10, and the results are shown in Figure 6A. As can be seen from the figure, the profit of sellers increases with the increase of selling price of agri-foods, while producers' profit increases first and then decreases with the increase of selling price. With the increase of unit value of agri-foods, supply chain profit gradually inclines to sellers.

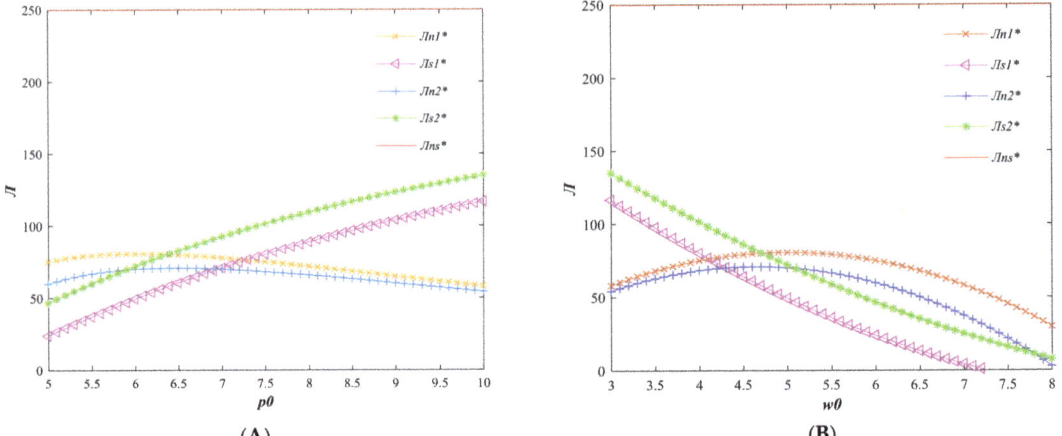

Figure 6. Variation of profits of supply chain participants with agricultural sales price p_0 (**A**) and wholesale price w_0 (**B**). (Variables with * are the optimal variable values).

Considering that the profit of sellers in the original parameters is low, in order to observe the change of profit of supply chain participants with the wholesale price of agri-foods, the retail price is assumed to be 10. When the wholesale price increases from 3 to 8, the change of profit of supply chain participants is shown in Figure 6B. From the figure, it can be seen that with the increase of wholesale price, the profit of sellers gradually decreases, and the profit of producers first increases and then decreases. The total profit of the centralized decision supply chain is the largest, mainly related to parameters such as total regional sales volume rather than prices.

6.3.2. The Impact of Cost-Sharing Factor

According to Proposition 8, the parameters $k = 1$ and $b = 0.375$ are reset to observe the changes of some variables, such as the quality of agri-foods, the level of price subsidy, and the profit of supply chain participants with the sharing coefficient β, under the cost-sharing contract. The results are shown in Figure 7. It is easy to find from Figure 7A that the adoption of cost-sharing contracts can significantly improve the quality of agri-foods, and also the price subsidy level is lower than that before the introduction of cost-sharing contracts, which means that the price of agri-foods is more "high-quality and low-price" at this time. From Figure 7B, it can be seen that as the cost-sharing coefficient of sellers increases, the supply chain profit keeps shifting to producers. The producer and seller can achieve the same profit at a specific moment.

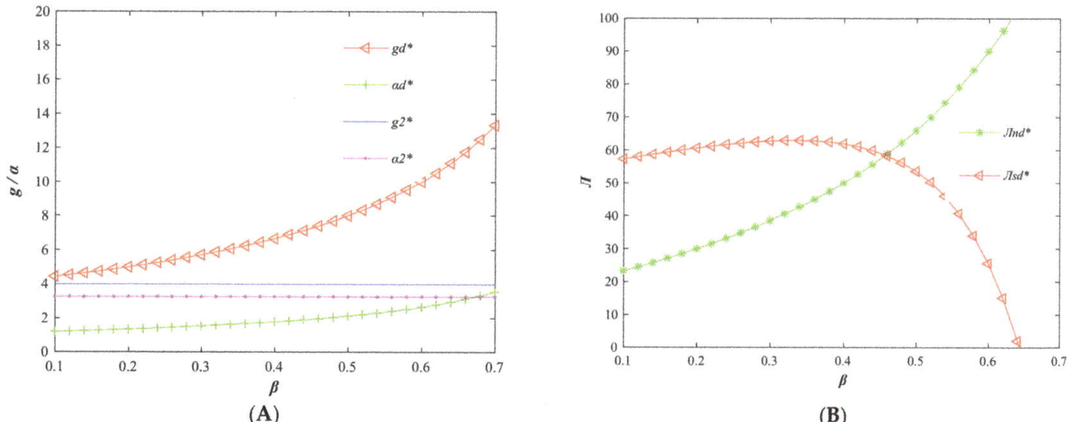

Figure 7. Variation of the main parameters (**A**,**B**) of the supply chain with the sharing ratio β after the introduction of the cost-sharing contract. (Variables with * are the optimal variable values).

7. Discussion

From the analysis above, we find a very interesting phenomenon. That is, the decision of the quality improvement initiative depends on the decision-making style, the value of agri-foods, and the consumers' consumption concept.

(1) In the decentralized decision-making model, the quality improvement of agri-foods led by sellers is better than that led by producers, but both are lower than the agricultural product quality level when the decision making is centralized to maximize the overall supply chain benefits. Therefore, from the consideration of improving the quality of agri-foods, it is the optimal strategy to improve the level of supply chain integration, establish supply chain alliances, and launch action plans of agri-food quality improvement jointly by supply chain members.

(2) Considering that the actual demand for agri-foods is affected by the wholesale price, the supply chain participants should choose decision-making strategies according to the characteristics of the products they operate when launching the action plan for agricultural produce quality improvement, from the perspective of opening up the market and increasing the market awareness of agricultural produce. Generally speaking, for general mass agri-foods, it is more favorable to expand sales volume by centralizing decisions for the supply chain to drive quality upgrades, while for high-end agri-foods with higher wholesale prices, it is more favorable to expand sales volume and open up the market for the supply chain to be led by sellers.

(3) The price and quality of agri-foods are affected by the price elasticity of demand and quality elasticity of demand. The more sensitive consumers are to price, the more popular low-quality and low-priced products will be, and the more sensitive consumers are to quality, the better quality and price mechanism is guaranteed when the overall benefit of the supply chain is maximized to improve the quality of agri-foods, and the quality improvement of agri-foods is more "good value for money" compared with the decentralized decision-making mode. Therefore, in order to promote the high-quality development of the agri-foods industry, raising the income level of agri-foods consumers and reducing the price elasticity of demand should be the first focus, and then, it is indispensable to promote the concept of food safety and improve the quality elasticity of demand so as to force the construction of a high-quality and high-quality price mechanism for agricultural produce.

(4) The use of cost-sharing contracts can significantly improve the quality of agri-foods in the supply chain, and the subsidized price of agri-foods under certain conditions is

lower than before the introduction of the contract, which helps to reduce the selling price and increase sales. Moreover, cost-sharing contracts facilitate the transfer of profit distribution in the supply chain to the producers. Therefore, from the perspective of protecting "small farmers" and increasing consumer surplus, sellers should be encouraged to promote quality upgrading initiatives in the form of cost-sharing.

8. Conclusions

Three basic decision models involving the decision-making behavior on the quality improvement of agri-foods in the supply chain, namely, the decentralized decision model dominated by producers, the decentralized decision model dominated by sellers, and the centralized decision model, were constructed using game theory. The market demand of agri-foods depends on their sales price and supply quality is considered, and two decision variables of "quality safety" and "price compensation factor" are introduced into the linear inverse demand function for the first time. A cost-sharing contract is designed to study the relationship between product price, quality, and market demand in the supply chain. The results show that the centralized model is better than the two other models from the perspective of quality improvement. However, how to make decisions to achieve quality and price of agri-foods varies. Given that the study only analyzed the decisions within the supply chain, that is, the quality improvement action is voluntary and endogenous, future research can further discuss the influence of quality improvement on the behavior of other stakeholders and make clear how quality works in the market. Moreover, the participants in the agricultural supply chain are assumed to be risk-neutral. To validate the findings further, studies can be conducted with risk-appetitive and risk-averse participants.

Author Contributions: Writing—original draft preparation, J.X.; writing—review and editing, J.C.; validation, P.D.; funding acquisition, G.Y. All authors have read and agreed to the published version of the manuscript.

Funding: National Natural Science Foundation of China: 72103178; Strategic Research and Consulting Project of Chinese Academy of Engineering: 2021-XZ-30; Postdoctoral Research Foundation of China: 2019M661960.

Data Availability Statement: Data is contained within the article.

Acknowledgments: This research was supported by the National Natural Science Foundation of China (72103178) and Strategic Research and Consulting Project of Chinese Academy of Engineering (2021-XZ-30).

Conflicts of Interest: The authors declare no conflict of interest.

References

1. Miao, Y.; Huang, J.; Liu, R.; Sun, J. Research hotspots and development trend of high quality and high price in China: Visualization analysis based on citespace. *Chin. J. Agric. Resour. Reg. Plan.* **2022**, *18*, 1–15.
2. Xu, J.; Yao, G.; Dai, P. Quality Decision-Making Behavior of Bodies Participating in the Agri-Foods E-Supply Chain. *Sustainability* **2020**, *12*, 1874.
3. Starbird, S.A. Moral Hazard, Inspection Policy, and Food Safety. *Am. J. Agric. Econ.* **2005**, *87*, 15–27. [CrossRef]
4. Chen, J. Game Analysis of Agri-food Quality Classification under Consumer Selection Behavior. *Oper. Res. Manag. Sci.* **2020**, *29*, 68–75.
5. Messer, K.D.; Costanigro, M.; Kaiser, H.M. Labeling food processes: The good, the bad and the ugly. *Appl. Econ. Perspect. Policy* **2017**, *39*, 407–427. [CrossRef]
6. Bonroy, O.; Constantatos, C. On the economics of labels: How their introduction affects the functioning of markets and the welfare of all participants. *Am. J. Agric. Econ.* **2015**, *97*, 239–259. [CrossRef]
7. Golan, E.; Kuchler, F.; Mitchell, L.; Greene, C.; Jessup, A. Economics of food labeling. *J. Consum. Policy* **2001**, *24*, 117–184. [CrossRef]
8. Hong, X.; Cao, X.; Gong, Y.; Chen, W. Quality information acquisition and disclosure with green manufacturing in a closed-loop supply chain. *Int. J. Prod. Econ.* **2020**, *232*, 107997. [CrossRef]
9. Li, S.; Shao, L. Upgrade Strategies in The Presence of Strategic Consumers. *Chin. J. Manag. Sci.* **2018**, *26*, 1–10.
10. Chen, J.; Liang, L.; Yao, D.Q.; Sun, S. Price and quality decisions in dual-channel supply chains. *Eur. J. Oper. Res.* **2017**, *259*, 935–948. [CrossRef]

11. Li, W.; Chen, J. Pricing and quality competition in a brand-differentiated supply chain. *Int. J. Prod. Econ.* **2018**, *202*, 97–108. [CrossRef]
12. Nie, W.; Bao, H.; Li, T. A Theoretical Basis for and Path Towards Quality Grading System of Agricultural Products in China-Welfare Evaluation Based on Choice Experiment Survey. *J. Agrotech. Econ.* **2021**, *7*, 65–78.
13. Nie, W.; Bao, H.; Li, T. Blocking Mechanism and Attributes Selection of Consumers' Searching Grading Information of Agricultural Products. *J. Huazhong Agric. Univ.* **2021**, *3*, 51–62.
14. Shao, Y.; Chen, G.; Yang, J. Potential Safety Hazard and Regulation Equilibrium of Primary Agricultural Products in Rural Areas. *Collect. Essays Financ. Econ.* **2020**, *9*, 104–112.
15. Kang, T.; Mu, Y. Asymmetry of Production and Marketing Information and Farmers' Behavior of Fertilizer Overuse. *J. Northwest A F Univ.* **2020**, *20*, 111–119.
16. Chen, Z.; Yuan, B. Labor Price, Information Asymmetry and Investment in Fine Management Technology—From the Perspective of Differences in Quality Recognition Ability. *J. Agrotech. Econ.* **2020**, *2*, 21–31.
17. Liu, X.; Dong, Y. Quantity, Quality or Price-performance Ratio—On the Impetus and Transformation of China's Agri-product Exports. *J. Int. Trade* **2019**, *11*, 100–115.
18. Abad, P.L. Supplier pricing and lot sizing when demand is price sensitive. *Eur. J. Oper. Res.* **1994**, *78*, 334–354. [CrossRef]
19. Sadjadi, S.J.; Oroujee, M.; Aryanezhad, M. Optimal production and marketing planning. *Comput. Optim. Appl.* **2005**, *2*, 195–203. [CrossRef]
20. Chan, C.K.; Kingsman, B.G. Coordination in a single-vendor multi-buyer supply chain by synchronizing delivery and production cycles. *Transp. Res. Part E* **2007**, *43*, 90–111. [CrossRef]
21. Dai, T.; Qi, X. An acquisition policy for a multi-supplier system with a finite-time horizon. *Comput. Oper. Res.* **2007**, *34*, 2758–2773. [CrossRef]
22. SeyedEsfahani, M.M.; Biazaran, M.; Gharakhani, M. A game theoretic approach to coordinate pricing and vertical co-op advertising in manufacturer-retailer supply chains. *Eur. J. Oper. Res.* **2011**, *211*, 263–273. [CrossRef]
23. Zhao, Y.; Wang, S.; Cheng, T.E.; Yang, X.; Huang, Z. Coordination of supply chains by option contracts: A cooperative game theory approach. *Eur. J. Oper. Res.* **2010**, *207*, 668–675. [CrossRef]
24. Leng, M.; Parlar, M. Game-theoretic analyses of decentralized assembly supply chains: Non-cooperative equilibriavs coordination with cost-sharing contracts. *Eur. J. Oper. Res.* **2010**, *204*, 96–104. [CrossRef]
25. Esmaeili, M.; Aryanezhad, M.B.; Zeephongsekul, P. A game theory approach in seller-buyer supply chain. *Eur. J. Oper. Res.* **2009**, *195*, 442–448. [CrossRef]
26. Ren, Y.; He, Z.; Luning, P.A. Luning, Performance of food safety management systems of Chinese food business operators in Tianjin. *Food Control* **2022**, *138*, 108980. [CrossRef]
27. Trienekens, J.; Zuurbier, P. Quality and safety standards in the food industry, developments and challenges. *Int. J. Prod. Econ.* **2008**, *113*, 107–122. [CrossRef]
28. Yao, G.X.; Xu, J. Agricultural supply chain decisions research under random yield. *Ind. Eng. Manag.* **2015**, *20*, 16–22.
29. Aftab, S.; Yaseen, M.R.; Anwar, S. Impact of rising food prices on consumer welfare in the most populous countries of South Asia. *Int. J. Social Econ.* **2017**, *44*, 1062–1077. [CrossRef]
30. Jiang, S.; Li, S. Green Supply Chain Game Models and Revenue Sharing Contract with Product Green Degree. *Chin. J. Manag. Sci.* **2015**, *23*, 169–176.
31. Yang, Y.; Yao, G. Fresh agricultural products supply chain coordination considering consumers' Dual preferences under carbon CAP-and-Trade mechanism. *J. Ind. Manag. Optim.* 2022. [CrossRef]

 foods MDPI

Article

"Food Village": An Innovative Alternative Food Network Based on Human Scale Development Economic Model

Giordano Stella [1], Biancamaria Torquati [1], Chiara Paffarini [1,*], Giorgia Giordani [1], Lucio Cecchini [1] and Roberto Poletti [2]

1 Department of Agricultural, Food and Environmental Sciences, University of Perugia, Borgo XX Giugno 74, 06121 Perugia, Italy; stellagiordano@libero.it (G.S.); bianca.torquati@unipg.it (B.T.); giorgia.giordani@studenti.unipg.it (G.G.); luciocecchini89@gmail.com (L.C.)
2 Sereco-Biotest Studi e Ricerche Ambientali, Via C. Balbo 7, 06121 Perugia, Italy; r.poletti@serecobiotest.it
* Correspondence: chiara.paffarini@unipg.it

Abstract: Although the different alternative food networks (AFNs) have experienced increases worldwide for the last thirty years, they are still unable to provide an alternative capable of spreading on a large scale. They in fact remain niche experiments due to some limitations on their structure and governance. Thus, this study proposes and applies a design method to build a new sustainable food supply chain model capable of realizing a "jumping scale". Based on the theoretical and value framework of the Civil Economy (CE), the Economy for the Common Good (ECG), and the Development on a Human Scale (H-SD), the proposed design model aims to satisfy the needs of all stakeholders in the supply chain. Max-Neef's Needs Matrix and Design Thinking (DT) tools were used to develop the design model. Applying the design method to the food chain has allowed us to develop the concept of the "Food Village", an innovative food supply network far from the current economic mechanisms and based on the community and eco-sustainability.

Keywords: alternative food networks (AFNs); human needs; food sovereignty; Civil Economy (CE); Economy for the Common Good (ECG)

Citation: Stella, G.; Torquati, B.; Paffarini, C.; Giordani, G.; Cecchini, L.; Poletti, R. "Food Village": An Innovative Alternative Food Network Based on Human Scale Development Economic Model. *Foods* **2022**, *11*, 1447. https://doi.org/10.3390/foods11101447

Academic Editor: Theodoros Varzakas

Received: 30 March 2022
Accepted: 14 May 2022
Published: 17 May 2022

1. Introduction

Overall, the food produced could feed about 10 billion people globally. However, the food system fails to achieve a fair allocation of resources [1]. Nearly 1.3 billion tons of food (about one-third of total production) are wasted every year, and approximately 56% of waste occurs in industrialised countries. However, they use better agricultural, transport, and conservation techniques [2].

In 2019, almost 690 million people were undernourished worldwide [3]. At the same time, by 2025, the prevalence of global obesity is predicted to reach an estimated 257 million adults, showing a rapid increase from 202 million in 2016 [4].

The agro-industrial system has enormous environmental impacts in biodiversity reduction, land degradation, aquifers' pollution, greenhouse gas emissions, and air pollution from the transport of foodstuffs, among others [5]. Moreover, the current food system has produced considerable social costs for small producers worldwide (product sales price lower than the production price, reduction in income, increased phenomena of failure, marginalisation, and loss of self-esteem, land abandonment, and migration to cities and other countries [6].

Previously, the global food problem was addressed, promoting agro-industry, free market, and the production methods of the so-called "Green Revolution" (use of pesticides, chemical fertilisers, mechanisation, genetically modified organisms, etc.) by the major international organisations [7].

The large-scale distribution represents in the industrialized world the way in which the agro-food system organizes the consumers' purchase of food. It is highly concentrated spatially and structurally and it is characterized by high levels of production and by long-distance import and export [8]; moreover, economic globalization and relatively cheap energy influence it [9]. This causes several negative environmental and social externalities [10].

However, recently, a change towards greater sustainability attributed to greater consumer sensitivity has been occurring towards socio-environmental issues [11,12] and numerous activist associations' actions have been implemented [13], such as those of the international association of producers "La Via Campesina" (LVC). LVC states that, to achieve food security, "Food Sovereignty", which is defined as "the right of people to healthy and culturally appropriate food produced through ecologically sound and sustainable methods, and their right to define their food and agriculture systems. It puts the aspirations and needs of those who produce, distribute and consume food at the heart of food systems and policies rather than the demands of markets and corporations", [14] must be promoted.

Food Sovereignty was analysed in terms of its history, meaning, and local experience application by several authors [15–17].

According to LVC (See http://www.viacampesina.org for background) (accessed on 1 October 2021), only through a change in the economic paradigm, capable of guaranteeing democratic, eco-sustainable exchange processes based on respect for others, can food security be achieved [14]. In this perspective, the direct participation of all stakeholders in the governance of the "common good" is central to an organized political level; this direct participation is also crucial to community processes emerging by an approach from the low earners.

Some authors have highlighted the importance of participatory democracy techniques for the construction of policies [18,19] and the role of food territorial planning [20,21] in building resilience, nourishing, and promoting rural development. These tools are important to facilitate territorial planning by the public decision makers and the destination of economic resources for local development.

In the last three decades, within the discussions on emerging new forms of food supply and distribution that involve different actors, first of all, producers, and consumers, the Alternative Food Networks (AFNs) represent those that are more spread and analysed [22].

AFNs are food supply chains based on a community mould. In social relationality, eco-sustainable agricultural practices, public health, and social equity are promoted [23]. However, although the current AFNs are experiments of the food system towards a new economic paradigm, the current AFNs have remained, over time, a limited phenomenon.

Some authors [24–26] have stated that AFNs do not make a change in scale and do not produce a transition in the agri-food system due to the strong structuring of the food system, their relatively recent birth, and their bottom-up approach, remaining mostly niche initiatives. They affirmed that a way to facilitate the move beyond the niche is to focus on the community-orientated schemes' role.

Agreeing with these authors, we argue that the reason AFNs have remained a niche phenomenon is that they fail to satisfy the needs of all stakeholders.

At the same time, the large-scale distribution creates environmental, social, and economic negative externalities.

From this perspective, both AFNs and large-scale distribution can be considered inefficient economic processes.

The aim of this paper is to propose an innovative AFN that can satisfy stakeholders' needs involved in the food chain.

Specifically, to create a new eco-sustainable agri-food chain based on collective well-being and social equity, which promotes food security and can be spread on a large scale, in our opinion, it is first required to identify an adequate theoretical economic framework.

Therefore, it became necessary to analyse the needs of all stakeholders involved in the food chain and the intimate connection between them. Aiming to design the economic processes, a theoretical framework based on the principles of Civil Economy (CE) [27,28], the Economy for the Common Good (ECG) [29,30], and the theory of Human Scale Development (H-SD) by Max-Neef [31] was adopted. All of these heterodox models promote a sustainable vision that is the base of both all AFNs and the new food network chain we want to propose.

Aiming to enlarge the scale of impact of AFNs, we referred in particular to Max-Neef's [31] model that theorizes that the economy has to respond to all needs of stakeholders.

All these considerations and methodological approaches were used to develop the "concept" of an innovative food network: "The Food Village".

The paper is organized in the following manner. In the Section 1.1 subparagraph we present the concepts of AFNs and the relevant literature on the topic, while in the Section 1.2 we compare four AFNs examples. In Section 2 "Materials and Methods", firstly, in Section 2.1 subparagraph, we describe the economic models underlying the new food supply chain model. In the Section 2.2 subparagraph, we present Max-Neef's matrix for needs analysis, while in the Section 2.3 subparagraph we describe a design method to conceive the innovative AFN for a new economic paradigm through Max-Neef's matrix. In the Section 2.4 subparagraph we describe the application of the proposed design process to develop the new food supply chain. In Section 3 the concept of the innovative food network that emerged from the designing process is discussed. A short discussion follows.

1.1. Alternative Food Networks: Past and Actuality

Following the first food scandals linked to large-scale distribution, and since the 1980s, there has been a radical change in consumer demand, the so-called "food turning-point" [32], that has caused greater importance to be attached to the transparency of the production processes [33]. Meanwhile, the first environmental movements, ambitious to overturn the modernisation paradigm in the food sector, also contributed to this change.

Considering this background, a new way of thinking about the supply chain has begun taking shape. The organic system was the substratum on which national and international networks of producers and consumers were created. However, farmers have begun revealing the economic unsustainability of the large-scale distribution system and agriculture productivity. The loss of power along the supply chain and large-scale production has been determined to crush small- and medium-sized farms [34].

By adopting more efficient production techniques and promoting high-quality products, the farmers redeemed their position within the agri-food system; this process, together with consumers' contribution in search of more sustainable food supply chains, has promoted the development of AFNs [35]. Among several definitions, Maye and Kirwan [36] (p. 1) described AFNs as "organised flows of food products connecting people (consumers) who care about the moral aspect behind their consumption practices. These people meet those (producers) who want a fair price for the way they produce food, far from the dominant (or conventional) logic of the market". Jarosz [37] (p. 1) affirmed that "alternative food networks represent efforts to re-spatialize and re-socialize food production, distribution and consumption in North America, Europe and Australia".

Terms such as AFNs and "Short Food Supply Chains" (SFSCs) are often used indiscriminately in such a way that the reduction in commercial nodes is the main feature of AFNs [38]. Instead, the "localisation" is given by an assortment of factors, and it is reductive to stop at the spatial conception alone. If taken individually as a criterion for evaluating the location of a given supply chain, the geographical configuration varies from radically local to radically global chains, with an infinity of intermediate cases in between. Therefore, for an exhaustive localisation analysis, several factors such as the product's identity (typicality, processing, tradition), management organisation of the supply chain, and technologies used [39] have to be considered.

From another perspective, Watts et al. [40] argued that AFNs are distinct from conventional supply chains based on their commitment and potential subordination to global chains (i.e., those supply chains that operate in a global neoliberal policy).

Another central aspect that outlines the boundary between conventional and alternative systems is the involvement of the consumer and the level of relationality established in the exchanges between the players of the supply chain [23]. Opitz et al. [41] stated that the interaction with producers is one reason that encourages consumers to choose AFNs.

Ilbery and Maye [42] affirmed that the boundary between conventional systems and AFNs is not clearly defined: neither operates completely autonomously and differently due to the economic motives pushing the alternative producers to operate in both systems. This is an increasingly widespread phenomenon of hybridisation and "conventionalisation" [43] within alternative supply chains [44,45]. "Conventionalisation" refers to the contamination of alternative supply chains, which take on some of the characteristics of conventional supply chains, from which they originally wanted to go away. This can also occur following the attempt of "conventional players" to expand their market by including some of the characteristics of alternative food supply chains.

Le Velly [46] recognised the AFNs' promise of diversity that a different organisation of the supply chain components should distribute benefits among producers, consumers, regions, and the environment. The difference is the characteristic that triggers the specific rules' definition interconnected with conventional rules. Therefore, Le Velly [46] proposed to address the question "from the perspective of the organisational innovation processes activated" (p. 9). These innovation processes could be implemented by adopting specific "alternative rules" that are new ways of relating between the producer and consumer as well as new methods of production, transport, and different contracts, among others. However, some rules adopted from conventional supply chain models are not excluded, such as the infrastructure and knowledge of wholesalers. Similar to others, Le Velly [46] also regarded AFNs as ongoing, both emergent and making, rather than already shaped systemic entities.

Even if the AFNs represent a concrete proposal for transition, the discussion regarding both the maximisation of the potential of these initiatives to spread their social, ecological, and economic innovations to transform food systems, and avoiding the erosion of their authenticity is open and animate [22,25,26].

Indeed, Rossi [47] asserts that AFNs' experiences are at a crucial point in their existence, also due to their growing interest in the demand side as well as the production; in fact, these experiences "on the one hand are both consolidating around their elements of alterity to the conventional food chains and, on the other, they are facing the challenge of growth and the interaction with the mainstream system" [22] (p. 4). Analysing five AFNs focused on community support agriculture, Rossi [47] argued whether the increase in AFNs could represent a way to enlarge the availability and affordability of the products by expanding the consumers' access to these initiatives; equally there is the issue of conventionalisation.

In the next subsection, we will analyse four examples of AFNs in order to highlight their characteristics and to clarify their differences to understand the mechanisms that can facilitate or hinder the change in scale of AFNs.

1.2. A Comparison of Four Relevant AFNs

The four relevant AFNs analysed are: Italian Solidarity Purchase Groups (SPGs), the Organised Group of Supply and Demand (OGSD, GODO in Italian), the Community Supported Agriculture (CSA), and the Food Coop Park Slope (FCPS) model. Highlighting their potential and limitations has contributed to developing the "concept" of the new innovative food network proposed in this paper.

Born in Italy in the mid-1990s, the SPGs are a collective food supply practice including consumers who cooperate by buying food products or common goods directly from local producers at a fair price for both parties. The group participants first define a list of products that they collectively intend to purchase. Based on this list, the different persons compile orders collected to define a group order, transmitted to the producer (almost always organic). Finally, goods delivered are divided among the group members, and each one pays for his share [48].

The OGSD is a particular SPG where producers and consumers are associated with the Italian Association of Organic Agriculture (AIAB in Italian) to encourage matching the demand and supply of local organic products. It promotes responsible consumption based on seasonality, producer visibility, and product exchange without intermediation [49]. The OGSD facilitates purchases from member farms, manages product deliveries, provides information on the organoleptic and nutritional qualities of products, and promotes visits to member farms and training on organic farming [49].

The CSA is a community that is committed to supporting agricultural activities by sharing the risks and benefits of production with the farmer. The community co-designs the production and purchases a share of the production before each growing season. Hence, the farmer receives working capital in advance, thus obtaining greater financial security and better prices. Depending on their contribution, in return, the members receive regular farm products throughout the season [50].

The FCPS model is inspired by one of the oldest consumer food cooperatives in the United States, born in 1973 in New York. Its goal is to be a purchasing agent for its members, the only ones who can shop food and household items in the store. To have the possibility to buy into the selling point, all of them contribute with 2 h and 45 min of work every four weeks to the Food Coop. The FCPS focuses on sustainability and prefers selling environmentally sustainable products. Usually, the mark-up is only 21% compared to the wholesale price (26–100% in large-scale distribution) (See https://www.grubstreet.com/2018/04/history-of-the-park-slope-food-coop.html) (accessed on 10 November 2021). Additionally, the members' work covers about 75% of the marketing costs associated with selling point employees (See http://foodcoop.film/) (accessed on 25 November 2021). Therefore, the Food Coop can be competitive and offer higher-quality products compared to large-scale distribution at lower prices. This model has also spread to Europe, where numerous cooperative supermarkets inspired by the FCPS experience have sprung up [51]. It has yielded significant results in creating social aggregation, a sense of community and solidarity, and promoting a fair and environmentally sustainable food supply.

As shown before, the literature on AFNs is vast and over time several authors have focused on different aspects. Recently, at the international level, there has been an open and growing debate on the social assumptions and on the characteristics of the economic processes necessary for the structuring, affirmation, and change in scale of the AFNs [52–54], among others.

Mount [52] wondered about the effects of the change in scale on the structure of AFNs, how this could affect the values that characterize them and the effectiveness with which they are able to translate them into coherent economic processes.

Wald, Hill [53] affirmed that reflecting on the scale concept helps both to perceive the development and the spread of food systems, and how certain alternative food system models could realize a "jumping scale".

Using a multi-actor perspective framework, Poças Ribeiro et al. [54] explored the limiting and facilitating factors impacting the emergence and consolidation of different types of AFNs in three different countries. They underlined the fundamental role of organizers concerning the development of AFNs and that, at the same time, a wide scope of actions by governmental and non-governmental stakeholders supporting the development of more AFNs are needed.

We considered the characteristics of the four AFNs in order to highlight both the elements that hinder their change in scale and the elements that could facilitate the construction of a new AFN model.

These four AFNs are characterised by being governed by a "sharing economy" system. Business models based on sharing can be very different. Still, social well-being issues and the positive effects on the sustainable use of goods are a priority for all. A comparison to highlight their features was made concerning five key components: (i) the localisation (or proximity); (ii) the consumer involvement degree; (iii) the reasons for joining; (iv) the effects on sustainability; (v) the limits (Table 1).

Table 1. Comparison between the AFNs.

		SPG	OGSD	CSA	FCPS
Localisation/proximity		**High**: It brings together local producers with consumers aiming to create a proximity exchange channel.	**Medium–High**: It is very important but not exclusive because OGSD supplies organic products exclusively.	**High**: Localisation is not an essential element but is strongly recommended.	**Medium**: The product's prevalence comes from a distance of 200–500 km and other continents at the members' wishes.
Consumer involvement		**High**: Consumers generally manage the organisational part of the group	**Medium–Low**: The products ordered are collected at distribution points or delivered home, so direct contact with the producers could be marginal.	**Medium–High**: Consumers support production before receiving products. Sometimes, they contribute to manual labour helping the producers.	**Very high**: Consumers work in it, decide which products should be selling and participate in assemblies by the directors' board.
Reasons for joining		High product quality; ethical–moral values; trust in producers; social interactions; socio-political values; support to producers; ecological sustainability; fair price. Acquisition of new knowledge is not strictly a reason for joining.	High product quality; ethical–moral values; trust in producers; social interactions; socio-political values; support to producers; ecological sustainability; fair price. Acquisition of new knowledge is not strictly a reason for joining.	High products quality; ethical–moral values; trust in producers; socio-political values; support to producers; fair price; ecological sustainability; acquisition of new knowledge. Social interaction is not strictly a reason for joining.	High product quality; socio-political values; social interactions; support to producers; fair price; acquisition of new knowledge. Ethical–moral values, trust in producers, ecological sustainability are not strictly reasons for joining.
Effects of sustainability	Healthy Eating:	Consumers are very interested in healthy eating and nutrition education.	It promotes organic agriculture and the critical consumption of healthy foods.	It usually promotes organic agriculture.	It arises from the consumers' need to find healthy foods in big cities.
	Use of natural resources:	Less impact of transport due to the high proximity supply–demand. Logistic system streamlined by governance. Low-input agriculture supported.	Less impact of transport due to the middle–high proximity supply–demand. Logistic system streamlined by governance. Organic agriculture supported.	Food waste can occur due to a wrong consumers' estimate of the product they need in advance. Producers are often organic.	Imported products from other continents, if members request, could determine an environmental impact.
Limits		A limited number of families can join. Limited temporal and logistic accessibility. Reduced product variability. The possible high involvement from all members could discourage many from joining.	Limited temporal and logistic accessibility. Reduced product variability. A limited number of consumers supplied by local producers	Products ordered in advance may not be satisfactory at the time of delivery, both qualitatively and quantitatively. Limited temporal and logistic accessibility. The possible high involvement from all members could discourage many from joining. Food waste could occur.	The high involvement from all members could discourage many from joining. A consolidated system requests a large number of adherents willing to collaborate periodically over the years.

Source: Authors own elaboration.

Localisation or proximity is a discriminating aspect between the different forms of AFNs. In SPGs, the proximity between producer and consumer is high, as they collaborate to enhance local production, favouring organic or sustainable ones. In the OGSD, localisation is important (up to 80% in some cases) but not exclusive; supplying exclusively organic products often also involves farms located throughout the country. For CSAs and Food Coops, the proximity between supply and demand is not determined. In both cases, producers and consumers may never come into direct contact, and the mutual trust is based on the sharing of some fundamental values, such as product quality and fair prices. Despite this, in CSAs, proximity between producer and consumer is strongly recommended and it is usually the norm; sometimes, consumers support the production through manual labour.

Consumer involvement in SPGs is a fundamental part of the group: they voluntarily manage orders, deliveries, and quality control, among others, and the relationship between producer and consumer is constant and without intermediaries. Instead, in the OGSD, the consumer–producer collaboration either does not occur or occurs partially: consumers decide to subscribe to a specific organised group and support the producers through a membership fee. Sometimes, consumers can provide voluntary help at the headquarters of the local OGSD quarter. The supply–demand interaction occurs mainly through web platforms: orders are placed periodically, and then consumers can collect their shopping at the designated logistics points.

In CSAs, the producer is economically supported by the consumer; usually, at the beginning of the production year, they meet to co-plan the productions, and sometimes, the consumers contribute to manual labour helping the producers.

In Food Coops, the consumers' involvement is active and high because they have to work within the coop to be able to buy from the selling point; they also decide which products they should be selling and participate in assemblies with the directors' board.

In the SPGs, OGSD, and CSAs, the quality of the product, attention to the environment, and ethical–moral values are priority aspects for the members [49,55]. Specifically, the trust in the producer within the SPGs is what distinguishes the relationships before the support. Instead, in OGSD, the trust and support to the producer are based on the membership fee that each member–consumer pays to become part of it. In Food Coops and, particularly in Park Slopes, the desire to eat quality food and the possibility of being able to decide the provenience of the food supply unite people.

We focused on nutrition and environmental impact regarding the effects of the AFNs on consumers. SPGs, OGSD, and CSA, promote the consumption of local and organic foods. In the FCPS, healthy eating is the reason that led to its creation due to the difficulty in finding good quality foods in New York; this is why an assembly decision-making system open to all members was set up so that all the goods represent the will of the consumers. The products are preferably, but not exclusively, local, organic, vegan, and non-GMO, among others.

Concerning the impact on the environment, a critical aspect of CSA is that the consumer often overestimates their needs during the advance order, leading to food waste. Additionally, selling non-local producers' goods, the FCPS model has a bigger impact on the environment because it uses more long-range transport with respect to the other AFNs.

Regarding the limits, the SPGs satisfy a restricted number of families' demands: generally, no more than 50 families per group based on the territoriality feature of the productions. Therefore, if the requests exceed this threshold, a spin-off or a new group is arranged. Moreover, it is characterized by reduced product variability.

In OGSD, consumers can access a more varied product portfolio but still lower than the large-scale distribution.

In CSAs, inconsistencies between the ordered product and the one received often occur both in quality and quantity.

In SPGs, OGSD, and CSAs, the temporal and logistic accessibility limits consumers because the products could be withdrawn only during limited times of the week.

In SPGs and CSAs, the possible high involvement from all members could discourage many from joining; in the FCPS model, the high degree of involvement and work required could discourage many from joining.

From this analysis, the four AFNs could not make a change in scale and produce a transition in the agri-food system due to some constitutive limits. In particular, we have individuated some issues, including the failure to respond fully to consumer's demand in terms of the assortment of food products, consumers requiring a medium–high involvement, the reduced time slots for accessibility to purchase, and uncompetitive prices.

It is necessary to develop an innovative model responding to all stakeholders involved in the supply chain, consumers, producers, and all operators, to surpass these critical issues and, thus, create a sustainable supply chain on a larger scale. Therefore, it is necessary to understand the economic principles and models that can form this innovative food chain structure. The next section reports the economic models that align with the principles and values underlying a new sustainable AFN.

2. Materials and Methods

2.1. The Economic Models Underlying the New Food Supply Chain Model

AFNs are characterised by human relatedness to distance themselves from the traditional supply chain model to accomplish the common good and environmental sensitivity.

To achieve these objectives on a large scale, we hypothesise that an innovative food supply chain based on an economic framework consistent with human values and ecology must be designed. Therefore, it seems necessary to extend Le Valley's "promise of diversity" [46] to the epistemology of the economic model. Following this perspective, the "diversity" of the innovative supply chain model is based on the change in the economic paradigm. From this, the "diversity" of production, organisational, and governance systems can be ideated and designed in a coherent approach.

Therefore, three alternative economic models have been identified, together constituting the economic paradigm of the "Food Village": the CE [27], ECG [29], and Economics of H-SD [31]. Starting from being different approaches, these models differ from the mainstream economic model because they focus on pursuing collective well-being and essential human needs. People are considered human beings rather than economic agents and are part of a single organism made up of the human community and the environment. From this perspective, the economy aims to achieve public happiness and to reach this result. Therefore, it must satisfy all essential human needs.

Born in Italy in the eighteenth century, the CE states that human beings act in terms of the market, friendship, reciprocity, gratuity, and fraternity in their economic actions. Hence, the market actor is seen as a person, not simply an individual, but a friendly economic agent. Public virtues replace the private interests of the political economy. In these terms, from Smith's "invisible hand" that considered the market regulated based on the individual's interests, it moves on to the "visible fabric" or to the civic virtues that each individual uses in the moment of exchange. Every economic operator should be endowed with moral qualities that can make a difference and enrich everyone. CE is based on five principles: (1) reciprocity, which makes the exchange personal and meaningful; (2) fraternity, which fosters diversity (cultural, religious, ethnic, etc.) and makes them compatible; (3) gratuity, open to others and treats them with respect, in reciprocity; (4) public happiness, arising from ethics and virtues as well as the common good and being the goal of society and the economy; (5) the plurality of economic actors, involving both public and private and profit and non-profit business, and overcoming the state–market duopoly, thereby making a more democratic economic system [27].

ECG is considered able to follow the necessary sustainable transformations on an economic, political, and social level across Europe (In 2016 an Opinion of the European Economic and Social Committee [56] was published concerning ECG where it was defined as a lever to the "transition towards a European Ethical Market which will foster social innovation, boost the employment rate, and benefit the environment".) and worldwide. Even though the ECG is based on several disciplinary approaches derived from ethics, ecology, political science, social psychology, neurobiology, pedagogy, discussed over a long time (since 2010), ECG has been developed as a practical economic model. Therefore, it needs to be explored in the study of practices and developed clearly and in a structured way. Recently, Dolderer et al. [30], based on Felber [57], defined the "Common Good Economics" as "the science of satisfying the needs of the present and future human generations, in alignment with democratic values and ecological planetary boundaries" [30] (p. 7). The Common Good is well-thought-out in its highest and broadest significance [57], and the ability to achieve it results from economic success. Practically, the ECG has implemented the Common Good Balance Sheet (CGBS). It is a scorecard that measures how much public, private, or third sector activity contributes to the Common Good, based on the preservation of five fundamental values: human dignity, cooperation and solidarity (which count as one), ecological sustainability, social justice, and democratic co-determination and transparency [58]. The CGBS comprises 17 indicators that emerged from a matrix, intersected by the five fundamental values with the stakeholders of the analysed activity: suppliers, lenders, employees and holders, customers, partner companies, and the social and civil context, understood as territory, population, future generations, other human beings, and nature globally [59]. The CGBS is compiled directly to measure the contribution of the economic activity to the Common Good, verified and certified by external auditing: the more the activities are structured socially, ecologically, democratically, and jointly, the higher the score. Companies adopting the CGBS will be encouraged through tax exemption and easier access to public contracts and funds.

The Economics of H-SD states that the economic system has to respond to human needs. Unlike Maslow [60], Max-Neef [61] argued that it was not possible to affix a hierarchy to human needs and classified them as existential and axiological [62]. Existential needs are distinguished according to the dimensions of "being", "having", "doing", and "interacting". In contrast, the axiological needs are distinguished in subsistence, protection, affection, knowledge, participation, creativity, identity, freedom, and free time. Surpassing the existential and axiological needs, Max-Neef developed a matrix of needs. Through the matrix, it is possible to identify the satisfiers of human needs (Table 2), which represent ways of fulfilling individual needs.

Table 2. Matrix of Needs and Satisfiers.

	Being	**Having**	**Doing**	**Interacting**
Subsistence	Health, Adaptability, Sense of humour	Food, Shelter, Work	Feed, Procreate, Rest, Work	Social setting, Environment
Protection	Care, Equilibrium, Solidarity	Rights, Social security, Family	Cooperate, Plan, Help	Living space, Dwelling
Affection	Self-esteem, Respect, Passion	Friendship, Family, Relation with nature	Make love, Share, Cultivate, Appreciate	Privacy, Intimacy, Home, Togetherness
Understanding	Critical conscience, Curiosity, Discipline	Literature, Education, Teachers	Investigate meditate experiment	Groups, Community, Schools, Family
Participation	Dedication, Respect, Receptiveness	Rights, Responsibility duties, Work	Cooperate, Dissent, Agree on, Interact	Associations, Churches, Family
Idleness	Curiosity, Tranquillity, Imagination	Peace of mind, Games, Parties	Day-dream, Relax, Remember, Brood	Privacy, Intimacy, Free time, Landscape
Creation	Passion, Intuition, Imagination	Abilities, Skills, Method, Work	Work, Invent, Build, Compose, Design	Productive settings, Workshops, Time
Identity	Sense of belonging, Self-esteem	Language, Symbols, Religion, Values	Commit oneself, Grow, Recognise	Social rhythms, Maturation stages
Freedom	Autonomy, Boldness, Passion	Equal rights	Dissent, Choose, Disobey, Run risks	Temporal/spatial plasticity

Source: Max-Neef et al. [61] (p. 33).

For example, the need for knowledge can be satisfied through literature, a satisfier (way), while the good (means) used for this purpose is—potentially—a book. A satisfier could also be represented by interacting with new people and discovering new places, among others. Satisfiers are characterised by intangibility, as these represent how society approaches a need, while materiality characterises goods.

The H-SD Economy departs from the ideology behind the current economic model, where needs are met through material goods and services, seen more as an end than a means. Instead, the introduction of the satisfiers uses a less materialistic approach. Unlike needs, which remain the same for Max-Neef even between different historical periods and cultures, the satisfiers are influenced and modified by several factors. These are the organisational structure of a certain society, political system, and social practices. Moreover, they vary for each individual according to subjective attitudes, such as the character or the ethical and moral values. Ultimately, the relationship between human needs, satisfiers, and material goods is concretised in the Max-Neef matrix.

The Max-Neef matrix is an integral part of the design method that we propose in this study and, therefore, will be more explored in the next subparagraph.

2.2. Max-Neef's Matrix for Needs Analysis

Max-Neef's matrix promotes H-SD, a notion based on satisfying basic human needs and increasing self-confidence levels. With this perspective, it is possible to construct optimal synergies between the human being and environment, man and technologies, global processes and local activities, personal interests and social interests, participatory planning and private initiative, and civil society and the state [31].

The economic paradigm of H-SD raises the quality of life through a holistic understanding of people's needs. Thus, each need is fulfilled by different sets of satisfiers that include all things that contribute to the satisfaction and well-being of the individual or the collective. In particular, the satisfiers corresponding to the existential needs of "being" refer to the individual or collective attributes, expressed with nouns, which refer to aptitudes—or particular inclinations—expressions of character, and personal values; the satisfiers corresponding to the needs of "having" represent the tools to contribute to these attributes. These tools can be norms, relational mechanisms, attitudes, or information. The satisfiers of "doing" are actions, individual or collective, expressed as verbs. These actions are aligned with the satisfiers of "being" and "having". Finally, regarding the satisfiers corresponding to the need for "interacting", places are used, in the spatial and temporal sense, in which people relate to and self-determine. Their articulation of the satisfiers is an essential element for realising total well-being, both individual and collective. If even one component of the matrix fails, the quality of life would inevitably be affected. Therefore, well-being arises at the level of the entire system once the right complementarities between the different dimensions are met: The well-being is not reduced to the accumulation of goods and services. The satisfiers are not exchanged and obtained through the market [63]; some have no exchange value and are neither exchanged nor exchangeable.

It should be remembered that the satisfiers do not respond univocally to needs. In the H-SD economic model, the needs are correlated, and a single satisfier can satisfy more than one. Conversely, a need may require more than one satisfier to be satisfied [31].

2.3. Designing Innovative AFN through the Max-Neef's Matrix

Based on Max-Neef's matrix, we defined a design method that we used to develop and to propose the innovative AFN "Food Village", which could potentially respond to all stakeholder's needs involved in the agri-food chain.

We based the design method starting from the Design Thinking (DT) approach. Today, DT represents a methodology of action that guides transformation, evolution, and innovation within the most varied economic sectors. DT was born to understand and identify spontaneous mental strategies of the designer, leading to the construction of a project [64].

Specifically, considering DT based on the designer's ability to integrate human needs, the available material and technical resources, and a project's constraints and opportunities, it is required that the designer is simultaneously analytical and emphatic, rational and emotional, methodical, and intuitive, oriented by plans and constraints, but spontaneous [65]. Some authors [66,67] call this attitude "abductive thinking", from the idea of Peirce [68] that affirmed that the deduction or induction could not elaborate a new idea using data acquired in the past. In this sense, "abductive thinking" considers feelings and emotions as important as rationality. In DT, "abductive thinking" is related to the "perceptive cognition", defined as "basic skill in the creation of new realities and artefacts" [64] (p. 3). In DT, "perceptual cognition" (feelings and emotions) and rationality are complementary to give full development to the "abductive reasoning", that is, to thought with no anchorage regarding past data and experiences. According to Tschimmel [69], this premise is the basis of DT.

Stickdorn and Schneider [70] argued that the first step in developing a DT model is the design of the process itself; it changes according to the context in which the good or service is created. Therefore, it differs from project to project [64]. This vision corresponds to "constructivism", the project's success depending on the social actors and interaction environment of constructivism.

As mentioned before, we decided to base the construction of the design process on the Max-Neef matrix in order to identify the needs of stakeholders in the economic process.

Moreover, aiming to display the results of this work, DT techniques were considered; in particular, both the principles of abductive reasoning and storytelling, a communication and dissemination tool, were used.

The method arranged can be summarised in the following stages represented in Figure 1.

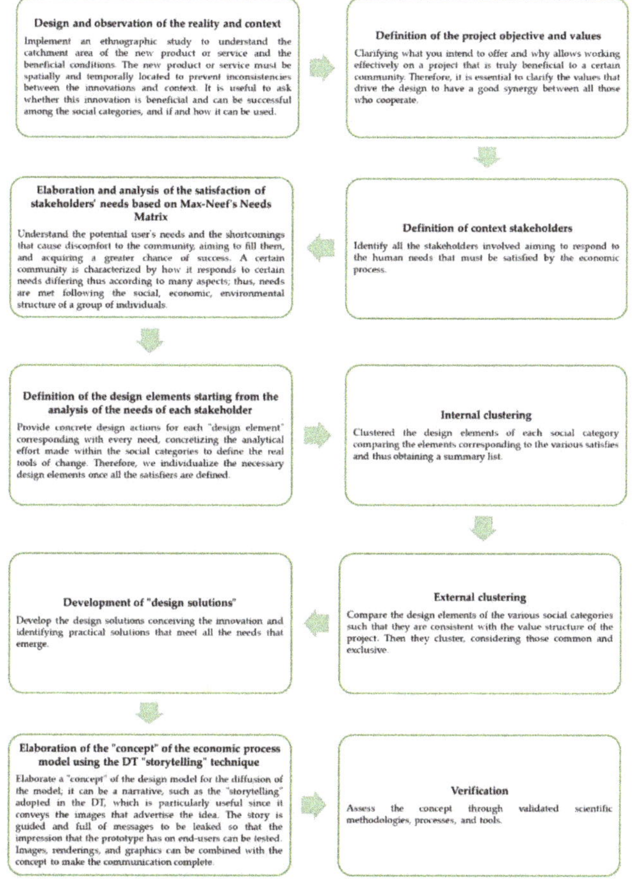

Figure 1. Design method.

2.4. A New Food Supply Chain Model for a New Economic Paradigm

The method presented previously aimed to define rational modus operandi for developing the new supply chain model that respects social, economic, and environmental sustainability. The various phases and a brief explanation of each are reported below. Furthermore, the matrices of the needs of five social categories (consumers, producers, owners or financial partners, employees, collectivity) are designed. Supplementary Materials reports the matrix of consumers (Table S1), producers (Table S2), holders (Table S3), employees (Table S4), and collectivity (Table S5) and the clustered design elements (Table S6).

The application of the design method is described as follows:

1. Definition and observation of the reality and context. As previously presented, a literature review on AFNs was conducted, clarifying the difference between them to understand the mechanisms that can facilitate or hinder the change in scale within the value chain.

2. Definition of the project objective and values. The new supply chain model represents a virtuous example of social, economic, and environmental sustainability through good practices and a specific value heritage. The values of H-SD, CE, and the Common Good are shared, such as those of reciprocity, fraternity, gratuitousness, happiness, and human dignity, solidarity, and social justice, environmental sustainability, transparency, and co-determination.

3. Definition of context stakeholders. This definition was partly made by referring to a Manual for drafting a report on the Common Good Sheet (https://www.ecogood.org/) (accessed on 2 December 2021) and applied to the following five categories:

- Customers or consumers: the end-users of the goods or services provided by the supply chain.
- Producers or suppliers: subjects who sell their products through the supply chain channel.
- Holders or financial partners: those who make their own or third-party capital available. Financial service providers also belong to this category, companies that deal with transactions, insurance, and asset or financial advice.
- Employees or collaborators as active operators within the supply chain space: people who perform duties in the sales or purchase point, seasonal and non-seasonal agricultural workers, and all those who provide their service in one of the production process phases.
- Collectivity or social context, all groups that indirectly experience entrepreneurial actions while focusing on residents close to the supply chain and potentially critical NGOs.

4. Elaboration and analysis of the satisfaction of stakeholders' needs based on Max Neef's Needs Matrix. The needs of the five categories chosen were classified into existential and axiological [62]. Therefore, the five matrices report on the abscissa the need for being, having, doing, interacting, and on the ordinate, subsistence, protection, affection, understanding, participation, leisure, creativity, identity, freedom, and spirituality. The latter was added as essential for achieving complete well-being: spirituality can be understood as how human beings experience transcendence or connection to a higher system or power [71]. This sense of connectedness may or may not have religious affiliations. According to Max-Neef [31], the satisfiers must be declined according to precise criteria and the existential and axiological needs they respond to. Hence, these can take different forms, such as an attribute, noun, verb, or condition. The satisfiers were also identified by considering the degree to which they improve or inhibit individuals' well-being concerning themselves, the community, or the environment [61]. Finally, their complementarity was considered to achieve complete well-being. The satisfiers' classification based on how each affects the different dimensions of well-being has made it possible to highlight how the freedom and autonomy of individuals can be significantly influenced by changing or shaping certain social and economic mechanisms. The satisfiers are reported in the first four columns of Tables S1–S5 in the Supplementary Materials.

5. Definition of the design elements starting from the analysis of the needs of each stakeholder. This phase represents the real innovative contribution to the Max-Neef method in studying the needs of a certain community. Once the satisfiers had been identified, they were matched with concrete actions or tangible tools for satisfying needs. These were named "design elements", representing the true answers of the applied analysis system. The design elements can be found in the last column of Tables S1–S5 in the Supplementary Materials.

6. Internal clustering. The clustering work allows a clearer view of the design elements for each category, thus having an operational purpose. In this study, the design elements of each social category are grouped into seven macro-areas: "Values/principles", "Governance/training model", "Training", "Spaces", "Communication", "Cooperation", "Characterizing elements". The clustered elements are shown in the first five columns of Table S6 in the Supplementary Materials.

7. External clustering. The comparison between the design elements of the various social categories ascertains their consistency concerning the subjects and the value heritage attributed to the supply chain. Furthermore, this allows us to understand the useful answers for several categories simultaneously and streamline the supply chain design. The overall design elements can be found in the sixth column of Table S6 in the Supplementary Materials.

8. Development of "design solutions". Some answers were elaborated. With these, we give substance to the design and organicity and completeness of the project work. The answers clearly outline the actions necessary for the realisation of the supply chain. The set of design solutions determines the concept of the supply chain model elaborated here.

9. Elaboration of the "concept" of the economic process model using the DT "storytelling" technique. Through this research work, a narrative form has been provided to the design work; then, the elaborated supply chain model has been described in words. Finally, in the following paragraph, a detailed discussion has been reported.

10. Verification of the design work. In DT, the innovative product or service is tested by submitting it to a panel of potential end-users to estimate its usefulness. Analyses were carried out on the propensity of consumers regarding the "The Food Village" concept using scientifically validated econometric tools. The results of this analysis will be presented in a future contribution.

3. Results

The "Food Villages": An Innovative Food Network Concept Proposal

The proposed "Food Villages" model promotes food resilience, health, environment care and defence, social aggregation, relationality, enhancement, and cultural biodiversity promotion, agroecology, and economic processes for the common good (increase in employment, fair compensation and rights of workers, appropriate production and services, etc.).

The model aims to establish a "Food Community" where the needs of all stakeholders can be satisfied. It is a prototype of an agri-food chain based on the ecological, civil, common good, and happiness economy principles to achieve the common good. The heart of the project is the "Community Pact for Food", a set of shared values and practices around food, its production, impact on the environment, the economy, and society.

The base innovation is established on the concept that consumers, local producers, and the Food Village's employees can be involved in the same legal entity, which combines supply, processing, and marketing to create a fair and ecological supply chain. The economic process could be ecological and achieve efficiency, redistribution, and relationality, thus becoming a tool for the development of the common good. Therefore, the "Community Cooperative" has been identified as the legal subject of the model.

Within this economic space, the needs of consumers and small–micro local producers are met and compared. The agricultural producers must create a stable income, receive fair compensation, operate in good working conditions, and improve the efficiency of production; at the same time, consumers need to constantly buy healthy and sustainable products at a fair price, and optimise their use of time, live spaces of relationship, and increase their awareness and self-determination.

The possibility to adhere to the "Community Cooperative" will be open and each kind of member will pay a membership fee to become part of it.

The products of the member farms conferring to this "Community Cooperative" will be sold within the Food Market.

Specifically, the food products' supply sold in the Food Market will include three levels:

(1) "ultra-local", characterized by the supply of members of the Food Village, located up to a maximum of 50 km away from the Food Market; this share of products will represent at least 20% of the total offer. All the ultra-local food products within the Food Market will be produced according to agroecological criteria or conferred by farms that are progressively in transition to agroecology, thus facilitating the involvement of local farms that would otherwise have been excluded.

(2) "local", characterized by the supply of non-associated farms located up to a maximum of 200 km away from the Food Market; this share of products will represent at least 50–60% of the total offer.

(3) "local extended" characterized by the supply of non-associated farms, located up to a maximum of 500 km away from the Food Market; this quota will represent at least 20–30% of the total offer and will concern all those products that cannot be produced in the local area. The products that cannot be cultivated in the area, such as coffee, tea, cocoa, etc. will also be sourced beyond 500 km away from the Food Market.

This supply structure was designed to increase the goods variability, according to the quality criteria expressed by the consumers.

Household economy products with specific eco-sustainability certifications would also be sold in the Food Market.

The Food Market will allow consumers to have constant access to various products purchased in bulk or using eco-sustainable packaging where this is not possible. For those interested in reducing the use of cars, the Cooperative will organise a shopping delivery system three times a week. For those interested or in need, but also to reduce the use of cars, the Cooperative will organise a shopping delivery system three times a week.

The Food Market was planned to allow consumers to have constant access to a wide variety of products, something that rarely happens in short-chain models such as SPG and OGSD.

A micro-transformation system and storage will be created through modules owned by the cooperative (See for example Self-Globe modular plants (https://www.selfglobe.com/) (accessed on 10 June 2021) to reinforce the cooperative member's local farmers' role in the food supply chain.

The micro transformation could involve different kinds of activities such as a mill, a pasta factory, cheese factory, oil mill, seed cleaning, slaughterhouse, fruit and vegetable processing and transformation, etc.

The transformation processes will be carried out by a dedicated staff of the Cooperative. It will allow small and micro local agricultural farms participating in the Cooperative to transform their production; in fact, usually, these farms are forced to sell to wholesalers and large-scale retailers because they do not produce adequate quantities, thus foregoing fair compensation to limited quantities of agricultural production.

The micro-transformation modules will allow producers to raise their net income per hectare conferred through an internal redistribution of the surplus achieved by processed products. Road transport is reduced when the processing and marketing sectors are located in a single place.

Beyond the products supplied by local producers, the Food Market will be supplied according to traditional methods, based on the quality criteria expressed by the consumers themselves. Although there will be particular attention on the eco-sustainability of the supply chains involved, the Food Market of the "Food Village" could also sell all the kinds of products that a traditional supermarket usually sells. Thus, the Food Market purchasing agents will also be able to source from both non-local producers and distributors.

Contiguously to the Food Market, spaces for participatory democracy are provided for social assemblies, co-planning of prices and production, and the participatory certification of the local production. All spaces will be built according to bio-ecological architecture for their autonomous energy requirements. Additionally, they will provide permanent training to local producers on agroecology, business management, production processes, agronomic best available technologies (BAT), crop accounts, and price formation. Moreover, consumers will be provided with courses on healthy lifestyles (balanced nutrition, physical activity/sports, self-awareness practices, facilitation and participation techniques, education in relationships, etc.); territorial, national, and international dissemination and enhancement of food cultures will also be organised.

Figure 2 shows a brief graphical summary of the principal characteristics of the "Food Village".

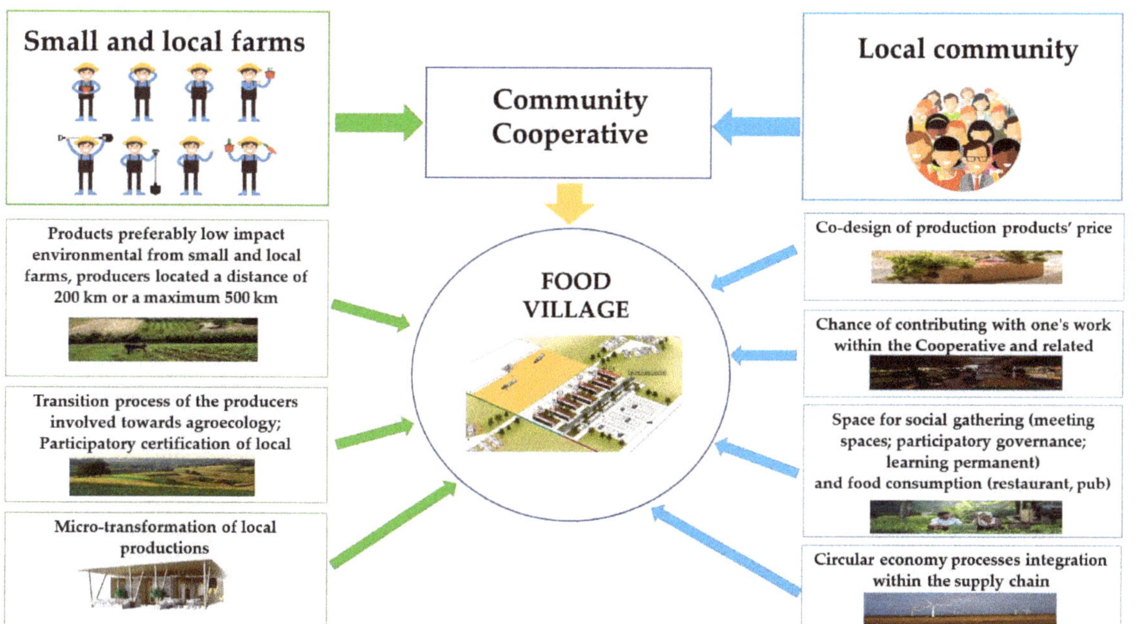

Figure 2. Food Village characteristics.

These spaces for participatory democracy could also be used to implement ecological transition projects (e.g., purchasing groups of green technologies such as solar and photovoltaic panels, electric bicycles, etc., as well as repairing and reusing objects).

Entertainment and catering sites are provided (such as a bar, restaurant, street food, theatre, etc.) to meet and attend artistic performances (music, presentations, books, readings, etc.) to facilitate the community aggregation. Restaurants mainly use products provided by members, thus creating another earning opportunity for the producers. Furthermore, the Cooperative could organise visits to the member farms to strengthen the sense of community, bond with the territory, and agricultural production.

"Food Village" is a replicable model; according to the needs and characteristics of a specific area, each "Food Village" Cooperative can open Food Markets separate from the headquarters to facilitate its increased usability. This is important in large cities where the space required to implement micro transformation and promote social aggregation and participatory democracy is unavailable in the city centres.

In each area, the Community Cooperative will aggregate the local offer and involve new farms based on the actual member's consumption. This, along with the reduction in food waste at the trade phase, will stabilise the income for producers who could have the guarantee of selling their products even before production due to supply contracts. Furthermore, the procurement contracts established with the producer members could be confirmed annually, making the farm's economic flows stable. A protocol could be defined within the cooperative regulations, governing the contractual relations between the cooperative and its producer members. The quality standards for production will be established for agroecological farmers and those in the agroecological transition process.

The product quality standards defined in the contracts will be verified through a participatory certification system in which the members are involved. If a producer fails to supply the Cooperative, the needs may be reallocated to other members in the same village or neighbouring villages. Software to coordinate operators and manage the compensation for production failures will be developed.

Moreover, the system will facilitate the work exchange and sharing of means of transport within the network to optimise resources, increase efficiency, and reduce costs of the production system. The Cooperative could facilitate the purchase, shared use of the production machinery, and reuse production waste within the farms involved or externally to implement circular productive and economic processes. This approach would raise the quality of the production system in ecological terms and also reduce production costs.

A "co-planning of production" model will be applied, stabilising producer members' income and cost-saving by consumers in terms of a discount. This process will regard the "ultra-local" farmers and it will be developed in two phases. In the first phase, two months before the start of the agricultural season, consumers must indicate their weekly food needs (expressed in kg) for each food class (bread, pasta, vegetables, fruit, meat, etc.); namely, their food preferences in terms of the type of food consumed. Due to a dedicated calculation system, consumers will compare their food needs with an average balanced diet based on the Mediterranean diet. Thus, consumers could analyse their consumption and modify it if they deem it appropriate. Consequently, a pre-order to the cooperative will be placed based on the consumer's food preferences using a matrix. Based on the previous year's prices, the system estimates the expense and consumers decide whether to continue the order. Then, they must indicate the supply period: three, six months, or one year to simplify logistics for producers.

The second phase develops into participation paths to decide the prices of the products together. A "commission of members" (producers and consumers) will be created to define the annual price of food produced by farmer members. The definitive price from the co-design process will consider (1) the production costs and an agreed percentage surplus concerning the average national unitary income for the crop or food supplied; (2) the processing and marketing costs; (3) the replication or dissemination costs of the project (opening new Food Villages); (4) the discounts extended to members based on their degree of participation.

Once the prices are defined, the consumers can confirm, cancel, or modify their order. Once the order is confirmed, consumers will receive the products on a weekly basis, from the moment of harvest.

This process of price formation, excluding profits, will enable increased accessibility to food (fair price), adequately pay producers (fair compensation), and disseminate environmentally sustainable agricultural practices.

The "commission of members" supports the Food Market purchasing agents in selecting products that will be bought from producers outside the Cooperative to ensure that the product prices respect the principles of fairness and accessibility. Every three months, the commission will check if there is a need to revise selling prices.

The Food Market will be structured following the model of FCPS that requires each member to work three hours a month within it to buy products.

This model is now widely tested. It has yielded significant results both in creating social aggregation, a sense of community and solidarity, and in promoting a fair and environmentally sustainable way of supplying food. In fact, it is possible to buy high-quality food, often organic and local, at affordable prices within the FCPS. In the Food Villages, there will be voluntary (non-compulsory) possibilities to participate in the FCPS model.

Members making their contributions to the Food Village system can have access to a dedicated discount. Specifically, those who participate in the "co-planning of production" will access up to 10% of discounts concerning the co-planned products, while a 20% discount on all products sold in the Food Market will be provided for those who participate in the FCPS. Consumers that adhere to both models will access up to 30% of discounts concerning the co-planned products and a 20% discount on all products sold in the Food Market.

Members that participate in none of these activities, will be allowed to purchase and participate in all initiatives (educational activities, social events, etc.), while the consumers that are not members will be allowed only to purchase.

The "Food Villages" could represent "solidarity communities", namely, social spaces where reciprocity is practised and one takes care of the other. Initiatives to meet the needs of the weaker social groups and emancipate those with difficulty in integration, including campaigns to satisfy fundamental human rights, promoting interculturality and inter-religiousness, work placement, right to housing, and food support (see the Last-Minute Market (https://www.lastminutemarket.it/) (accessed on 15 July 2021) experience and Banco Alimentare (https://www.bancoalimentare.it/it) (accessed on 15 July 2021), will be supported.

The community dimension and values expressed by the Food Villages and the technical, logistic, and governance models on which they are based on make Food Villages a real CE and ECG prototype. Therefore, to measure the impact of this model on the socio-economic and environmental fabric, the Food Villages will adhere to the guidelines of the ECG. Moreover, they will carry out the "Common Good Balance Sheet" annually. The "Common Good Balance Sheet" will also allow the Cooperative to foresee actions to improve Common Good and redirect the production processes of the partner farms of the Food Villages for the common good.

4. Discussion

This study proposed a new food supply chain model by reviewing four AFN models highlighting their characteristics, potential, and limits. Although AFNs promote an eco-sustainable paradigm change in the food chain, the current examples cannot provide an alternative capable of large scale spreading due to some constitutive limitations on their structure and governance explained previously. Therefore, we argue that an alternative should have the ability to "jump the scale".

Our proposal is based on considering all the needs of all stakeholders in the supply chain simultaneously, as neither the current large-scale distribution nor the AFNs do that. Therefore, the Needs Matrix developed by Max-Neef within the economic model of H-SD was used to identify the needs (and their satisfiers) of all stakeholders in the supply chain. In contrast, the methodological framework of DT was used to develop a systematic and comprehensive design procedure. Therefore, the design model was built based on the criteria of the H-SD economic model and was intended to be a tool for spreading the vision and values, starting with the restructuring of economic processes.

Through the elaborate design method, the "Food Village" model was shaped and proposed in this study as a food supply network far from the current economic mechanisms and based on the community. Starting from the needs of the stakeholders, the design has allowed the strength of the community in creating economic processes that simultaneously allow sustainability, equity, reciprocity, and freedom as envisaged by the vision of the CE and the ECG to be enhanced. Basing governance on the community has favoured constructing a logistic and economic system capable of systemically incorporating all the system elements that favour sustainability (circular economic processes, agroecology, participatory governance, fair price or fair compensation, bio-architecture, etc.). Finalising the elements of the system to respond to current needs has opened up space for new processes within the economic exchange. An example is the co-design of prices. Establishing the Food Village on a community cooperative invests micro-transformation modules of local production on-site as a venture for all the stakeholders. Thus, this technology reduces the intermediary costs, leaving a greater margin for producers and consumers to compare and identify a fair price or fair compensation in a participatory process. Such a structured economic process promotes social interactions based on reciprocity by opening the space to a more cohesive society. Moreover, the model of the Food Village can be a powerful tool in promoting the rural economy, increasing the food resilience of the area, and reducing environmental impacts on the territory.

Following its application to the food supply chain, the proposed design system effectively identifies the satisfiers and design solutions capable of simultaneously promoting the different stakeholders' satisfiers. Therefore, the Max-Neef Needs Matrix is an excellent analytical tool. Furthermore, the proposed method is useful and adaptable to economic contexts other than that of the Food Chain. Therefore, this procedure was considered a useful tool for the design of new economic processes capable of responding to the needs of all stakeholders.

The main limitation of this system is the length of time required for the need analysis process and a certain redundancy of the initial design elements. Therefore, in this model, these latest approaches are subsequently clustered.

This study aimed to provide a way of "moving alternative food networks beyond the niche" [25] (p. 1). Thus, further research and exchange of opinions are expected to arise from our food for thought.

Supplementary Materials: The following supporting information can be downloaded at: https://www.mdpi.com/article/10.3390/foods11101447/s1, Table S1: matrix of consumers, Table S2: matrix of producers, Table S3: matrix of holders, Table S4: matrix of employees, Table S5: matrix of collectivity, Table S6: the clustered design elements.

Author Contributions: Conceptualization, methodology, writing—original draft preparation, G.S. Methodology, paper administration, supervision, B.T. Writing—original draft preparation, visualization, writing—reviewing, editing, C.P. Investigation, G.G. Writing—reviewing, L.C. Conceptualization, visualization, R.P. All authors have read and agreed to the published version of the manuscript.

Funding: This research was funded by both the project "Community Project Cibo Nostrum: the added value for agricultural products", financed by Sub-Measure 16.2.2 of the Rural Development Programme (RDP) 2014–2020 for Umbria Region (Italy) (Unique Identification Code I68I18000200002) and of the project "Entrepreneurial choices in a context of AgriSocial Business Model and of Welfare Society", financed by University of Perugia's Basic Research Programme.

Institutional Review Board Statement: Not applicable.

Informed Consent Statement: Not applicable.

Data Availability Statement: The data are available on request from the authors.

Conflicts of Interest: The authors declare no conflict of interest.

References

1. Holt-Giménez, E.; Shattuck, A.; Altieri, M.; Herren, H.; Gliessman, S. We Already Grow Enough Food for 10 Billion People ... and Still Can't End Hunger. *J. Sustain. Agric.* **2012**, *36*, 595–598. [CrossRef]
2. FAO. *Global Food Losses and Food Waste—Extent, Causes and Prevention*; FAO: Rome, Italy, 2011.
3. FAO; IFAD; UNICEF; WFP; WHO. *The State of Food Security and Nutrition in the World 2020. Transforming Food System for Affordable Healthy Diets, 2020*; FAO: Rome, Italy, 2020; Available online: https://doi.org/10.4060/ca9692en (accessed on 12 November 2021).
4. Abarca-Gómez, L.; Abdeen, Z.A.; Hamid, Z.A.; Abu-Rmeileh, N.M.; Acosta-Cazares, B.; Acuin, C.; Cho, Y. Worldwide trends in body-mass index, underweight, overweight, and obesity from 1975 to 2016: A pooled analysis of 2416 population-based measurement studies in 128.9 million children, adolescents, and adults. *Lancet* **2017**, *390*, 2627–2642. [CrossRef]

5. Altieri, M.A.; Nicholls, C.I.; Ponti, L. *Agroecologia: Una via Percorribile per un Pianeta in Crisi. (Agroecology: A Feasible Path for a Planet in Crisis)*, 1st ed.; Edagricole: Bologna, Italy, 2015; p. 328.
6. Pimbert, M. *Towards Food Sovereignty: Another World Is Possible for Food and Agriculture (Part I: Chap. 1–3)*; IIED: London, UK, 2009; Available online: https://pubs.iied.org/g02268 (accessed on 10 July 2021).
7. Holt-Giménez, E.; Altieri, M.A. Agroecology, Food Sovereignty and the New Green Revolution. *Agroecol. Sustain. Food Syst.* **2013**, *37*, 90–102. [CrossRef]
8. Saltmarsh, N.; Wakeman, T. Mapping food supply chains and identifying local links in the broads and rivers area of Norfolk. *East Angl. Food Link Proj. Rep.* 2004. Available online: http://www.eafl.org.uk/downloads/LocalLinksMainWeb.pdf (accessed on 25 November 2021).
9. Martinez, S. *Local Food Systems; Concepts, Impacts, and Issues*; Diane Publishing: Collingdale, PA, USA, 2010.
10. Friedmann, H. Feeding the empire: The pathologies of globalized agriculture. *Soc. Regist.* **2005**, *41*. Available online: https://socialistregister.com/index.php/srv/article/view/5828 (accessed on 10 November 2021).
11. Cecchini, L.; Torquati, B.; Chiorri, M. Sustainable agri-food products: A review of consumer preference studies through experimental economics. *Agric. Econ.* **2018**, *64*, 554–565. [CrossRef]
12. Govindan, K. Sustainable consumption and production in the food supply chain: A conceptual framework. *Int. J. Prod. Econ.* **2018**, *195*, 419–431. [CrossRef]
13. Martinez-Alier, J.; Shmelev, S.; Anguelovski, I.; Bond, P.; Del Bene, D.; Demaria, F.; Gerber, J.-F.; Greyl, L.; Haas, W.; Healy, H.; et al. Between activism and science: Grassroots concepts for sustainability coined by Environmental Justice Organizations. *J. Political Ecol.* **2014**, *21*, 19–60. [CrossRef]
14. LVC (La Via Campesina). Sustainable Peasant and Family Farm Agriculture Can Feed the World. Via Campesina Views No. 6. p. 15, Jakarta, Indonesia, September 2010. Available online: https://viacampesina.org/downloads/pdf/en/paper6-EN-FINAL.pdf (accessed on 25 November 2021).
15. Schiavoni, C. The global struggle for food sovereignty: From Nyéléni to New York. *J. Peasant Stud.* **2009**, *36*, 682–689.
16. Van der Ploeg, J.D. Peasant-driven agricultural growth and food sovereignty. *J. Peasant Stud.* **2014**, *41*, 999–1030. [CrossRef]
17. Dunford, R. Peasant activism and the rise of food sovereignty: Decolonising and democratising norm diffusion? *Eur. J. Int. Relat.* **2017**, *23*, 145–167. [CrossRef]
18. Núñez-Ríos, J.E.; Aguilar-Gallegos, N.; Sánchez-García, J.Y.; Cardoso-Castro, P.P. Systemic Design for Food Self-Sufficiency in Urban Areas. *Sustainability* **2021**, *12*, 7558. [CrossRef]
19. Menconi, M.E.; Grohmann, D.; Mancinelli, C. European farmers and participatory rural appraisal: A systematic literature review on experiences to optimize rural development. *Land Use Policy* **2017**, *60*, 1–11. [CrossRef]
20. Peters, C.J.; Picardy, J.; Darrouzet-Nardi, A.F.; Wilkins, J.L.; Griffin, T.S.; Fick, G.W. Carrying capacity of U.S. agricultural land: Ten diet scenarios. *Elem. Sci. Anthr.* **2016**, *4*, 116. [CrossRef]
21. Stella, G.; Coli, R.; Maurizi, A.; Famiani, F.; Castellini, C.; Pauselli, M.; Tosti, G.; Menconi, M. Towards a National Food Sovereignty Plan: Application of a new Decision Support System for food planning and governance. *Land Use Policy* **2019**, *89*, 104216. [CrossRef]
22. Berti, G. Sustainable agri-food economies: Re-territorialising farming practices, markets, supply chains, and policies. *Agriculture* **2020**, *10*, 64. [CrossRef]
23. Corsi, A.; Barbera, F.; Dansero, E.; Peano, C. Alternative Food Networks. In *An Interdisciplinary Assessment*; Palgrave Macmillan: London, UK, 2018. [CrossRef]
24. Renting, H.; Marsden, T.K. Understanding alternative food networks: Exploring the role of short food supply chains in rural development. *Environ. Plan.* **2003**, *35*, 393–411. [CrossRef]
25. Maye, D. Moving Alternative Food Networks beyond the Niche. *Int. J. Sociol. Agric. Food* **2013**, *20*, 383–389. [CrossRef]
26. Goodman, D.; DuPuis, E.M.; Goodman, M.K. *Alternative Food Networks: Knowledge, Practice, and Politics*; Routledge: London, UK, 2012; p. 320. [CrossRef]
27. Bruni, L.; Zamagni, S. *Civil Economy. Efficiency, Equity, Public Happiness*, 1st ed.; Peter Lang: Bern, Switzerland, 2007; p. 284.
28. Martino, M.G. Civil Economy: An Alternative to the Social Market Economy? Analysis in the Framework of Individual versus Institutional Ethics. *J. Bus. Ethics* **2020**, *165*, 15–28. [CrossRef]
29. Felber, C. *Die Gemeinwohl-Ökonomie: Das Wirtschaftsmodell der Zukunft*; Deuticke: Wien, Austria, 2010; p. 160.
30. Dolderer, J.; Felber, C.; Teitscheid, P. From Neoclassical Economics to Common Good Economics. *Sustainability* **2021**, *13*, 2093. [CrossRef]
31. Max-Neef, M. Development and human needs. In *Real Life Economics*; Ekins, P., Max-Neef, M., Eds.; Routledge: London, UK, 1992; pp. 197–214.
32. Goodman, D.; DuPuis, E.M. Knowing food and growing food: Beyond the production-consumption debate in the sociology of agriculture. *Sociol. Rural.* **2002**, *42*, 5–22. [CrossRef]
33. Goodman, M.K. Reading fair trade: Political ecological imaginary and the moral economy of fair trade foods. *Political Geogr.* **2004**, *23*, 891–915. [CrossRef]
34. Davidova, S.; Thomson, K. Family farming in Europe: Challenges and prospects. In *Brussels: DG for Internal Policies, Policy Department B: Structural and Cohesion Policies*; European Parlament Pubblications Office: Bruxelles, Belgium, 2014; Available online: http://www.europarl.europa.eu/studies (accessed on 10 September 2021).

35. Kneafsey, M.; Venn, L.; Schmutz, U.; Balázs, B.; Trenchard, L.; Eyden-Wood, T.; Blackett, M. *Short Food Supply Chains and Local Food Systems in the EU. A State of Play of Their Socio-Economic Characteristics*; EUR 25911; Publications Office of the European Union: Luxembourg, 2013; Available online: https://publications.jrc.ec.europa.eu/repository/handle/JRC80420 (accessed on 5 October 2021).
36. Maye, D.; Kirwan, J. *Alternative Food Networks, Sociology of Agriculture and Food Entry for Sociopedia. Isa*; Universidad Complutense Madrid: Madrid, Spain, 2010; Available online: http://www.sagepub.net/isa/resources/pdf/AlternativeFood-Networks.pdf (accessed on 16 November 2021).
37. Jarosz, L. The city in the country: Growing alternative food networks in Metropolitan areas. *J. Rural Stud.* **2008**, *24*, 231–244. [CrossRef]
38. Bazzani, C.; Canavari, M. Alternative Agri-Food Networks and Short Food Supply Chains: A review of the literature. *Econ. Agro-Aliment.* **2013**, *2*, 11–34. [CrossRef]
39. Brunori, G.; Rossi, A.; Malandrin, V. Co-producing Transition:Innovation Processes in Farms Adhering to Solidarity-based Purchase Groups (GAS) in Tuscany, Italy. *Int. J. Sociol. Agric. Food* **2011**, *18*, 28–53. [CrossRef]
40. Watts, D.C.; Ilbery, B.; Maye, D. Making reconnections in agro-food geography: Alternative systems of food provision. *Prog. Hum. Geogr.* **2005**, *29*, 22–40. [CrossRef]
41. Opitz, I.; Specht, K.; Piorr, A.; Siebert, R.; Zasada, I. Effects of consumer-producer interactions in alternative food networks on consumers' learning about food and agriculture. *Morav. Geogr. Rep.* **2017**, *25*, 181–191. [CrossRef]
42. Ilbery, B.; Maye, D. Food supply chains and sustainability: Evidence from specialist food producers in the Scottish/English borders. *Land Use Policy* **2005**, *22*, 331–344. [CrossRef]
43. Guthman, J. Agrarian Dreams: The Paradox of Organic Farming in California. Ph.D. Dissertation, University of California, Berkeley, CA, USA, 2004.
44. Sonnino, R.; Marsden, T. Beyond the divide: Rethinking relationships between alternative and conventional food networks in Europe. *J. Econ. Geogr.* **2006**, *6*, 181–199. [CrossRef]
45. Le Velly, R.; Dufeu, I. Alternative food networks as "market agencements": Exploring their multiple hybridities. *J. Rural Stud.* **2016**, *43*, 173–182. [CrossRef]
46. Le Velly, R. Allowing for the projective dimension of agency in analysing alternative food networks. *Sociol. Rural.* **2019**, *59*, 2–22. [CrossRef]
47. Rossi, A. Beyond food provisioning: The transformative potential of grassroots innovation around food. *Agriculture* **2017**, *7*, 6. [CrossRef]
48. Brunori, G.; Rossi, A.; Guidi, F. On the new social relations around and beyond food. Analysing consumers' role and action in Gruppi di Acquisto Solidale (Solidarity Purchasing Groups). *Sociol. Rural.* **2012**, *52*, 1–30. [CrossRef]
49. Torquati, B.; Taglioni, C.; Viganò, E. Construction of Alternative Food Networks for organic products: A case study of Organized Groups of Supply and Demand. *New Medit* **2016**, *15*, 53.
50. Cone, C.A.; Myhre, A. Community-supported agriculture: A sustainable alternative to industrial agriculture? *Hum. Organ.* **2000**, *59*, 187–197. [CrossRef]
51. Giacchè, G.; Retière, M. The "promise of difference" of cooperative supermarkets: Making quality products accessible through democratic sustainable food chains? *Redes. Rev. Do Desenvolv. Reg.* **2019**, *24*, 35–48. [CrossRef]
52. Mount, P. Growing local food: Scale and local food systems governance. *Agric. Hum. Values* **2012**, *29*, 107–121. [CrossRef]
53. Wald, N.; Hill, D.P. 'Rescaling' alternative food systems: From food security to food sovereignty. *Agric. Hum. Values* **2016**, *33*, 203–213. [CrossRef]
54. Poças Ribeiro, A.; Harmsen, R.; Feola, G.; Rosales Carréon, J.; Worrell, E. Organising alternative food networks (AFNs): Challenges and facilitating conditions of different AFN types in three EU countries. *Sociol. Rural.* **2021**, *61*, 491–517. [CrossRef]
55. Barbera, F.; Dagnes, J.; Di Monaco, R. Participation for what? Organizational roles, quality conventions and purchasing behaviors in solidarity purchasing groups. *J. Rural Stud.* **2020**, *73*, 243–251. [CrossRef]
56. European Economic and Social Committee. Economy for the Common Good. ECO/378-EESC-2015-02060-00-00-ac-tra. 2016. Available online: https://www.eesc.europa.eu/en/our-work/opinions-information-reports/opinions/economy-common-good (accessed on 7 October 2021).
57. Felber, C. *This Is Not Economy: Aufruf zur Revolution der Wirtschaftswissenschaft*; Paul Zsolnay Verlag: Wien, Austria, 2019; p. 258.
58. Frémeaux, S.; Michelson, G. The common good of the firm and humanistic management: Conscious capitalism and economy of communion. *J. Bus. Ethics* **2017**, *145*, 701–709. [CrossRef]
59. Felber, C.; Campos, V.; Sanchis, J.R. The common good balance sheet, an adequate tool to capture non-financials? *Sustainability* **2019**, *11*, 3791. [CrossRef]
60. Maslow, A. *Towards a Psychology of Being*, 2nd ed.; Van Nostrand Reinold: New York, NY, USA, 1968; p. 240.
61. Max-Neef, M.; Elizalde, A.; Hopenhayn, M.; Herrera, F.; Zemelman, H.; Jataba, J.; Weinstein, L. Human Scale Development. In *Development Dialogue*; Cepaur-Dag Hammarskjöld Foundation: Uppsala, Sweden, 1989.
62. Max-Neef, M. *Human Scale Development: Conception, Application and Further Reflections*; The Apex Press: New York, NY, USA; London, UK, 1991; Available online: http://www.areanet.org/fileadmin/user_upload/papers/Max-neef_Human_Scale_development.pdf (accessed on 5 September 2021).

63. Cruz, I.; Stahel, A.; Max-Neef, M. Towards a systemic development approach: Building on the Human-Scale Development paradigm. *Ecol. Econ.* **2009**, *68*, 2021–2030. [CrossRef]
64. Tschimmel, K. Design Thinking as an effective Toolkit for Innovation. In Proceedings of the XXIII ISPIM Conference: Action for Innovation: Innovating from Experience, Barcelona, Spain, 17–20 June 2012; ISBN 978-952-265-243-0.
65. Pombo, F.; Tschimmel, K. Sapiens and Demens in Design Thinking–perception as core. In Proceedings of the 6th International Conference of the European Academy of Design EAD, Madrid, Spain, 29–31 March 2005; Volume 6, p. 62.
66. Martin, R.; Martin, R.L. *The Design of Business: Why Design Thinking Is the Next Competitive Advantage*; Harvard Business Press: Boston, MA, USA, 2009; p. 256.
67. Cross, N. *Design Thinking: Understanding How Designers Think and Work*; Berg Publishers: Oxford, UK, 2011.
68. Peirce, C.S. *Collected Papers of Charles Sanders Peirce*; Harvard University Press: Cambridge, MA, USA, 1974; Volume 7.
69. Tschimmel, K. Training Perception—The Heart in Design Education. In *Design Education: Tradition and Modernity*; Papers from the International Conference DETM'05; National Institute of Design: Ahmedabad, India, 2007; pp. 120–127. ISBN 8186199578.
70. Stickdorn, M.; Schneider, J. (Eds.) *This is Service Design Thinking: Basics, Tools, Cases*; John Wiley & Sons Inc.: Hoboken, NJ, USA, 2012; p. 384.
71. Aly, D.; Hassan, D.K.; Kamel, S.M.; Hamhaber, J. Human Needs as an Approach to Designed Landscapes. *J. Nat. Resour. Dev.* **2018**, *8*, 15–26. [CrossRef]

Article

Food from the Depths of the Mediterranean: The Role of Habitats, Changes in the Sea-Bottom Temperature and Fishing Pressure

Porzia Maiorano [1,2], Francesca Capezzuto [1,2,*], Angela Carluccio [1,2], Crescenza Calculli [3], Giulia Cipriano [1,2], Roberto Carlucci [1,2], Pasquale Ricci [1,2], Letizia Sion [1,2], Angelo Tursi [1,2] and Gianfranco D'Onghia [1,2]

1 Department of Biology, University of Bari Aldo Moro, Via E. Orabona 4, 70125 Bari, Italy; porzia.maiorano@uniba.it (P.M.); angela.carluccio@uniba.it (A.C.); giulia.cipriano@uniba.it (G.C.); roberto.carlucci@uniba.it (R.C.); pasquale.ricci@uniba.it (P.R.); letizia.sion@uniba.it (L.S.); angelo.tursi@uniba.it (A.T.); gianfranco.donghia@uniba.it (G.D.)

2 Consorzio Nazionale Interuniversitario per le Scienze del Mare (CoNISMa), Piazzale Flaminio 9, 00196 Roma, Italy

3 Department of Economy and Finance, University of Bari Aldo Moro, Largo Abbazia S. Scolastica, 70124 Bari, Italy; crescenza.calculli@uniba.it

* Correspondence: francesca.capezzuto@uniba.it; Tel.: +39-3486521418

Abstract: As part of the "*Innovations in the Food System: Exploring the Future of Food*" Special Issue, this paper briefly reviews studies that highlight a link between deep-sea fishery resources (deep-sea food resources) and vulnerable marine ecosystems (VME), species, and habitats in the Mediterranean Sea, providing new insights into changes in commercial and experimental catches of the deep-sea fishery resources in the central Mediterranean over the last 30 years. About 40% of the total landing of Mediterranean deep-water species is caught in the central basin. Significant changes in the abundance of some of these resources with time, sea-bottom temperature (SBT), and fishing effort (FE) have been detected, as well as an effect of the Santa Maria di Leuca cold-water coral province on the abundance of the deep-sea commercial crustaceans and fishes. The implications of these findings and the presence of several geomorphological features, sensitive habitats, and VMEs in the central Mediterranean are discussed with respect to the objectives of biodiversity conservation combined with those of management of fishery resources.

Keywords: seafood; fisheries resources; vulnerable marine ecosystem; environmental change; conservation; Mediterranean

Citation: Maiorano, P.; Capezzuto, F.; Carluccio, A.; Calculli, C.; Cipriano, G.; Carlucci, R.; Ricci, P.; Sion, L.; Tursi, A.; D'Onghia, G. Food from the Depths of the Mediterranean: The Role of Habitats, Changes in the Sea-Bottom Temperature and Fishing Pressure. *Foods* **2022**, *11*, 1420. https://doi.org/10.3390/foods11101420

Received: 4 April 2022
Accepted: 9 May 2022
Published: 13 May 2022

1. Introduction

Seafood is a fundamental source of proteins and nutrients for human nutrition. Its global consumption has increased since 1960 at rates per year higher than that of all land animals, combined and individually (i.e., bovine, ovine, porcine, etc.), except for poultry [1]. However, with the increasing demand for sea products, most of the fishery resources of the world's oceans have been overfished, and many are in a condition of variable levels according to the vulnerability of the different species. In addition, the overfishing condition of continental shelf fish resources has pushed the fishing activities to move towards the exploitable living resources of the deep sea [2–4]. In this regard, deep waters (i.e., beyond the continental shelf, and deeper than approximately 200 m) have acted as a refuge for several stocks with an extensive vertical distribution, where no fishing was occurring until the first decades of the last century [5]. With the expansion of fishing to deeper waters, favoured by the development of new technologies, muddy deep bottoms and other deep-water refuges—such as soft-bottom coral gardens (CGs), cold-water coral (CWC) reefs, submarine canyons (SCs), and seamounts (SEs)—have been affected by this activity, and may no longer play their role of providing structure, food, and shelter for fishery resources. All of these deep-water ecosystems have been identified as hotspots of biodiversity (e.g., [6–8]), and since they can be valuable fishing sites, due to the occurrence of large sizes and high abundances of commercial species, they are often impacted by commercial fishing [9–20]. Fisheries for some deep-sea fish stocks have also collapsed due to heavy pressure on species whose life history traits do not make them well-suited to an intensive harvest [3,4,21]. In fact, deep-sea fish species have a longer lifespan, slow growth, and later sexual maturity, and are consequently more vulnerable and less resilient to overfishing [22–27].

The EU began to deal with deep-sea fisheries in 1992, as a result of assessments carried out by the International Council for the Exploration of the Sea (ICES), who stated that most of the deep-water species of commercial interest were overfished [28]. The current scientific evidence suggests that many deep-sea fish stocks are being exploited beyond sustainable levels [23,29–32], thus emphasising the need to improve the management of these species [3,32–35].

It is well-known that fishing can impact harvested populations directly by excessive removal of individuals, and indirectly by reducing habitat complex structures that guarantee their bioecological activities and provide protection from predators, as well as from adverse physical factors (e.g., [36–40]). Globally, around 40% of trawling fishing grounds are on waters deeper than the continental shelf [41], and several deep-sea habitats and ecosystems have been impacted by fishery activities, with consequent depletion of or reduction in economically important species (e.g., [3,9,21,42–45]).

According to the FAO [46], vulnerable marine ecosystems (VMEs) are groups of species, communities, or habitats that may be vulnerable to impacts from fishing activities. Corals, together with sponges, echinoderms, molluscs, and other epibenthic species, play a significant role in the formation of VMEs. Furthermore, most coral taxa are included in relevant lists of protected species, such as the Red List drawn up by the International Union for the Conservation of Nature (IUCN) [47–50]. Their vulnerability is linked to their likelihood of experiencing substantial alterations from short-term or chronic disturbances, as well as to their difficulty in recovering (for example, slow growth rate, late age of maturity, low or unpredictable recruitment, and long life expectancy) [51–55]. In addition, there is sufficient evidence that VMEs act as essential fish habitats (EFHs), defined as "those waters and substrates necessary to fish for spawning, breeding, feeding, or growth to maturity" [56–58]. Therefore, there is an international consensus in favour of the protection of VMEs in order to combine conservation and fisheries-management objectives according to the ecosystem approach to fisheries (EAF) [46,59–64].

Fish consumption has always been an important part of people's diets around the Mediterranean Sea. The annual production of about 788,000 tonnes, for a total revenue of USD 3.4 billion, offers employment opportunities to several hundred thousand people, supplies seafood products for human consumption to local and regional markets, and creates many other indirect benefits, although fisheries and related activities also produce a large amount of marine litter (including plastics) as a global increasing threat [1]. In the Mediterranean Sea the exploitation of deep-water resources only started in the first decades of the last century. In particular, the deep-water red shrimps *Aristaeomorpha foliacea* (giant red shrimp) and *Aristeus antennatus* (blue and red shrimp) were the target species for deep-water bottom trawling in the 1930s in the Ligurian Sea, where catches of these two shrimps were between 100 and 200 kg/day and, after the Second World War, they could be up to 1000 kg/day per boat [65,66]. In the 1940s, these resources began to be exploited in the Catalan and Balearic seas, and subsequently in other Mediterranean areas [67]. In the Mediterranean Basin, the development of deep-water commercial fisheries was due to the narrowness of the shelf, crossed by several submarine canyons, as well as the growth of the human population along the basin and high demand for a much-appreciated food. However, considering the multispecies nature of the Mediterranean fisheries, the deep-water red shrimps are mostly caught in bathyal muddy bottoms, between 400 and 800 m, together with many other valuable demersal species, such as the European hake (*Merluccius merluccius*), the deep-water rose shrimp (*Parapenaeus longirostris*), the Norway lobster (*Nephrops norvegicus*), the blackspot seabream (*Pagellus bogaraveo*), the greater forkbeard (*Phycis blennoides*), and the wreckfish (*Polyprion americanus*). Other deep-water species of commercial interest, such as the blue whiting (*Micromesistius poutassou*), European conger (*Conger conger*), blackbelly rosefish (*Helicolenus dactylopterus*), angler (*Lophius piscatorius*), blackbellied angler (*Lophius budegassa*), blackmouth catshark (*Galeus melastomus*), bluntnose sixgill shark (*Hexanchus griseus*), golden shrimp (*Plesionika martia*), southern shortfin squid (*Illex coindetii*), European flying squid (*Todarodes sagittatus*), and lesser flying squid (*Todaropsis eblanae*), represent additional components of deep catches. Most of the fish species are commonly caught with trawl nets, longlines, and gillnets, while crustaceans and cephalopods are fished mainly with trawl nets, and sometime with pots.

Most of the abovementioned species are distributed across a wide depth range, between shelf and slope; some of them are typically distributed in deep waters. Moreover, past and recent investigations in deep-sea habitats with a complex and heterogeneous structure—such as coral ecosystems, submarine canyons, and seamounts (now considered VMEs)—have proven that most of the abovementioned species use these types of habitats for shelter, feeding, spawning, and nursery (e.g., [20,58,68–71]). Indeed, habitat features play an important role in determining the structure of species assemblages, and habitat loss may impact—albeit in different ways—on all life stages and critical phases of the different species [38,72]. While adult individuals may not be strictly affected by habitat as juveniles, the detrimental effects of habitat loss on juvenile survival may have longer-term impacts on adult populations [73].

The innovation and sustainability of the food (in the present case, the fishery resources represented by deep-sea fish, crustaceans, and molluscs) used by humanity are mainly based on the systems where it is produced (i.e., the deep-sea sensitive habitats and VMEs), and on the systems that allow its harvest (i.e., the fishing techniques). As part of the "*Innovations in the Food System: Exploring the Future of Food*" Special Issue, this paper aims (1) to briefly review studies that highlight a link between deep-sea fishery resources and VMEs in the Mediterranean Sea; (2) to provide new insights into commercial and experimental catches of the deep-sea fishery resources in the central Mediterranean for the past 30 years; (3) to evaluate changes in the abundance of these resources with time, sea-bottom temperature (SBT), fishing effort (FE), and depth; and (4) to reveal an effect of the Santa Maria di Leuca cold-water coral province on the abundance of the deep-sea fishery resources. The implications of these findings and the presence of several geomorphological features, sensitive habitats, and VMEs in the central Mediterranean are discussed in terms of conservation of biodiversity, combined with the sustainable management of the fishery resources.

2. Materials and Methods

Data on food resources from deep-sea sensitive habitats and VMEs in the whole Mediterranean basin refer to international volumes and publications, and references therein (e.g., [11,55,67,68,74–77]).

In order to define the contribution of the most important commercial deep-sea species as food resources from the depths of the Mediterranean Sea, landing data from official FAO statistics were explored (https://www.fao.org/fishery/statistics-query/en/gfcm_capture/gfcm_capture_quantity, accessed on 12 January 2022), focusing on the central part of the basin (Figure 1).

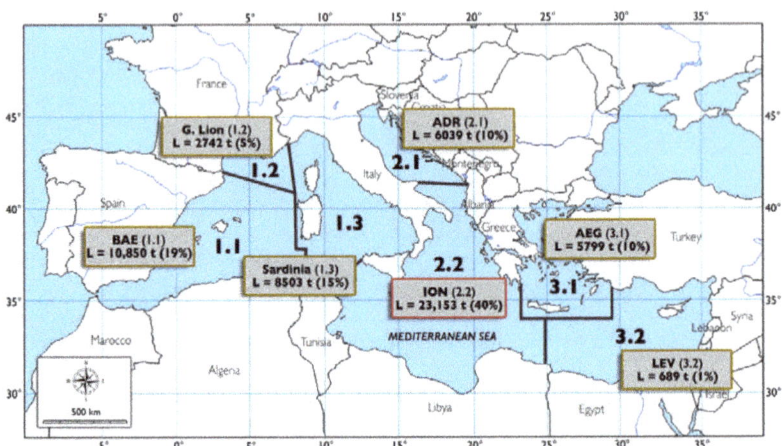

Figure 1. Average total landing (L) of deep-sea commercial species for each FAO Mediterranean subdivision, expressed in tonnes (t) and percentages (%), calculated for the period 1994–2019. FAO subdivisions are coded as Balearic (BAE, 1.1), Gulf of Lion (G. Lion, 1.2), Sardinia (SAR, 1.3), Adriatic (ADR, 2.1), Ionian (ION, 2.2), Aegean (AEG; 3.1), and Levant (LEV, 3.2).

The temporal trends over a period of 25 years (1994–2019) were analysed by means of Spearman's non-parametric correlation.

New observations from the central Mediterranean, the southwestern Adriatic, and the northwestern Ionian, as well as on muddy bottoms of the northwestern Ionian Sea, were derived from data collected as part of national and international study projects carried out in the last two decades by the ecology team from the Department of Biology at the University of Bari Aldo Moro. In particular, data from deep-sea sensitive habitats and VMEs were taken using different low-impact fishing techniques [12,71,78–85], while data from muddy bottoms were collected during the Mediterranean Trawl Surveys (MEDITS) programme, included in the EU Data Collection Framework to date [86]. The MEDITS surveys are carried out in the Mediterranean from late spring to summer every year, according to a standardised protocol that includes gear characteristics, haul duration, and sampling procedures, following a depth-stratified random design, from 10 to 800 m in depth [86–89].

Using MEDITS data, the abundances in weight (expressed as biomass index kg/km^2) and numbers (expressed as density index N/km^2) of the deep-sea species distributed on deep muddy bottoms (200–800 m) of the northwestern Ionian Sea (Figure 2) were evaluated for the period 1994–2020, and their changes over time were tested using Spearman's non-parametric correlation. Data on the sea-bottom temperature (SBT) were recorded using a probe at the start and the end of different MEDITS hauls carried out from 1998 to 2020, and an average value of SBT was computed per year. The fishing pressure of bottom trawling fleets on the resources of the northwestern Ionian Sea was analysed using fishing effort (FE) data, in terms of number of vessels and gross tonnage (GT). Data were obtained from the European Fishing Fleet Register (http://ec.europa.eu/fisheries/fleet/index.cfm, accessed on 12 January 2022) for the period 1994–2020. The relationship between abundance of the deep-sea species and SBT was evaluated using linear regression analysis. Spearman's non-parametric correlation was also applied to the abundance data in weight (kg/km^2) of the deep-sea species that showed a significant trend over time versus the FE—expressed as the total annual number of vessels—throughout the study period.

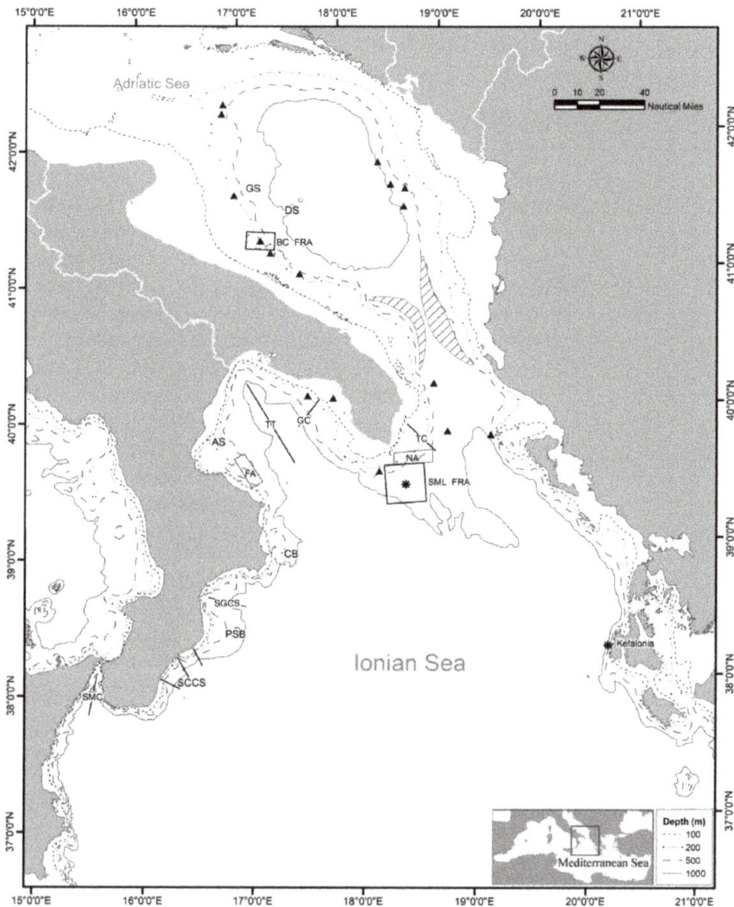

Figure 2. Topographic features, highs, banks, canyons, sensitive habitats, and VME species and habitats along the southwestern Adriatic Sea and northwestern Ionian Sea (central Mediterranean). GS = Gondola Slide; BC FRA = Bari Canyon Fisheries Restricted Area; SML FRA = Santa Maria di Leuca Fisheries Restricted Area; TC = Tricase Canyon; GC = Gallipoli Canyon; TT = Taranto Trench; AS = Amendolara Shoal; CB = Crotone Bank; SGCS = Squillace Gulf Canyon System; PSB = Punta Stilo Bank; SCCS = South Calabria Canyon System; SMC = Strait of Messina Canyon; triangle = hard bottom corals; asterisk = hard- and soft-bottomed corals [58,81,90–97]; lined areas = *Isidella elongata* facies [98]; NA = sampling area near the SML FRA; FA = sampling area far from the SML FRA.

Furthermore, for the shrimps *A. foliacea* and *P. longirostris*, due to correlation between environmental covariates ($\rho = -0.73$, $p < 0.001$), and in order to avoid multicollinearity problems, linear regression models were estimated to investigate the dependence between log-transformed abundances and environmental drivers (i.e., SBT and FE). Transformed responses ensure, in these cases, that the model assumptions are met. For both species, s = {*A. foliacea*, *P. longirostris*}, and for each y_z with z = {abundance in weight, abundance in number}, the linear model is specified as follows:

$$\log(y_z^{(s)}) = \beta_0^{(s)} + \beta_1^{(s)} x_i^{(s)} + \epsilon^{(s)} \quad (1)$$

where x_i represents the independent covariate, with i = {FE, SBT}.

Using linear regression analysis, the relationship between length and depth was evaluated for *M. merluccius*, *P. bogaraveo*, *P. blennoides*, *H. dactylopterus*, and *Galeus melastomus*, collected both on muddy bottoms with the MEDITS trawl net and in VMEs with an experimental longline (e.g., [81]), and the boxplots of the length were represented for all of these species in both habitats.

In order to detect an effect of the presence of a VME on fishery resources, MEDITS abundance data on the weight and number of the species *A. foliacea*, *P. martia*, *M. merluccius*, *P. bogaraveo*, *P. blennoides*, and *H. dactylopterus*, for an area near the Santa Maria di Leuca (SML) cold-water coral (CWC) province (NA), were compared with those of other species far from this coral province (FA) (Figure 2). Data from 54 trawl hauls, carried out between 200 and 800 m, were used for each area. Relative boxplots of the abundances in weight and number were produced, and the differences in the abundances between the two areas were tested using the Kruskal–Wallis non-parametric test. The pressure of fishing activity in these two areas was assessed in order to exclude effects due to this activity on the results of the comparison between the two areas. Specifically, FE was calculated for NA and FA by aggregating the fleets operating close to the two areas. In particular, the fleets of Gallipoli, Leuca, and Otranto were considered for NA, while those of Corigliano Calabro and Cirò Marina for FA. The differences in the species abundances between the two areas were tested using the Kruskal–Wallis non-parametric test.

3. Review of the Mediterranean Studies on the Link between Deep-Sea VMEs and Fishery Resources

In the Mediterranean Sea, the deep-water resources are mainly exploited by trawl fishing on the soft bottoms of the bathyal grounds. However, there are areas characterised by the occurrence of VME species on soft and hard bottoms, canyons, and seamounts, where the fishing is carried out using different types of gears.

3.1. Open Slope, Soft Bottoms

Between the shelf break and descent to bottoms deeper than 1000 m, soft corals can be found that can form dense aggregations on soft bottoms—called sea pen fields, sea fan corals, and arborescent corals—which build up coral gardens or coral forests. Coral gardens can develop on soft or hard substrata, depending on the habitat-forming species. Both sea pen fields and coral gardens and/or coral forests contribute to making more heterogeneous and complex habitats, attracting mobile and swimming fauna [55].

The sea pen fields built up by the octocoral *Funiculina quadrangularis* are mostly distributed on the upper slope, generally at less than 400 m in depth, on soft muddy habitats characterised by noticeable bottom currents. These habitats are commonly inhabited by commercially valuable crustaceans, such as the deep-water rose shrimp (*P. longirostris*) and the Norway lobster (*N. norvegicus*). Soft-bottomed coral gardens, structured by the gorgonian *Isidella elongata* (bamboo coral), can be found from the shelf break down to 1600 m. The valuable deep-water red shrimps (*A. antennatus*, *A. foliacea*) and the golden shrimp (*P. martia*) are frequently associated with these coral gardens [11,54,66,99–101]. Bamboo coral seems to play a role in habitat formation, increasing the three-dimensional habitat complexity on flat bathyal bottoms with its candelabrum-like shape. As a passive feeder, its occurrence is often associated with plankton-rich currents which, in turn, favours a high density of prey, such as pandalid shrimps and other crustaceans [102–104]. These prey animals attract predators of different trophic levels, such as the abovementioned deep-water red shrimps, bony fishes (e.g., *M. merluccius*, *P. blennoides*, *H. dactylopterus*, *P. bogaraveo*, *L. boscii*), and sharks, such as the lesser spotted dogfish (*Scyliorhinus canicula*) and the blackmouth catshark (*G. melastomus*), along with cephalopods (e.g., *I. coindetii*, *T. sagittatus*, *T. eblanae*), all of commercial interest [70,98,105–107]. In addition to the important implications as a feeding area for bentho-pelagic species, the arborescent complexity of the colonies could further act as shelter and spawning/nursery sites for several species that can grow to greater sizes than in areas where bamboo coral does not occur [98,104,107,108]. The commercial species associated with bamboo coral account for about 5% of all of the income of the professional fisheries in the Mediterranean [34], with increasing landings—especially in Italy and Spain—the main producers in Europe [55,109].

3.2. Open Slope, Hard Bottoms

Coral gardens or coral forests between the shelf break and upper slope are mainly made up of antipatharians and alcyonaceans. Among the former, the most widespread species are *Antipathes dichotoma*, *Parantipathes larix*, *Leiopathes glaberrima*, and *Antipathella subpinnata*, which form monospecific or multispecific forests [54,55]. Several commercial fish species are often associated with these antipatharians (e.g., [52,108,110]). Alcyonaceans are present on Mediterranean Sea with hard bottoms with several species, covering a wide bathymetric range. The whip-like gorgonian *Viminella flagellum* and the fan-shaped gorgonian *Callogorgia verticillata* are present in the bathyal zone at depths from 100 to 500 m and from 150 to 1000 m in depth, respectively [54,55]. Off the southwestern coasts of Sardinia, several fish species—some of commercial interest—have been observed hiding among the colonies of *L. glaberrima*, while egg capsules of the shark *S. canicula* have been found on the branches of this antipatharian at depths between 186 and 210 m [52,111]. The presence of egg capsules of this shark on *L. glablerrima* colonies had previously been observed on El Idrissi Bank (Alboran Sea) at 647 m and 452 m [112]. Cau et al. [111] suggested that the coral forest from a representative for southwestern Sardinia represent nursery grounds for *S. canicula*.

In the eastern Ionian Sea, the shark *G. melastomus* and the teleost fish *H. dactylopterus* were the most common fish species caught in the area, characterised by the presence of black coral (*L. glaberrima*) and bamboo coral (*Isidella elongata*) [108]. The shark seems to use the its habitat as a feeding area, and the teleost as a refuge area [108]; *H. dactylopterus* was also found together with other commercial fish species, such as the silver scabbardfish (*Lepidopus caudatus*), and the wreckfish (*P. americanus*) in a coral forest dominated by *L. glaberrima* on the Malta Escarpment (310–315 m) [113].

The fishes that coral zooxanthellate reefs produce in tropical waters account for 17% of animal protein consumed [114]. On the open slope of the Mediterranean Sea, there are still hard bottoms characterised by cold-water coral (CWC) communities, including solitary and colonial zooxanthellae cnidarians [94]. Colonial species have a complex branching morphology, and are habitat formers. The main species, known as white corals, are the colonial species *Madrepora oculata* and *Lophelia pertusa* (recently renamed as *Desmophyllum pertusum*), as well as the solitary coral *Desmophyllum dianthus*. These species have a broad frame-building ability, being able to deposit calcium carbonate and build up durable biogenic substrata. CWCs, as passive suspension feeders, depend on the supply of current-transported particulate organic matter and zooplankton for their trophic requirements. They are preferentially distributed on topographic irregularities on the slope, in canyons and on seamounts, where there are strong currents and the sedimentation rate is low [53,91,115–117].

CWC habitats are impacted by fishing due to the occurrence of large sizes and high abundances of commercial species [9,10,13,16,18,58,70,118,119] and references therein]. The presence of corals is generally known to the local fishers, who experience gear damage and losses, although they often fish close to these areas with the aim of obtaining a greater catch and larger specimens of valuable commercial species, such as the deep-water red shrimps (*A. antennatus* and *A. foliacea*) and the European hake (*M. merluccius*). In fact, side-scan sonar and underwater video images show the characteristic seabed scars of otter trawls ploughing through the coral banks [12,82]. Longline is also used in these areas of complex bottom topography, and is not accessible to trawling, so as to catch wreckfish, greater forkbeard, blackbelly rosefish, blackspot seabream, and bluntnose sixgill shark [71,78,81].

CWC habitats provide a suitable ground for larval settlement and juvenile growth of benthic species. They are spawning and nursery areas for vagile and swimming fauna, acting as an EFH for several commercial and non-commercial fish and invertebrate species [12,17,57,71,78–81,83,84,111].

3.3. Canyons

The fishing targeting the deep-water shrimps in the northwestern Mediterranean is carried out on both the slope and the walls of the submarine canyons [11,14,67,120–124].

In the western Mediterranean, *C. conger* and *P. blennoides* are the fish species captured with the highest biomass at the head of the Blanes Canyon [124]. The juveniles of some deep-sea shrimps (e.g., *Plesionika heterocarpus, P. edwardsii, P. gigliolii*, and *P. martia*) and fish (such as *P. blennoides, Mora moro*, and *Lepidion lepidion*) are found to be distributed in the benthic intermediate nepheloid layers of the Blanes Canyon, which seems to act as a nursery area for these species [125,126]. From the canyons in the eastern part of the Gulf of Lions, blackmouth catshark and European hake are among the most abundant species [15]. Spawning females of angler and European hake have been more commonly observed within the submarine canyons of Petit-Rhône and Grand-Rhône than on the adjacent open slope [15]. *G. melastomus, H. dactylopterus*, and *P. blennoides* are the most frequently observed fish species in French Mediterranean submarine canyons characterised by the presence of CWC species [17]. Discarded fishing gear, including entangled nets and lines, has been observed in the canyons of the western Mediterranean where CWC species thrive [17,127–129].

The shark *G. melastomus* and the teleost fish *H. dactylopterus* and *P. blennoides* are the most frequently captured species in the Quirra Canyon (Sardinian waters), where valuable species—such as the European hake and the deep-water red shrimps—are also collected [130].

Eastward, in the Bari Canyon where CWC species thrive, the blackmouth catshark and the teleost fish European conger, blackbelly rosefish, European hake, blackspot seabream, and greater forkbeard have been found to be more abundant in the canyon than in the adjacent area [71,81]. In particular, greater forkbeard showed significantly greater abundance and biomass in the canyon than outside. Blackspot seabream are exclusively caught inside the canyon. A greater number of both smaller and larger individuals of European conger and greater forkbeard are found in the Bari Canyon than on the open slope. Mature females and males are mostly observed in the canyon in all of the most abundant species, indicating the role of the Bari Canyon as a refuge area and an EFH for fish species exploited in the neighbouring fishing grounds [71,83]. Longline remains have been observed in this canyon, which has a complex bottom topography and is not accessible to trawling [97].

Indeed, canyons seem to benefit and support fisheries [14,71,83,131,132], providing spawning and nursery sites as well as refuges for several commercial species [15,17,69,79, 80,84,97,122,126,133].

3.4. Seamounts

Mediterranean seamounts also host habitat-former species, such as the deep-water glass sponges, sea fans and sea pens, antipatharians—which form large forests up to epibathyal depths—and cold-water corals, such as *Dendrophyllia cornigera* or the white corals *M. oculata* and *D. pertusum*, which thrive at bathyal depths (e.g., [94,134,135]). Due to the topographic and hydrographic conditions allowing the presence of habitat-former species, the seamounts act as biodiversity hotspots, and attract bentho-pelagic fish and migratory species, such as tuna, swordfish, sharks, and cetaceans, as well as several demersal fishery resources [68,76].

The Ulisse Seamount (Ligurian Sea) is a fishing ground for semi-professional and recreational fishermen targeting *P. bogaraveo, M. merluccius, P. americanus*, pink spiny lobster (*Palinurus mauritanicus*), swordfish (*Xiphias gladius*), and red seabream (*Pagellus acarne*) [136]. In the 1970s, according to the number of hooks employed on the fishing line, the catches were up to several hundreds of kilos, represented by wreckfish, blackspot seabream, bluntnose sixgill sharks, and European conger. A total of 120 wreckfish were caught between 1972 and 1975, before their complete disappearance from the catch data [20].

Accidental by-catch, mainly represented by the large arborescent primnoid anthozoan *Callogorgia verticillata*, provided the first evidence of the existence of coral forests on the summit of the Ulisse Seamount [20].

A. foliacea and *A. antennatus* have been reported for other Mediterranean seamounts and banks [68], including the Baronie Seamount located off the northeastern coast of Sardinia [137], representing a site of particular biological and economic interest. Several commercial species—such as the common squid (*Loligo vulgaris*), *P. edwardsii*, *H. dactylopterus*, *A. foliacea*, *A. antennatus*, *G. melastomus*, and *P. blennoides*—have been caught on this seamount [137].

The occurrence of deep-water shrimps (e.g., *A. foliacea*, *A. antennatus*, and *P. martia*), scleractinian corals (e.g., *Caryophyllia calveri*, *Desmophyllum dianthus*), and high densities of other invertebrates was reported in [138] from the Eratosthenes Seamount in the Levantine Basin. In this basin, the presence of sharks (e.g., *G. melastomus* and the spiny dogfish, *Squalus acanthias*), the greater forkbeard, and commercially important deep-sea shrimps (e.g., *P. martia*, *A. foliacea*, and *A. antennatus*) has also been reported for the Turgut Reis Bank [139].

The occurrence of several shark species (e.g., *Prionace glauca*, *H. griseus*, *Cetorhinus maximus*, *Carcharodon carcharias*, *Isurus oxyrhincus*, *Carcharhinus brevipinna*, *Lamna nasus*, *Odontaspis ferox*, and *Sphyrna lewini*) has been detected in the sea area close to the Alcione and Casoni seamounts in the South Tyrrhenian Sea [140].

In the Seco de los Olivos Seamount (western Mediterranean), several commercial species are caught with different fishing techniques, impacting benthic habitats and species [141]: the blue whiting, blackbelly rosefish, silvery pout (*Gadiculus argenteus*), and European hake with otter trawl; *Pagellus* spp., blue whiting, red scorpionfish (*Scorpaena* spp.), and mullets (*Mullus* spp.) using set gillnet; soldier shrimp (*Plesionika* spp.) with traps; and blackspot seabream using bottom longline. In addition, recreational fishing is also carried out on this seamount—mostly on the steeper slopes of the surrounding ridges, targeting the grey grouper (*Epinephelus caninus*) [141].

4. Results

4.1. Data from FAO Official Statistics

During the period 1994–2019, the bulk of the deep-water catches in the Mediterranean Sea was due to the species *M. merluccius*, *P. longirostris*, *A. foliacea*, *A. antennatus*, *N. norvegicus*, *Lophius* spp., *G. melastomus*, *C. conger*, *H. dactylopterus*, *P. bogaraveo*, *Phycis blennoides*, and *P. americanus*. The average total landing for the whole basin was 57,774 (±1857) tonnes per year (7% on average of the Mediterranean total landing; 788,000 tonnes). In the Ionian Sea (FAO subdivision 37.2.2), the average total landing of deep-water species was equal to 23,153 (±1859) tonnes (40% on average of the total landing of Mediterranean deep-water species) (Table 1), thus being the most important area for the exploitation of deep-water resources (Figure 1). A marked fluctuation in landings of the deep-water resources over time has been detected both in the whole Mediterranean and in the Ionian Sea, without any significant trends (Figure 3).

Table 1. Total landing (in tonnes and %), with mean and standard error (SE), in the Mediterranean (MED) and Ionian Sea (ION), and by species, in the period 1994–2019. ρ = Spearman's rank correlation coefficient; p = p-value; n.s.: non-significant values.

	Landing Tonnes			
	Mean ± SE	**%**	**Spearman's ρ**	p
MED	57,774 ± 1857	-	−0.059	n.s
ION	23,153 ± 1754	40%	−0.210	n.s
M. merluccius	9865 ± 1445	43%	−0.637	<0.001
P. longirostris	7597 ± 521	33%	0.426	<0.05
A. foliacea and *A. antennatus*	1921 ± 177	8%	0.446	<0.05
N. norvegicus	1673 ± 117	7%	−0.588	<0.01
Lophius spp.	1330 ± 207	6%	−0.462	<0.05
C. conger	296 ± 41	1%	0.079	n.s
G. melastomus	373 ± 101	2%	0.853	<0.001
H. dactylopterus	52 ± 26	>0.2%	0.584	<0.01
P. blennoides	17 ± 7	>0.2%	0.542	<0.01
P. bogaraveo	9 ± 3	>0.2%	0.681	<0.001
P. americanus	20 ± 4	>0.2%	−0.060	n.s

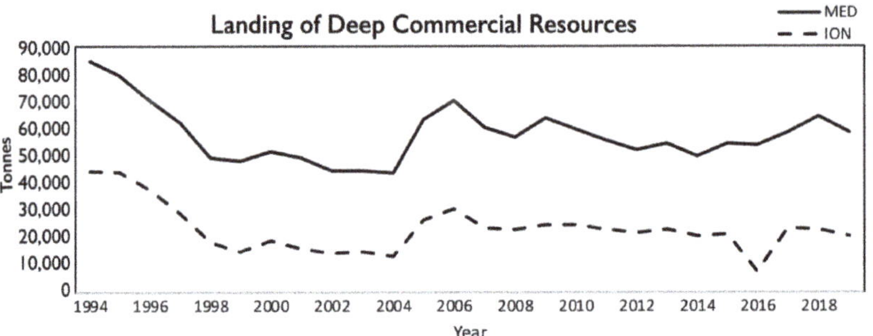

Figure 3. Landings (in tonnes) of deep-sea commercial species in the Mediterranean Sea (black line) and the Ionian Sea (dashed line) in the period 1994–2019, based on official FAO statistics.

The main harvested species in the Ionian Sea are the European hake (9865 ± 1445 tonnes and 43% on average), the deep-water rose shrimp (7597 ± 521 tonnes, 33%), the deep-water red shrimps (*A. foliacea* and *A. antennatus*, 1921 ± 177 tonnes, 8%), the Norway lobster (1673 ± 117 tonnes, 7%), and anglers (*Lophius* spp., 1330 ± 207 tonnes, 6%). For *M. merluccius*, *N. norvegicus*, and *Lophius* spp. significant negative temporal trends have been detected ($p < 0.001$, $p < 0.01$, and $p < 0.05$, respectively). In contrast, the landings of the shrimp *P. longirostris* and deep-water red shrimps showed significant positive trends ($p < 0.05$) (Table 1, Figure 4a,b). For the European conger, a clear landing decrease was shown from 2006 (991 tonnes) to 2019 (172 tonnes) (Figure 4c). The exploitation of other deep-water commercial species has increased in this period, as is the case of the blackmouth catshark, with an increase in the landings from a minimum of 193 tonnes in 2009 to a maximum of 1465 tonnes in 2018 ($p < 0.001$). Similarly, *H. dactylopterus* showed a landing increase in the last five years of the time series ($\rho = 0.584$; $p < 0.01$), as did *P. bogaraveo* ($\rho = 0.681$; $p < 0.001$) and *P. blennoides* ($\rho = 0.542$; $p < 0.01$) (Table 1, Figure 4c,d).

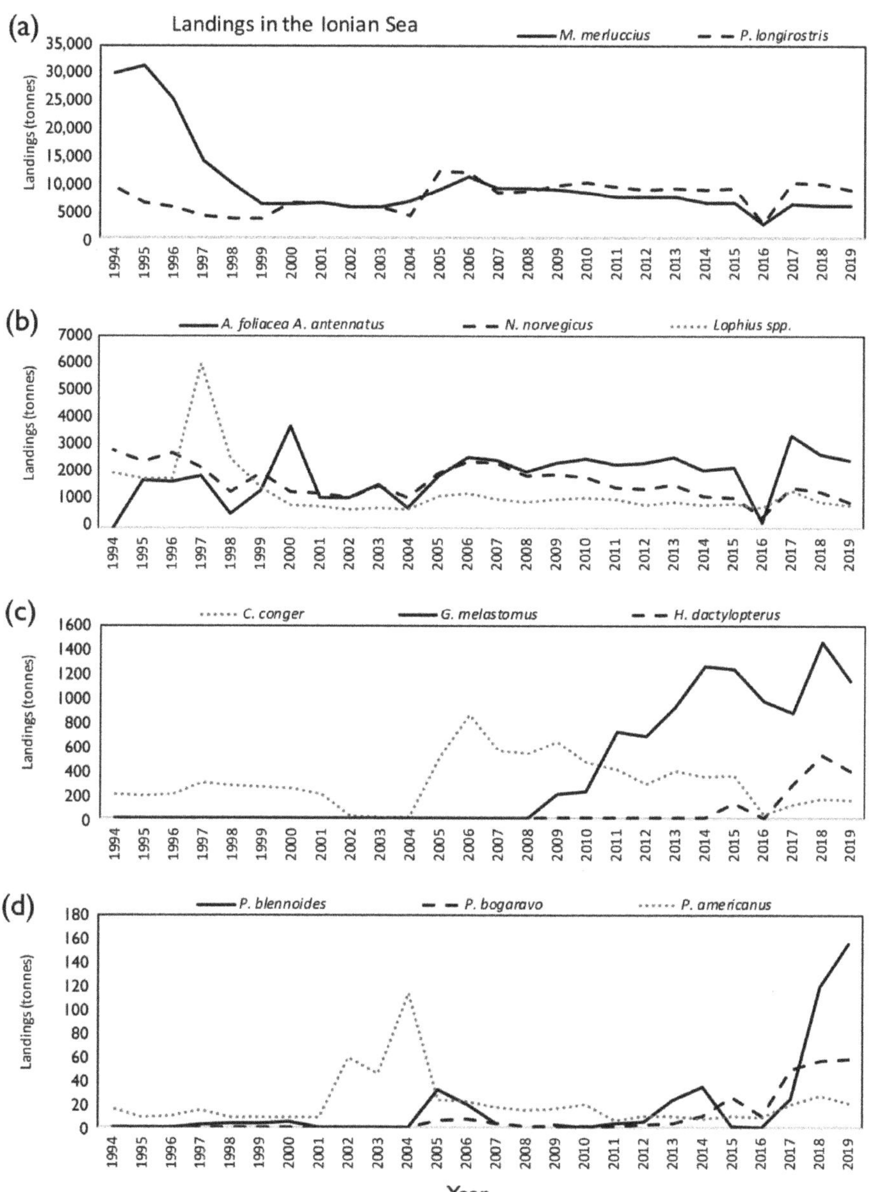

Figure 4. Landings (in tonnes) by deep-sea commercial species—(**a**) *M. merluccius* and *P. longirostris*; (**b**) *A. foliacea*, *A. antennatus*, *N. norvegicus*, and *Lophius* spp.; (**c**) *C. conger*, *G. melastomus*, and *H. dactylopterus*; (**d**) *P. bogaraveo*, *P. blennoides*, and *P. americanus*—caught in the Ionian Sea (FAO subdivisions 37.2.2) during the period 1994–2019.

In the Ionian Sea, the Italian fleet shows the highest exploitation of the deep-water resources, with an average landing value of 19,504 tonnes per year in the period 1994–2019, equal to an average of 84% of the total landing from this basin, followed by Tunisia (1418 tonnes; 6%), Albania (1109 tonnes; 5%), and Greece (1062 tonnes; 5%). Marked fluctuations have been observed in the Italian annual landing, with a non-significant decrease, while significant increases have been observed in landings in Tunisia and Albania ($p < 0.001$). A stable trend of landing was shown for Greece (Table 2, Figure 5).

Table 2. Total landing (in tonnes and %), with mean and standard error (SE), by Ionian countries in the period 1994–2019. ρ = Spearman's rank correlation coefficient; p = p-value; n.s.: non-significant p-values.

	Landing (Tonnes)			
Country	**Mean ± SE**	**%**	**Spearman ρ**	**p**
Italy	19,504 ± 1792	84%	−0.353	n.s.
Greece	1062 ± 67	5%	−0.332	n.s.
Malta	48 ± 7	<0.2%	0.760	<0.001
Tunisia	1418 ± 192	6%	0.902	<0.001
Albania	1109 ± 215	5%	0.856	<0.001

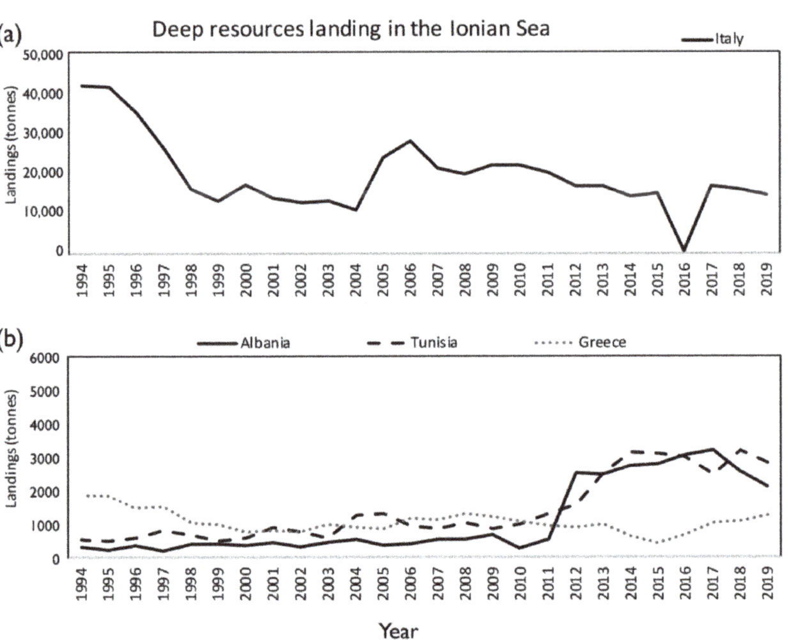

Figure 5. Landings by country ((**a**) Italy; (**b**) Albania, Tunisia and Greece) in the Ionian Sea (FAO subdivisions 37.2.2).

4.2. *Data from MEDITS Trawl Surveys (NorthWestern Ionian Sea, GSA 19)*

The European hake and the deep-water rose shrimp are the most abundant species in weight and number, respectively. Highly significant increases in abundance in weight and number over time were detected for the deep-water rose shrimp (*P. longirostris*) ($p < 0.001$, $\rho = 0.737$ and 0.804, respectively), the giant red shrimp (*A. foliacea*) ($p < 0.001$, $\rho = 0.692$ and 0.599, respectively), and the blackspot seabream (*P. bogaraveo*) ($p < 0.01$, $\rho = 0.538$ and $p < 0.001$, 0.634, respectively). A highly significant negative trend for abundance in both weight and number was detected for *N. norvegicus* ($p < 0.001$, $\rho = -0.766$ and −0.785 respectively). A biomass decrease was only shown for *A. antennatus* ($p < 0.05$, $\rho = -0.391$). Fluctuating abundances with no significant trends were observed for the European hake, greater forkbeard, anglers, and blackmouth catshark over the study period (Table 3, Figure 6a,b).

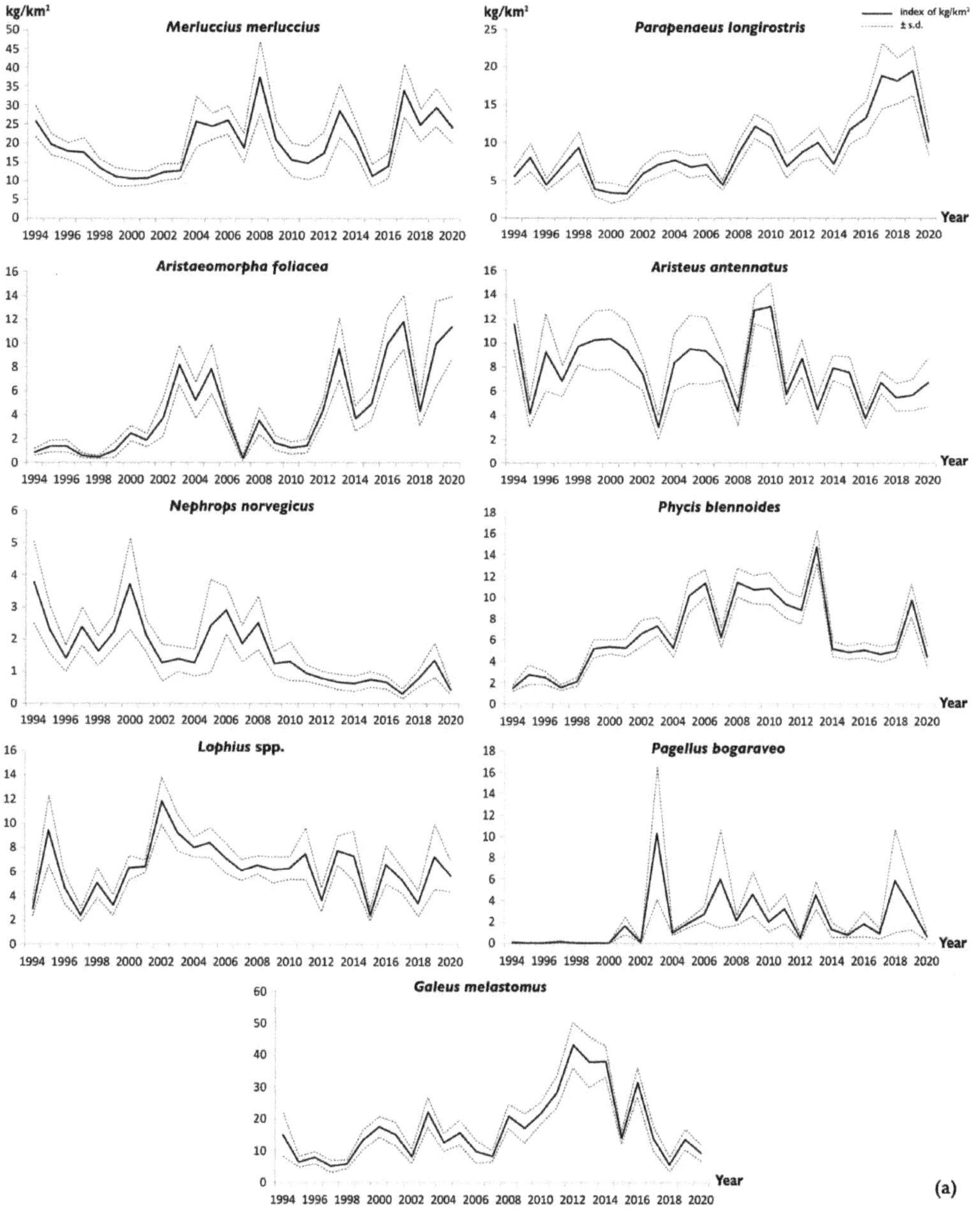

Figure 6. *Cont.*

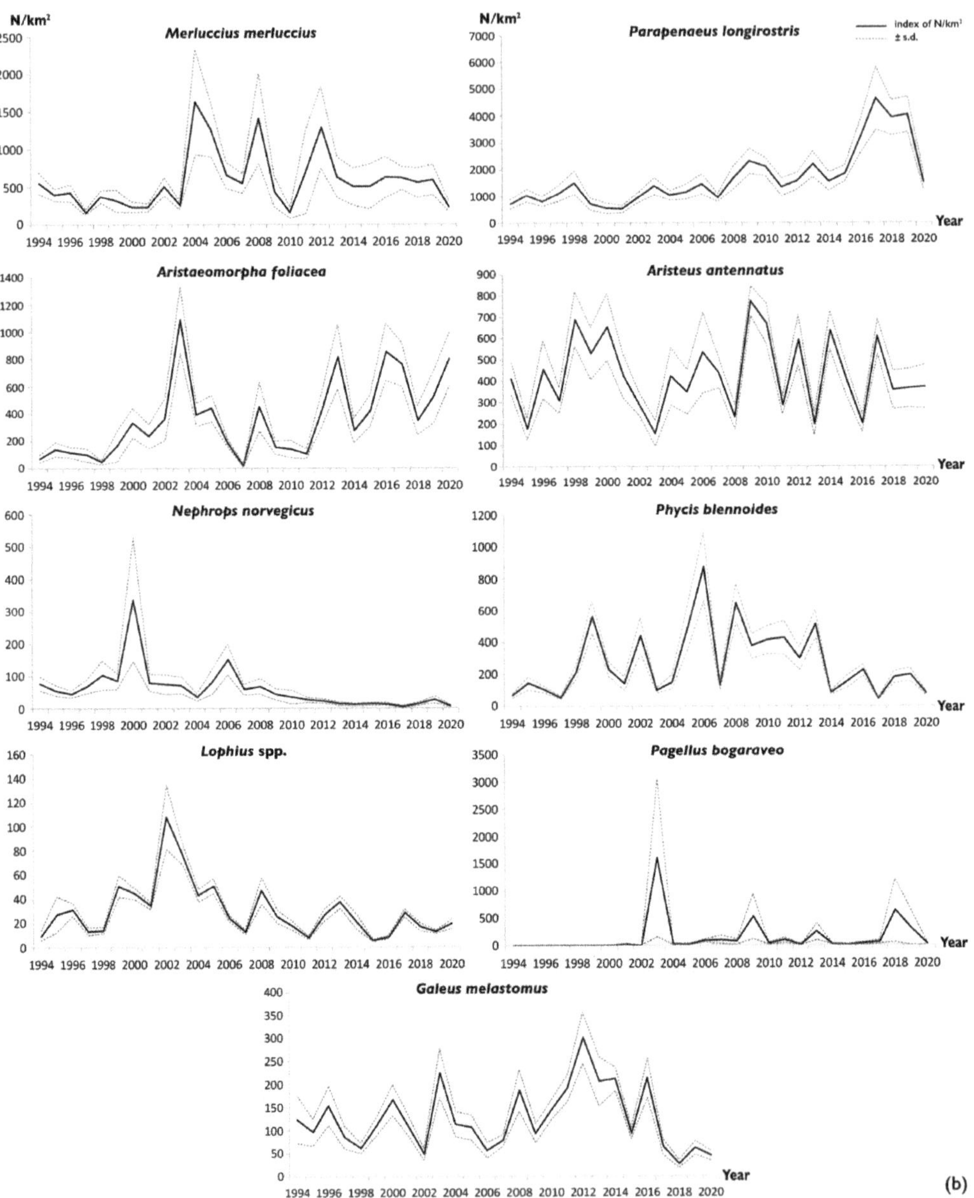

Figure 6. (**a**) Time series of biomass (kg/km^2) index by species sampled during experimental trawl surveys carried out in the northwestern Ionian Sea from 1994 to 2020. (**b**) Time series of abundance (N/km^2) index by species sampled during experimental trawl surveys carried out in the nortwestern Ionian Sea from 1994 to 2020.

Table 3. Mean values of biomass (kg/km^2) and density (N/km^2) indices with standard deviation (s.d.), computed by species on the 1994–2020 time series of experimental trawl surveys carried out in the northwestern Ionian Sea. ρ = Spearman's rank correlation coefficient; p = p-value; n.s.: non-significant p-value.

	kg/km^2			**N/km^2**		
	Mean ± s.d.	**Spearman ρ**	p	**Mean ± s.d.**	**Spearman ρ**	p
Merluccius merluccius	19.98 ± 7.40	0.338	n.s.	585 ± 385	0.336	n.s.
Parapenaeus longirostris	8.87 ± 4.50	0.737	<0.001	1697 ± 1086	0.804	<0.001
Aristaeomorpha foliacea	4.34 ± 3.63	0.692	<0.001	362 ± 284	0.599	<0.001
Aristeus antennatus	7.77 ± 2.67	−0.391	<0.05	427 ± 173	−0.034	n.s.
Nephrops norvegicus	1.59 ± 0.90	−0.766	<0.001	61 ± 65	−0.785	<0.001
Phycis blennoides	6.59 ± 3.47	0.313	n.s.	272 ± 211	0.052	n.s.
Lophius spp.	7.77 ± 2.67	0.023	n.s.	427 ± 173	−0.326	n.s.
Pagellus bogaraveo	2.06 ± 2.44	0.538	<0.01	145 ± 333	0.634	<0.001
Galeus melastomus	16.96 ± 10.54	0.344	n.s.	125 ± 67	−0.085	n.s.

A significant positive relationship was shown between the abundance and SBT in *A. foliacea* ($p < 0.01$, $\rho = 0.580$) and *P. longirostris* ($p < 0.05$, $\rho = 0.464$), while there was a significant negative relationship for *N. norvegicus* ($p < 0.01$, $\rho = -0.642$).

Furthermore, a significant increase in abundance in weight (kg/km^2) in relation to the decreasing FE was observed for *A. foliacea* and *P. longirostris* ($p < 0.001$, $\rho = -0.736$ and -0.746 respectively), as well as for *P. bogaraveo* ($p < 0.005$, $\rho = -0.520$).

The estimated regression coefficients β_0 and β_1 for each model, applied to the log-transformed abundances of *A. foliacea* and *P. longirostris*, are reported in Table 4.

Table 4. Effect of fishing effort and temperature covariates on *P. longirostris* and *A. foliacea* abundance indices in biomass (kg/km^2) and density (N/km^2). Estimate = Estimated coefficient; SE = Standard Error; t = t-value; p = p-value.

		Effect	**Estimate**	**SE**	t	p
P. longirostris	Biomass (kg/km^2)	Intercept	4.33	0.42	10.29	<0.001
		Fishing effort	−0.01	0.01	−5.30	<0.001
		Intercept	−3.36	2.16	−1.56	0.136
		Temperature	0.36	0.14	2.55	0.02
	Density (N/km^2)	Intercept	9.9	0.52	19.10	<0.001
		Fishing effort	−0.01	0.01	−4.97	<0.001
		Intercept	0.97	2.57	0.38	0.712
		Temperature	0.42	0.17	2.49	0.022
A. foliacea	Biomass (kg/km^2)	Intercept	5.04	0.72	6.97	<0.001
		Fishing effort	−0.02	0.01	−5.14	<0.001
		Intercept	−9.48	3.39	2.80	0.011
		Temperature	0.72	0.23	3.21	0.005
	Density (N/km^2)	Intercept	8.39	0.81	10.32	<0.001
		Fishing effort	−0.01	0.003	−3.18	0.005
		Intercept	−1.61	3.38	−0.48	0.639
		Temperature	0.5	0.22	2.21	0.04

Results highlight the significant effects of both covariates on the abundances of the two species. In particular, for both species, increased abundances significantly depend on a decrease in FE, while significant positive effects are estimated for abundances in relation to the increase in SBT.

Regardless of the type of tool used and the type of habitat investigated, the relationships between the sizes and depths show the occurrence of the largest individuals at the greatest depths, with these results being highly significant for all species on soft bottoms, and only for *H. dactylopterus* in VMEs (Figure 7). The sizes were greater in VMEs than on soft bottoms, but this cannot be properly compared due to the different sampling methods and tools (Figure 8).

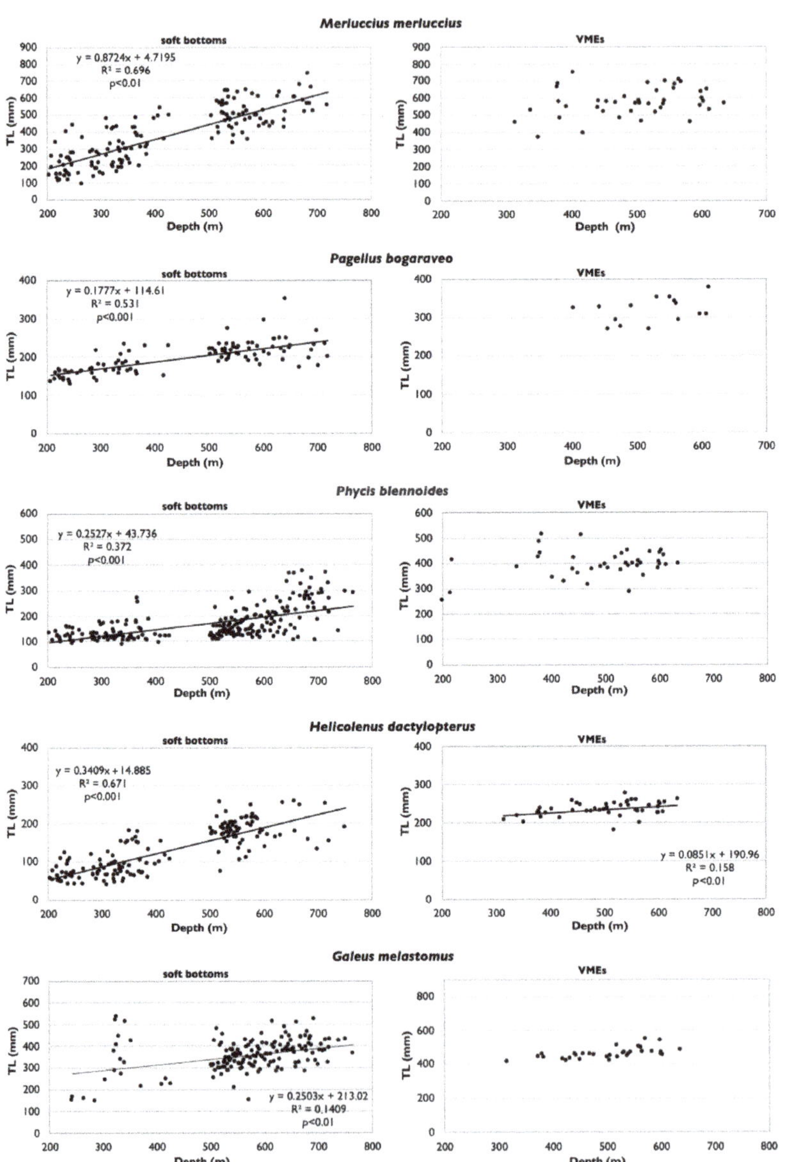

Figure 7. Relationships of total length (TL) with depth of deep-sea species collected on soft bottoms (**left**) and in vulnerable marine ecosystems (VMEs) (**right**) of the central Mediterranean.

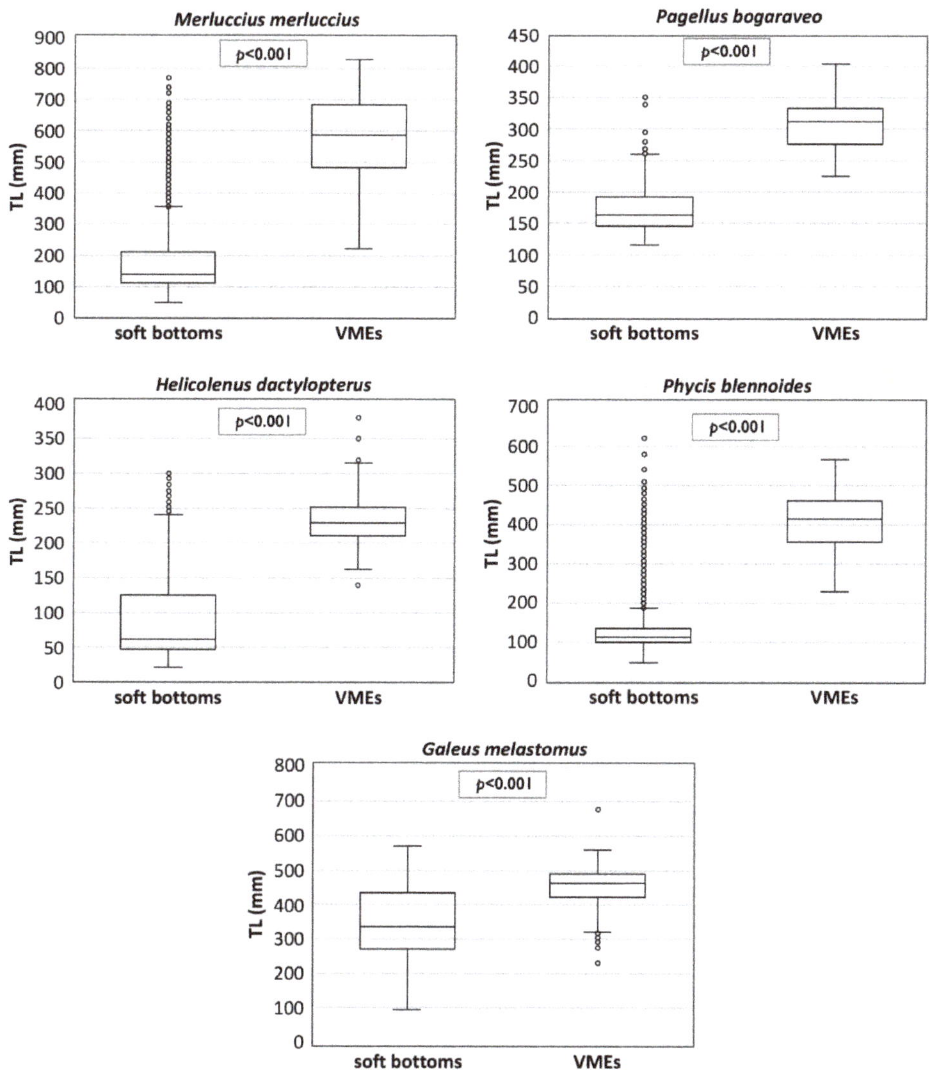

Figure 8. Boxplots of the total length (TL) by species collected on soft bottoms and in vulnerable marine ecosystems (VMEs) of the central Mediterranean.

For all of the deep-sea species considered, the abundances in weight and number for an area near the SML CWC province were greater than those for the area far from this coral province (Figure 9). However, the differences were significant for the abundance in both weight and number of *A. foliacea*, *H. dactylopterus*, and *P. bogaraveo*. The differences between the two areas were significant for the abundance in number of *P. blennoides* and for the abundance in weight of *M. merluccius* (Figure 9). All of these differences are even more significant due to the fact that the fishing effort is significantly greater near the SML CWC province than in the area far from this province in terms of both the number ($p < 0.001$) and the GT of vessels ($p < 0.05$) (Figure 10).

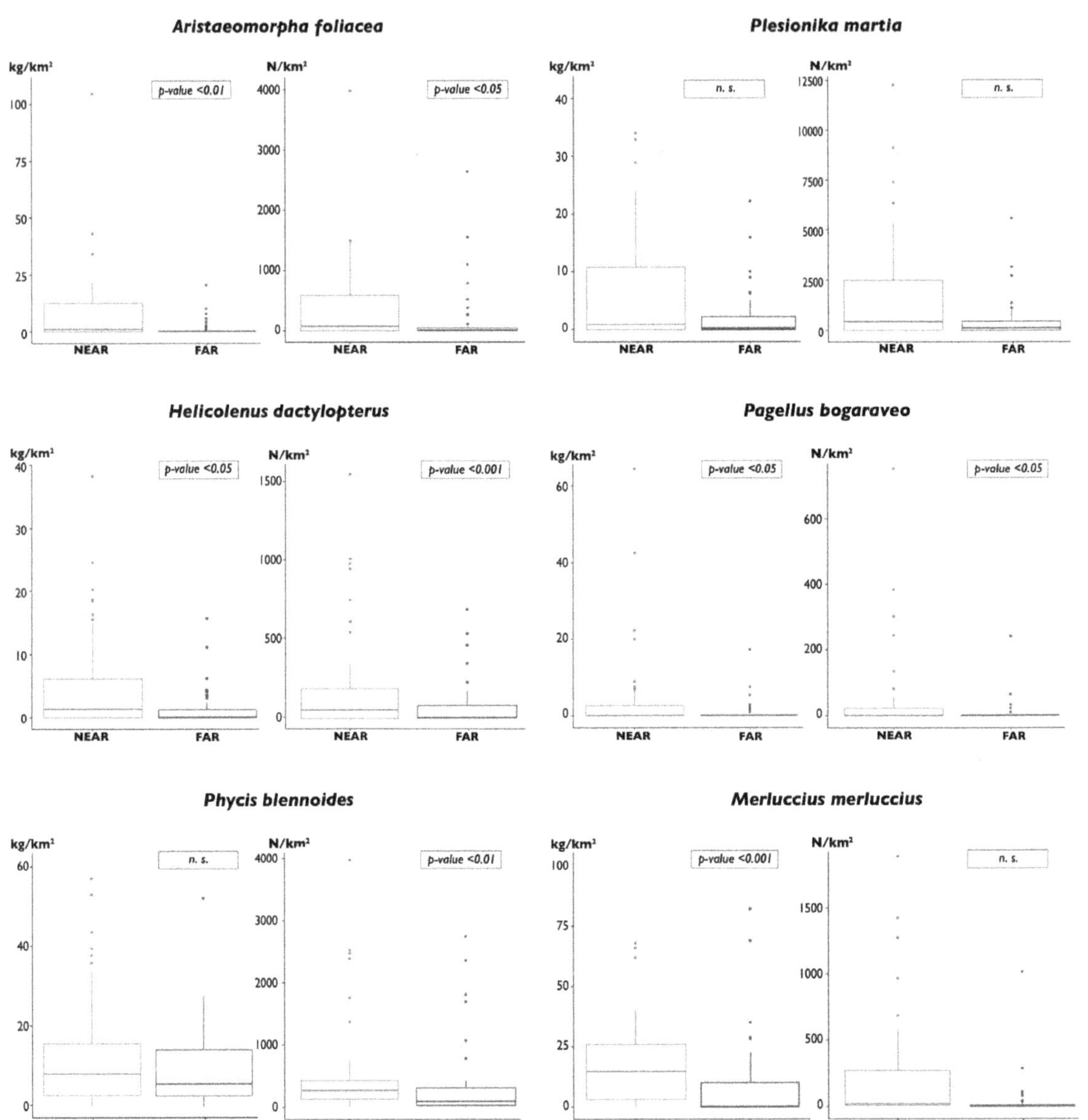

Figure 9. Boxplots of biomass (kg/km^2) and abundance (N/km^2) by species collected in two areas of the northwestern Ionian Sea, near and far from the SML fishery restricted area, from 1994 to 2020.

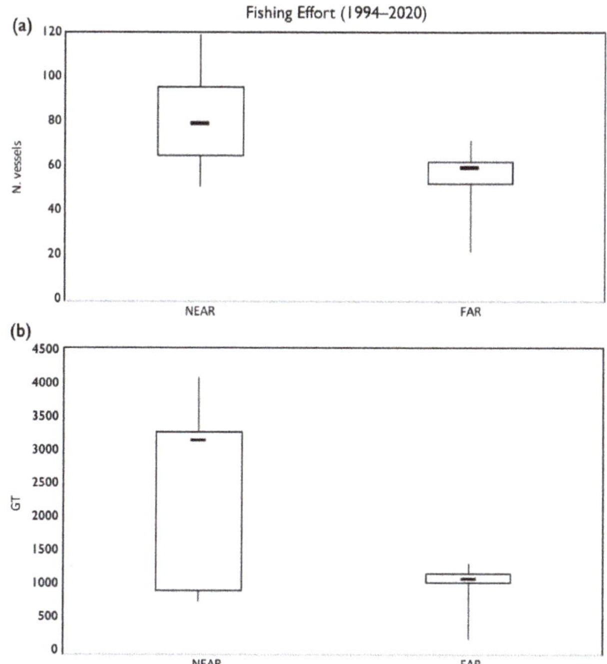

Figure 10. Boxplots of the fishing effort in terms of number (N) (**a**) and gross tonnage (GT) (**b**) of trawl vessels operating near to and far from the SML CWC province during the period 1994–2020.

5. Discussion

Macrobenthic invertebrates, such as soft and hard corals, contribute to the formation of heterogeneous and complex habitats on slopes, in canyons, and on seamounts, which constitute VMEs. Deep VMEs are hotspots of biodiversity, due to the variety of species they host, and are valuable fishing areas due to the occurrence of large individuals of commercial species and higher catches than in adjacent areas (e.g., [20,55,58,69,70]).

The deep-sea species distributed in the deep-sea habitats and VMEs of the Mediterranean make up 7% of the total quantified landings of approximately 788,000 tonnes [1]. The Ionian Sea is the basin where the largest fraction (about 40%) of the Mediterranean's deep resources is captured. This basin, in addition to being the deepest in the entire Mediterranean, presents an articulated and complex hydrography and geomorphology of the seabed, with canyons, banks, rocky bottoms, and VMEs where soft and hard coral species thrive (e.g., [81,92–94,142–144]).

Catches from official FAO statistics show consistency with catches from MEDITS trawl surveys. In both cases, abundance fluctuations were observed for most species. The most abundant deep-sea food resources in the Ionian Sea are the European hake (*M. merluccius*), the deep-water rose shrimp (*P. longirostris*), and the deep-water red shrimps (*A. foliacea* and *A. antennatus*). Other deep-sea food resources are represented by the crustacean *N. norvegicus*, the teleost fish *Lophius* spp., *C. conger*, *H. dactylopterus*, *P. bogaraveo*, *P. blennoides*, and *P. americanus*, and the shark *G. melastomus*.

For some species, significant increases in abundances have been observed, and this may be related to the reduction in fishing effort (e.g., *A. foliacea*, *P. longirostris*, *P. bogaraveo*), the increase in SBT (e.g., *A. foliacea*, *P. longirostris*), and the presence of refuge areas scarcely accessible to fishing, which constitute VMEs and sites of resource renewal (e.g., *A. foliacea*, *H. dactylopterus*, *P. bogaraveo*, and, to a lesser extent, *P. blennoides* and *M. merluccius*). In particular, adult individuals of *P. bogaraveo* were almost exclusively collected in VMEs between the southwestern Adriatic and northwestern Ionian seas (e.g., [71,78–80,145]). The golden shrimp (*P. martia*) was also found to be associated with the presence of corals [145].

Time-series data for the period 1985–2005 reveal a positive relationship of *A. foliacea* and *P. longirostris* abundances with a rise in the water temperature. Inverse relationships of *A. foliacea*, *P. longirostris*, *N. norvegicus*, and *P. blennoides* abundances with FE were also detected [78].

The increase in the abundance of *P. longirostris* observed in the northwestern Ionian Sea has also been reported in all European Mediterranean waters, demonstrating that the abundance of some stock is closely linked to climate change [146]. The increase in temperature observed in recent years may have produced an increase in suprabenthos, e.g., *Lophogaster typicus*, which represents the main prey of the deep-water rose shrimp [146,147]. In addition, as observed in the northern Tyrrhenian, high temperatures and low wind circulation negatively affect the recruitment of *M. merluccius*. Hake juveniles prey upon *P. longirostris* juveniles [148], so the lower predation pressure could have further enhanced the recruitment success of the shrimp [146,149]. The significant increase in abundance detected for *A. foliacea* is consistent with observations in the eastern Ionian Sea and southern Adriatic Sea [150]. However, for both *P. longirostris* and *A. foliacea*, the reduction in FE may have influenced the increase in their abundances.

Regarding *A. antennatus*, in the western Mediterranean it was observed that this deep-water shrimp seems to prefer relatively cold temperatures (13.1–13.2 °C) and relatively salty waters (>38.5) with low currents and moderate variability [123]). However, the occurrence of *A. antennatus* appears to be driven in a nonlinear manner by environmental conditions, including local temperature [123].

Times-series data regarding the abundance of *M. merluccius* in the northwestern Ionian for the period 1985–2005 reveal a positive relationship with the NAO index [78]. Recently, it has been observed that environmental factors can affect the spatiotemporal distribution pattern of the European hake throughout the Mediterranean basin [151]. In particular, high predicted biomass levels were observed especially at 200 m and between 14 and 18 °C, highlighting the preference of the species for colder waters. Moreover, the high biomass of this teleost fish has been correlated with the presence of nursery areas in many Mediterranean areas, some of which have been identified along the northwestern Ionian and in the southwestern Adriatic [152–155], connected to VMEs [81,83]. In this respect, the effect of the presence of the SML VME has been detected for *M. merluccius* and other deep-water species, in agreement with previous investigations in the central Mediterranean (e.g., [12,71,81,83,85,98]). Indeed, the SML CWC province has an effect on the abundance of deep-water resources, while also contributing to the spillover of individuals of commercial species exploited in the surrounding fishing grounds, subjected to a greater fishing pressure than in other areas of the northwestern Ionian Sea.

The most abundant species captured on the muddy bottoms, both by commercial fishing (FAO data) and during trawl surveys (MEDITS data), are also those most frequently caught in sensitive habitats and VMEs. Most of these species exhibit a bigger–deeper pattern, both on muddy bottoms—where larger individuals are generally caught by trawling—and in sensitive habitats and VMEs, partially protected from fishing due to the roughness of the seabed and the presence of rocks and hard corals. The largest individuals distributed in deep waters and in VMEs represent the adult fraction of the stocks. These are breeder individuals that often concentrate in VMEs, making them EFHs, since these ecosystems act as spawning and nursery areas [12,57,71,81,98], contributing to the renewal of the stocks on fishing grounds [83]. Both the spillover (i.e., active movement of juveniles and adults into fished areas) and larval seeding (i.e., the dispersal of eggs and larvae in fished areas) may contribute to the renewal of stocks in neighbouring fishing grounds. All of the species examined in this study carry out their life cycles on the bathyal bottoms by distributing themselves between muddy bottoms and refuge areas represented by VMEs, but only in these ecosystems is their vulnerability to fishing activities greatly reduced. In fact, models developed using data from trawl surveys carried out on muddy bottoms provide indications of overfishing status for most of the species examined in the present study (e.g., [156]).

Species with a wide bathymetric distribution, such as anglers, have not shown any significant trend over time. Norway lobster, for both types of data, showed a decrease in abundance over time. This may be due to the fact that this species needs soft bottoms to dig burrows in the mud, which are also the most subjected to trawling. Another explanation deals with the negative relationship between Norway lobster biomass and SST, with the lowest indices associated with high temperatures [146]. The decreasing abundance related to higher temperatures is due to the reduction in the organic matter flux resulting from the decreased rainfall and river discharge, which influences benthic feeders and predators such as the Norway lobster [146,147].

Several deep-water species also represent an important feeding resource for several species of odontocetes inhabiting the offshore areas of the northwestern Ionian Sea [157–160]. Bathyal bentho-pelagic squids are fundamental prey of the Risso's dolphin (*Grampus griseus*) and the sperm whale (*Physeter macrocephalus*). Myctophids and demersal species, such as *M. merluccius*, are hunted by the striped dolphin (*Stenella coeruleoalba*) [161–164]. The trophic interactions between cetaceans and deep-water prey contribute to the recycling of energy and matter in the pelagic domain and coastal areas. Odontocetes can increase the CO_2 absorption capability of phytoplankton, thus playing a critical role in climatic regulation [165]. Thus, the management of deep sensitive habitats and their species contributes to ensuring the stability of several biological components that play a critical role in ecosystem functioning [164].

The increase in landings observed for Tunisia and Albania could be due to the fact that there are still deep areas that are largely unexploited by fishing activity. Moreover, here, the fishing pressure is lower than that along the Italian coasts of the central Mediterranean.

The VMEs are often impacted by commercial fishing (e.g., [17,20,64,68,69,82,98,103,104,106,107,127,129]). An overview of available information on the incidental catch of VME indicator taxa from fishery-dependent and fishery-independent surveys has been recently provided by Chimienti et al. [55]. Bottom trawls represent the most impactful fishing practice to deep soft-bottom VMEs, followed by longlines, gillnets, pots, and traps.

Other human activities—including oil and gas exploration and exploitation, pollution, marine litter, ocean acidification, and climate change—are also harmful to VMEs and their biodiversity on the open slopes, canyons, and seamounts (e.g., [14,64,69,126,166–168]). In addition, biodiversity loss affects ecosystem services (ESs) and impairs the ocean's capacity to provide food, maintain water quality, and recover from disturbance [169–172]. ESs are the benefits that humans derive, either directly or indirectly, from the functions of ecosystems [173]. The Millennium Ecosystem Assessment (MA) [174] estimates that 60% of global ecosystem services are degraded or are being managed unsustainably. As biodiversity is lost and ecosystems are degraded, the biocapacity of the planet to support living organisms decreases. As biocapacity decreases, there are diminishing resources available to support a growing human population [175]. These considerations also concern the Mediterranean and its VMEs that provide several ESs [54] and, as noted above, many of them are in danger due to anthropogenic impacts.

The habitat structured by macrobenthic invertebrates in VMEs is a supporting ES that provides organisms—including those of commercial species—with suitable physical and chemical features, food and spatial resources, places for courtship, mating, and spawning, breeding sites and nurseries, places to hide from predators, and refuges to escape from adverse environmental conditions (e.g., [38,45,58,72,176,177]). The decline of the habitat can cause negative effects on the species that use it for bioecological processes, affecting community composition and ecosystem functioning [178,179].

A strategy based on ecosystem-based fishery management (EBFM) was adopted by the EU Common Fishery Policy (CFP) for fishery management, with the overall objective of sustaining healthy marine ecosystems and the fisheries that they support [180]. This implies sustainable management not only of the commercial stocks, but also of the whole environmental system that supports their production, including the importance of the economic and social dimensions.

Worldwide, fish consumption has grown enormously since 1961, surpassing even that of several species of the most common terrestrial animals in the human diet. Thinking about the fact that food innovation and sustainability are mainly based on systems where it is produced and systems by which food is collected, there is an urgent need to protect deep-sea habitats and VMEs by promoting the use of sustainable fishing systems.

Many governments have made international commitments to the conservation of marine biodiversity. In particular, since the adoption of the Convention on Biological Diversity (CBD) in 1992, biodiversity considerations in relation to the management of fisheries and aquaculture have been focused on policies and actions for the conservation of threatened species and vulnerable habitats. Within the CBD context, the scientific criteria to identify ecologically and biologically significant marine areas have been established [181], with the aim of defining management measures that ensure the conservation of the biodiversity of these areas. The process in the Mediterranean was started by a regional workshop in 2014 [182], and culminated at the CBD COP 12 with the endorsement of 15 ecologically and biologically significant marine areas, recognising the biological and ecological significance of deep-sea habitats in the Mediterranean [64]. Following specific UNGA resolutions [183–185], similar criteria were adopted by the FAO [46] to identify VMEs and to develop international guidelines for the management of deep-sea fisheries in the high seas, in order to ensure the protection of certain groups of species and habitats from significant adverse impacts (SAIs) caused by fisheries.

The Habitats Directive (92/43/EEC) [186] considers the CWC biotope as the habitat type "1170 Reefs", for which measures should be taken to maintain and restore a conservation status for this type of habitat. As part of the Barcelona Convention, the protocol concerning the specially protected areas and biological diversity adopted in 1995 is a tool for implementing the CBD, since it aims to protect and conserve biodiversity in valuable areas and species in the Mediterranean Sea. More recently, the Marine Strategy Framework Directive [187] aims to ensure that the collective pressure of human activities on the environment is kept within levels compatible with the achievement of good environmental status. This can also be achieved through the creation of marine protected areas (MPAs) focused on achieving a balance between sustainable fisheries and other human activities and habitat conservation. Although several conservation initiatives have been developed with the aim of protecting threatened hotspots of marine biodiversity, the marine protected areas (MPAs) that have been designated in deep waters are very limited [64,188,189].

During the last few decades, the Mediterranean deep-sea habitats have been protected through the institution of fishery restricted areas (FRAs), established by the General Fishery Commission for the Mediterranean and Black Sea (GFCM) with the aim of protecting VMEs and/or essential fish habitats (EFHs). To date, 10 FRAs have been established by the GFCM, including 1 large deep-water FRA below 1000 m. Trawling is forbidden in areas deeper than 1000 m in depth throughout the Mediterranean Sea. However, this conservation measure is not enough, considering that most of the coral habitats known so far are present within this depth limit. For this reason, the limitation of trawling up to 800 m depth would be more effective for the conservation of deep-sea coral habitats [54,64]. The FRAs aim to protect EFHs and/or sensitive habitats of high ecological value, such as VMEs, from any SAIs of fishing activities [46]. In particular, only two of the existing FRAs in the Mediterranean Sea have been created to target the conservation of a CWC habitat—namely, the *Lophelia* Reef off Santa Maria di Leuca (Italy, Ionian Sea) [82] and, more recently, the Bari Canyon in the southern Adriatic Sea (recommendation GFCM/44/2021/3) [133]. Three FRAs in the Strait of Sicily (northeast and northwest Malta) and one in the Gulf of Lion include some CWC sites, but they have been created to manage fishing stocks; thus, trawling is present there, albeit somewhat regulated. The Jabuka/Pomo Pit FRA aims to protect EFHs and an unquantified sea pen field [190], while the Eratosthenes Seamount FRA targets the protection of peculiar geological formations (with only a few specimens of solitary scleractinians recorded) [138], and the Nile Delta FRA is characterised by the presence of chemosynthetic fauna [54]. Trawling and dredging are forbidden in the FRAs for the conservation of VMEs, while they are regulated in those for the management of EFHs. Bottom longlining can be allowed—often in a buffer zone and under authorisation—depending on the regulations of the single FRA, while artisanal fishing practices are usually not performed in offshore areas, such as in the existing FRAs. These FRAs are currently isolated, while a desirable network of FRAs is a long way from being created. This network should be established in the pathway of the Mediterranean water mass circulation in order to connect the different FRAs all over the basin by means of larval dispersal [53,83,129].

As part of goal 14 (life below water) of the 2030 Sustainable Development Agenda [191], all of the targets regard VMEs, and specifically targets 14.2 (protect and restore ecosystems), 14.3 (reduce ocean acidification), and 14.4 (sustainable fishing). More recently, the EU Biodiversity Strategy for 2030 [192] reports that biodiversity is also seen as essential for safeguarding EU and global food security; its loss threatens our food systems, putting our food security and nutrition at risk.

6. Conclusions and Recommendations

In the EU, at least 30% of the land surface and 30% of the sea should be protected, and for areas with high value or potential for biodiversity and, therefore, greater vulnerability, more stringent protection is needed.

Regarding the deep-sea sensitive habitats and VMEs, the implementation of a network of protected or fishing-restricted areas represents a fundamental measure to guarantee the proper conservation of the sites known in the Mediterranean Sea. A network of protected areas (mostly MPAs and FRAs) would satisfy both the conservation of vulnerable habitats and the management of fishery resources according to the EBFM [63]. In this respect, after the past and recent establishment of the FRAs in the SML CWC province and Bari Canyon, respectively, appropriate spatial measures aimed at preserving the ecological function of *I. elongata facies* identified in the southern Adriatic should be adopted [98]. The Bari Canyon, *I. elongata facies*, and SML CWC province are in the path of the flow of the dense-water masses that pour from the southern Adriatic into the northern Ionian [53,81,92,94,97,129,193]. This stream of water masses connects the deep-sea coral communities distributed along the Apulian margin, which represent a network of refuge/renewal areas of fishery resources [81] that needs coherent conservation measures and management strategies according to the EBFM [53–55].

In the southern Adriatic, the activity of trawler fleets is more concentrated on the continental shelf and on the upper part of the slope, with target species including the European hake, red mullet, spottail mantis shrimp, deep-water rose shrimp, and Norway lobster [194]. This explains the persistence of *I. elongata* in the southern Adriatic [98], and would suggest that the protection measure to be adopted would be accepted by the fishermen. If not accepting an area closure measure, other management approaches—such as encounter protocols with associate thresholds—can be developed and implemented according to the FAO's International Guidelines for the Management of Deep-Sea Fisheries [46,195], in order to ensure the protection of *I. elongata* from SAIs. In addition, the use of onboard observers and the correct adoption of digital logbooks could be applicable to trawl fishing vessels, as well as to most of the deep-sea benthic longlining vessels, which must be equipped with vessel monitoring by satellite systems (VMSs) and/or automated identification systems (AISs). The management and control of the many small artisanal fishing boats, which use gillnets and trammel nets in shallower waters, could be achieved through the designation of landing points, obligations of notice of arrival in port, and control of landings [54].

Through blue growth (BG), the European Union seeks to meet human needs—such as food and energy—in a sustainable way, by creating new jobs and new sources of growth while safeguarding biodiversity and protecting the marine environment, thus preserving the services that healthy and resilient marine and coastal ecosystems provide [196]. In this respect, marine spatial planning [187] consists of a public process of analysing and allocating the spatial and temporal distribution of human activities in marine areas to achieve ecological, economic, and social objectives, reducing conflicts and creating synergies between different activities, while protecting the environment by assigning protected areas, calculating impacts on ecosystems, and identifying opportunities for multiple uses of space. The use of an ecosystem for economic returns and social benefits must be done in a way that minimises negative impacts. If an ecosystem is degraded, the ESs that are derived from it will also be modified, including those concerning the availability and safety of food.

The good governance of an ecosystem requires, first of all, knowledge and monitoring of its condition. The central Mediterranean is an area with a complex hydrology and geomorphology, rich in biodiversity and biological resources, requiring greater knowledge and continuous monitoring of sensitive habitats and VMEs, as well as their associated biotic components, which can represent food resources [64,76]. Danovaro et al. [197] used expert elicitation (1155 deep-sea scientists consulted and 112 respondents) to indicate a wide consensus that monitoring should prioritise large organisms (that is, macro- and megafauna) living in deep waters and in benthic habitats, whereas monitoring of ecosystem functioning should focus on trophic structure and biomass production. They suggested that deep-sea conservation efforts should focus primarily on VMEs and habitat-forming species.

This is of particular importance in relation to the growth of human populations and the expansion of activities regarding the deep-sea environment, and for which the involvement of stakeholders with an interest in the deep-sea will be necessary, together with mechanisms that promote wide participation at the national and international levels, and that ensure conservation and long-term effective ecosystem-based fishery management measures.

Author Contributions: Conceptualisation: G.D. and P.M.; investigation: G.D., P.M., L.S., F.C., A.C., G.C., R.C., P.R. and A.T.; contributed materials/analyses/tables/figures: C.C., F.C., L.S., P.M. and P.R.; writing—original draft preparation: G.D. and P.M.; references: A.C.; corresponding author: F.C. All authors have read and agreed to the published version of the manuscript.

Funding: The data used in this study were collected under the Data Collection Framework (DCF), supported by the Italian Ministry of Agriculture, Food and Forestry Policy (MiPAAF) and by the European Commission (EU Regulations 1004/2017).

Institutional Review Board Statement: All specimens analysed in this study were collected from the fishery (Data Collection Framework [DCF]; EU Reg. 199/2008). Therefore, this study does not comply with the European Commission recommendations (Directive 2010/63/EU of the European Parliament and of the Council of 22 September 2010) or with Italian National Law (Decree Law n. 26 of 4 March 2014) on the protection of animals used for scientific experiment.

Informed Consent Statement: Not applicable.

Data Availability Statement: The data used in this study were collected under the Data Collection Framework (DCF), supported by the Italian Ministry of Agriculture, Food and Forestry Policy (MiPAAF) and by the European Commission (EU Regulations 1004/2017). Data provided are owned by the Italian Ministry of Agriculture, Food and Forestry Policy (MiPAAF).

Acknowledgments: The authors thank their colleague Richard Lusardi, a native English-speaking expert, for manuscript review and Daniela Potenza, graphics expert, for the editing of the Figures.

Conflicts of Interest: The authors declare no conflict of interest.

References

1. FAO. *The State of Mediterranean and Black Sea Fisheries 2020*; FAO, General Fisheries Commission for the Mediterranean: Rome, Italy, 2020; ISBN 978-92-5-133724-0.
2. Gordon, J.D.M. Deep-Water Fisheries at the Atlantic Frontier. *Cont. Shelf Res.* **2001**, *21*, 987–1003. [CrossRef]
3. Morato, T.; Watson, R.; Pitcher, T.J.; Pauly, D. Fishing down the Deep. *Fish Fish.* **2006**, *7*, 24–34. [CrossRef]
4. Villasante, S.; Morato, T.; Rodriguez-Gonzalez, D.; Antelo, M.; Österblom, H.; Watling, L.; Nouvian, C.; Gianni, M.; Macho, G. Sustainability of Deep-Sea Fish Species under the European Union Common Fisheries Policy. *Ocean. Coast. Manag.* **2012**, *70*, 31–37. [CrossRef]

5. Caddy, J.F. Toward a Comparative Evaluation of Human Impacts on Fishery Ecosystems of Enclosed and Semi-enclosed Seas. *Rev. Fish. Sci.* **1993**, *1*, 57–95. [CrossRef]
6. Rogers, A.D. The Biology of Seamounts. In *Advances in Marine Biology*; Elsevier: Amsterdam, The Netherlands, 1994; Volume 30, pp. 305–350, ISBN 978-0-12-026130-7.
7. Koslow, J.A. Seamounts and the Ecology of Deep-Sea Fisheries. *Am. Sci.* **1997**, *85*, 168–176.
8. De Leo, F.C.; Smith, C.R.; Rowden, A.A.; Bowden, D.A.; Clark, M.R. Submarine Canyons: Hotspots of Benthic Biomass and Productivity in the Deep Sea. *Proc. Biol. Sci.* **2010**, *277*, 2783–2792. [CrossRef] [PubMed]
9. Fosså, J.H.; Mortensen, P.B.; Furevik, D.M. The Deep-Water Coral *Lophelia Pertusa* in Norwegian Waters: Distribution and Fishery Impacts. *Hydrobiologia* **2002**, *471*, 1–12. [CrossRef]
10. Husebø, Å.; Nøttestad, L.; Fosså, J.H.; Furevik, D.M.; Jørgensen, S.B. Distribution and Abundance of Fish in Deep-Sea Coral Habitats. *Hydrobiologia* **2002**, *471*, 91–99. [CrossRef]
11. Cartes, J.E.; Maynou, F.; Sardà, F.; Lloris, D.; Tudela, S. *The Mediterranean Deep-Sea Ecosystems: An Overview of Their Diversity, Structure, Functioning and Anthropogenic Impacts*; WWF, IUCN, Eds.; WWF Mediterranean Program; IUCN Centre for Mediterranean Cooperation: Málaga, Spain, 2004; 64p.
12. D'Onghia, G.; Maiorano, P.; Sion, L.; Giove, A.; Capezzuto, F.; Carlucci, R.; Tursi, A. Effects of Deep-Water Coral Banks on the Abundance and Size Structure of the Megafauna in the Mediterranean Sea. *Deep. Sea Res. Part II Top. Stud. Oceanogr.* **2010**, *57*, 397–411. [CrossRef]
13. Muñoz, P.D.; Murillo, F.J.; Sayago-Gil, M.; Serrano, A.; Laporta, M.; Otero, I.; Gómez, C. Effects of Deep-Sea Bottom Longlining on the Hatton Bank Fish Communities and Benthic Ecosystem, North-East Atlantic. *J. Mar. Biol. Assoc.* **2011**, *91*, 939–952. [CrossRef]
14. Company, J.B.; Ramirez-Llodra, E.; Sardà, F.; Aguzzi, J.; Puig, P.; Canals, M.; Calafat, A.; Palanques, A.; Solé, M.; Sanchez-Vidal, A.; et al. Submarine canyons in the Catalan Sea (NW Mediterranean): Megafaunal biodiversity patterns and anthropogenicthreats. In *Mediterranean Submarine Canyons: Ecology and Governance*; Würtz, M., Ed.; IUCN: Gland, Switzerland; Málaga, Spain, 2012; pp. 133–144.
15. Farrugio, H. A Refugium for the Spawners of Exploited Mediterranean Marine Species: The Canyons of the Continental Slope of the Gulf of Lion. In *Mediterranean Submarine Canyons: Ecology and Gorvernance*; Wurtz, M., Ed.; IUCN: Gland, Switzerland; Malaga, Spain, 2012; pp. 45–49.
16. Henry, L.-A.; Moreno Navas, J.; Roberts, J.M. Multi-Scale Interactions between Local Hydrography, Seabed Topography, and Community Assembly on Cold-Water Coral Reefs. *Biogeosciences* **2013**, *10*, 2737–2746. [CrossRef]
17. Fabri, M.-C.; Pedel, L.; Beuck, L.; Galgani, F.; Hebbeln, D.; Freiwald, A. Megafauna of Vulnerable Marine Ecosystems in French Mediterranean Submarine Canyons: Spatial Distribution and Anthropogenic Impacts. *Deep. Sea Res. Part II Top. Stud. Oceanogr.* **2014**, *104*, 184–207. [CrossRef]
18. Kutti, T.; Bergstad, O.A.; Fosså, J.H.; Helle, K. Cold-Water Coral Mounds and Sponge-Beds as Habitats for Demersal Fish on the Norwegian Shelf. *Deep. Sea Res. Part II Top. Stud. Oceanogr.* **2014**, *99*, 122–133. [CrossRef]
19. Hinz, H. Impact of Bottom Fishing on Animal Forests: Science, Conservation, and Fisheries Management. In *Marine Animal Forests: The Ecology of Benthic Biodiversity Hotspots*; Rossi, S., Ed.; Springer International Publishing: Cham, Switzerland, 2017; pp. 1041–1059.
20. Bo, M. Unveiling the Deep Biodiversity of the Janua Seamount (Ligurian Sea): First Mediterranean Sighting of the Rare Atlantic Bamboo Coral Chelidonisis Aurantiaca Studer, 1890. *Deep. Sea Res. Part I Oceanogr. Res. Pap.* **2020**, *156*, 103186. [CrossRef]
21. Lack, M.; Traffic Oceania; Short, K.; Willock, A. *Managing Risk and Uncertainty in Deep-Sea Fisheries: Lessons from Orange Roughy*; WWF Traffic Oceania: Sydney, Australia, 2003; 73p.
22. Jennings, S.; Greenstreet, S.P.; Reynolds, J.D. Structural Change in an Exploited Fish Community: A Consequence of Differential Fishing Effects on Species with Contrasting Life Histories. *J. Anim. Ecol.* **1999**, *68*, 617–627. [CrossRef]
23. Koslow, J. Continental Slope and Deep-Sea Fisheries: Implications for a Fragile Ecosystem. *ICES J. Mar. Sci.* **2000**, *57*, 548–557. [CrossRef]
24. Clark, M. Are Deepwater Fisheries Sustainable?—The Example of Orange Roughy (*Hoplostethus Atlanticus*) in New Zealand. *Fish. Res.* **2001**, *51*, 123–135. [CrossRef]
25. Dulvy, N.K.; Sadovy, Y.; Reynolds, J.D. Extinction Vulnerability in Marine Populations. *Fish Fish.* **2003**, *4*, 25–64. [CrossRef]
26. Dulvy, N.K.; Ellis, J.R.; Goodwin, N.B.; Grant, A.; Reynolds, J.D.; Jennings, S. Methods of Assessing Extinction Risk in Marine Fishes. *Fish Fish.* **2004**, *5*, 255–276. [CrossRef]
27. Cheung, W.W.L.; Pitcher, T.J.; Pauly, D. A Fuzzy Logic Expert System to Estimate Intrinsic Extinction Vulnerabilities of Marine Fishes to Fishing. *Biol. Conserv.* **2005**, *124*, 97–111. [CrossRef]
28. International Council for the Exploration of the Sea (ICES). *Report of the ICES Advisory Committee on Fishery Management ICES Cooperative Research Report No. 246*; ICES: Copenhagen, Denmark, 2001.
29. Watson, R.; Morato, T. Exploitation Patterns in Seamount Fisheries: A Prelimary Analysis. In *Fisheries Centre Research Reports*; Appendix 1; Morato, T., Pauly, D., Eds.; University of British Columbia: Vancouver, BC, Canada; Volume 12, pp. 61–66.
30. Devine, J.A.; Baker, K.D.; Haedrich, R.L. Deep-Sea Fishes Qualify as Endangered. *Nature* **2006**, *439*, 29. [CrossRef] [PubMed]
31. International Council for the Exploration of the Sea (ICES). *Report of the ICES Advisory Committee on Fishery Management, Advisory Committee on the Marine Evironment and Advisory Committee on Ecosystems*; Book 8; International Council for the Exploration of the Sea: Copenhagen, Denmark, 2007; ISBN 978-87-7482-000-0.

32. Bailey, D.M.; Collins, M.A.; Gordon, J.D.; Zuur, A.F.; Priede, I.G. Long-Term Changes in Deep-Water Fish Populations in the Northeast Atlantic: A Deeper Reaching Effect of Fisheries? *Proc. R. Soc. B Biol. Sci.* **2009**, *276*, 1965–1969. [CrossRef] [PubMed]
33. Sadovy, Y.; Cheung, W.L. Near Extinction of a Highly Fecund Fish: The One That Nearly Got Away. *Fish Fish.* **2003**, *4*, 86–99. [CrossRef]
34. European Commission. *Communication from the Commission to the Council and the European Parliament on Review of the Management of Deep-Sea Fish Stocks*; European Commission: Brussels, Belgium, 2007.
35. Norse, E.A.; Brooke, S.; Cheung, W.W.; Clark, M.R.; Ekeland, I.; Froese, R.; Gjerde, K.M.; Haedrich, R.L.; Heppell, S.S.; Morato, T. Sustainability of Deep-Sea Fisheries. *Mar. Policy* **2012**, *36*, 307–320. [CrossRef]
36. Jennings, S.; Kaiser, M.J. The Effects of Fishing on Marine Ecosystems. In *Advances in Marine Biology*; Elsevier: Amsterdam, The Netherlands, 1998; Volume 34, pp. 201–352, ISBN 978-0-12-026134-5.
37. Auster, P.J.; Langton, R.W. The Effects of Fishing on Fish Habitat. *Am. Fish. Soc. Symp.* **1999**, *22*, 150–187.
38. Caddy, J.F. Marine Habitat and Cover: Their Importance for Productive Coastal Fishery Resources. In *Monographs on Oceanographic Methodlogy*; UNESCO Publishing: Paris, France, 2007; Volume 11.
39. Caddy, J.F. The Importance Of "Cover" In The Life Histories Of Demersal And Benthic Marine Resources: A Neglected Issue in Fisheries Assessment And Management. *Bull. Mar. Sci.* **2008**, *83*, 46.
40. Juanes, F. Role of Habitat in Mediating Mortality during the Post-Settlement Transition Phase of Temperate Marine Fishes. *J. Fish Biol.* **2007**, *70*, 661–677. [CrossRef]
41. Roberts, C.M. Deep Impact: The Rising Toll of Fishing in the Deep Sea. *Trends Ecol. Evol.* **2002**, *17*, 242–245. [CrossRef]
42. Koenig, C.C.; Coleman, F.C.; Grimes, C.B.; Fitzhugh, G.R.; Scanlon, K.M.; Gledhill, C.T.; Grace, M. Protection of Fish Spawning Habitat for the Conservation of Warm-Temperate Reef-Fish Fisheries of Shelf-Edge Reefs of Florida. *Bull. Mar. Sci.* **2000**, *66*, 24.
43. Reed, J.K.; Shepard, A.N.; Koenig, C.C.; Scanlon, K.M.; Gilmore, R.G. Mapping, Habitat Characterization, and Fish Surveys of the Deep-Water Oculina Coral Reef Marine Protected Area: A Review of Historical and Current Research. In *Cold-Water Corals and Ecosystems*; Freiwald, A., Roberts, J.M., Eds.; Springer: Berlin/Heidelberg, Germany, 2005; pp. 443–465, ISBN 978-3-540-24136-2.
44. Hall–Spencer, J.; Allain, V.; Fosså, J.H. Trawling Damage to Northeast Atlantic Ancient Coral Reefs. *Proc. R. Soc. Lond. B* **2002**, *269*, 507–511. [CrossRef]
45. Du Preez, C.; Tunnicliffe, V. Shortspine Thornyhead and Rockfish (Scorpaenidae) Distribution in Response to Substratum, Biogenic Structures and Trawling. *Mar. Ecol. Prog. Ser.* **2011**, *425*, 217–231. [CrossRef]
46. FAO. *Report of the Technical Consultation on International Guidelines for the Management of Deep-Sea Fisheries in the High Seas: Rome, 4–8 February and 25–29 August 2008*; FAO Fisheries and Aquaculture Report; Organisation des Nations Unies pour L'Alimentation et L'Agriculture, Ed.; FAO: Rome, Italy, 2009; ISBN 978-92-5-006190-0.
47. General Fisheries Commission for the Mediterranean. *Criteria for the Identification of Sensitive Habitats of Relevance for the Management of Priority Species*; General Fisheries Commission for the Mediterranean: Malaga, Spain, 2009; Volume 3.
48. General Fisheries Commission for the Mediterranean. *Report of the First Meeting of the Working Group on Vulnerable Marine Ecosystems (WGVME)*; General Fisheries Commission for the Mediterranean: Malaga, Spain, 2017. Available online: http://www.fao.org/gfcm/technical-meetings/detail/en/c/885358/ (accessed on 8 March 2022).
49. General Fisheries Commission for the Mediterranean. *Report of the Second Meeting of the Working Group on Vulnerable Marine Ecosystems (WGVME)*; FAO Headquarters: Rome, Italy, 2018. Available online: http://www.fao.org/gfcm/technical-meetings/detail/en/c/1142043 (accessed on 8 March 2022).
50. Otero, M.D.M.; Numa, C.; Bo, M.; Orejas, C.; Garrabou, J.; Cerrano, C.; Kružic, P.; Antoniadou, C.; Aguilar, R.; Kipson, S. *Overview of the Conservation Status of Mediterranean Anthozoa*; International Union for Conservation of Nature: Malaga, Spain, 2017; ISBN 2-8317-1845-7.
51. Bo, M.; Bava, S.; Canese, S.; Angiolillo, M.; Cattaneo-Vietti, R.; Bavestrello, G. Fishing Impact on Deep Mediterranean Rocky Habitats as Revealed by ROV Investigation. *Biol. Conserv.* **2014**, *171*, 167–176. [CrossRef]
52. Bo, M.; Bavestrello, G.; Angiolillo, M.; Calcagnile, L.; Canese, S.; Cannas, R.; Cau, A.; D'Elia, M.; D'Oriano, F.; Follesa, M.C.; et al. Persistence of Pristine Deep-Sea Coral Gardens in the Mediterranean Sea (SW Sardinia). *PLoS ONE* **2015**, *10*, e0119393. [CrossRef] [PubMed]
53. Chimienti, G.; Bo, M.; Taviani, M.; Mastrototaro, F. Occurrence and Biogeography of Mediterranean Cold-Water Corals. In *Mediterranean Cold-Water Corals: Past, Present and Future*; Springer: Berlin/Heidelberg, Germany, 2019; pp. 213–243.
54. Chimienti, G.; Mastrototaro, F.; D'Onghia, G. Mesophotic and Deep-Sea Vulnerable Coral Habitats of the Mediterranean Sea: Overview and Conservation Perspectives. In *Advances in the Studies of the Benthic Zone*; Soto, L.A., Ed.; IntechOpen: London, UK, 2019; pp. 1–20. [CrossRef]
55. Chimienti, G.; De Padova, D.; Adamo, M.; Mossa, M.; Bottalico, A.; Lisco, A.; Ungaro, N.; Mastrototaro, F. Effects of Global Warming on Mediterranean Coral Forests. *Sci. Rep.* **2021**, *11*, 20703. [CrossRef] [PubMed]
56. Rosenberg, A.; Bigford, T.E.; Leathery, S.; Hill, R.L.; Bickers, K. Ecosystem Approaches to Fishery Management through Essential Fish Habitat. *Bull. Mar. Sci.* **2000**, *66*, 8.
57. Capezzuto, F.; Sion, L.; Ancona, F.; Carlucci, R.; Carluccio, A.; Cornacchia, L.; Maiorano, P.; Ricci, P.; Tursi, A.; D'Onghia, G. Cold-Water Coral Habitats and Canyons as Essential Fish Habitats in the Southern Adriatic and Northern Ionian Sea (Central Mediterranean). *Ecol. Quest.* **2018**, *29*, 9–23. [CrossRef]

58. D'Onghia, G. Cold-Water Corals as Shelter, Feeding and Life-History Critical Habitats for Fish Species: Ecological Interactions and Fishing Impact. In *Mediterranean Cold-Water Corals: Past, Present and Future*; Springer: Berlin/Heidelberg, Germany, 2019; pp. 335–356.
59. Garcia, S.M. Ecosystem Approach to Fisheries: Issue, Terminology, Principles, Institutional Foundations, Implementation and Outlook. *FAO Fish Tech. Pap.* **2003**, *443*, 1–71.
60. de Juan, S.; Lleonart, J. A Conceptual Framework for the Protection of Vulnerable Habitats Impacted by Fishing Activities in the Mediterranean High Seas. *Ocean. Coast. Manag.* **2010**, *53*, 717–723. [CrossRef]
61. de Juan, S.; Moranta, J.; Hinz, H.; Barberá, C.; Ojeda-Martinez, C.; Oro, D.; Ordines, F.; Ólafsson, E.; Demestre, M.; Massutí, E. A Regional Network of Sustainable Managed Areas as the Way Forward for the Implementation of an Ecosystem-Based Fisheries Management in the Mediterranean. *Ocean. Coast. Manag.* **2012**, *65*, 51–58. [CrossRef]
62. Tursi, A.; Maiorano, P.; Sion, L.; D'Onghia, G. Fishery Resources: Between Ecology and Economy. *Rend. Fis. Acc. Lincei* **2015**, *26*, 73–79. [CrossRef]
63. Grehan, A.J.; Arnaud-Haond, S.; D'Onghia, G.; Savini, A.; Yesson, C. Towards Ecosystem Based Management and Monitoring of the Deep Mediterranean, North-East Atlantic and Beyond. *Deep. Sea Res. Part II Top. Stud. Oceanogr.* **2017**, *145*, 1–7. [CrossRef]
64. Otero, M.D.M.; Marin, P. 46 Conservation of Cold-Water Corals in the Mediterranean: Current Status and Future Prospects for Improvement. In *Mediterranean Cold-Water Corals: Past, Present and Future*; Springer: Berlin/Heidelberg, Germany, 2019; pp. 535–545.
65. Relini, G.; Orsi Relini, L. The Decline of Red Shrimps Stocks in the Gulf of Genoa. *Investig. Pesq.* **1987**, *51*, 254–260.
66. Relini, G. La Pesca Batiale in Liguria. *Biol. Mar. Mediterr.* **2007**, *14*, 190–244.
67. Sardà, F.; D'Onghia, G.; Politou, C.Y.; Company, J.B.; Maiorano, P.; Kapiris, K. Deep-Sea Distribution, Biological and Ecological Aspects of *Aristeus Antennatus* (Risso, 1816) in the Western and Central Mediterranean Sea. *Sci. Mar.* **2004**, *68*, 117–127. [CrossRef]
68. Würtz, M.; Rovere, M. *Atlas of the Mediterranean Seamounts and Seamount-like Structures*; IUCN Gland: Gland, Switzerland, 2015; ISBN 2-8317-1750-7.
69. Fernandez-Arcaya, U.; Ramirez-Llodra, E.; Aguzzi, J.; Allcock, A.L.; Davies, J.S.; Dissanayake, A.; Harris, P.; Howell, K.; Huvenne, V.A.I.; Macmillan-Lawler, M.; et al. Ecological Role of Submarine Canyons and Need for Canyon Conservation: A Review. *Front. Mar. Sci.* **2017**, *4*, 5. [CrossRef]
70. Rueda, J.L.; Urra, J.; Aguilar, R.; Angeletti, L.; Bo, M.; García-Ruiz, C.; González-Duarte, M.M.; López, E.; Madurell, T.; Maldonado, M. 29 Cold-Water Coral Associated Fauna in the Mediterranean Sea and Adjacent Areas. In *Mediterranean Cold-Water Corals: Past, Present and Future*; Springer: Berlin/Heidelberg, Germany, 2019; pp. 295–333.
71. Sion, L.; Calculli, C.; Capezzuto, F.; Carlucci, R.; Carluccio, A.; Cornacchia, L.; Maiorano, P.; Pollice, A.; Ricci, P.; Tursi, A.; et al. Does the Bari Canyon (Central Mediterranean) Influence the Fish Distribution and Abundance? *Prog. Oceanogr.* **2019**, *170*, 81–92. [CrossRef]
72. Armstrong, C.W.; Falk-Petersen, J. Habitat–Fisheries Interactions: A Missing Link? *ICES J. Mar. Sci./J. Cons.* **2008**, *65*, 817–821. [CrossRef]
73. Komyakova, V.; Munday, P.L.; Jones, G.P. Comparative Analysis of Habitat Use and Ontogenetic Habitat-Shifts among Coral Reef Damselfishes. *Environ. Biol. Fishes* **2019**, *102*, 1201–1218. [CrossRef]
74. Würtz, M. Submarine Canyons and Their Role in the Mediterranean Ecosystem. In *Mediterranean Submarine Canyons*; Würtz, M., Ed.; IUCN: Gland, Switzerland; Málaga, Spain; Mediterranean Submarine Canyons: Ecology and Governance: Gland, Switzerland; Málaga, Spain, 2012; pp. 11–26.
75. Rossi, S.; Bramanti, L.; Gori, A.; Orejas, C. *Marine Animal Forests: The Ecology of Benthic Biodiversity Hotspots*; Springer: Berlin/Heidelberg, Germany, 2017; ISBN 3-319-21012-2.
76. IUCN. *Thematic Report—Conservation Overview of Mediterranean Deep-Sea Biodiversity: A Strategic Assessment*; IUCN: Gland, Switzerland; Malaga, Spain, 2019; 122p.
77. Orejas, C.; Jiménez, C. 1 An Introduction to the Research on Mediterranean Cold-Water Corals. In *Mediterranean Cold-Water Corals: Past, Present and Future*; Orejas, C., Jiménez, C., Eds.; Coral Reefs of the World; Springer International Publishing: Cham, Switzerland, 2019; Volume 9, pp. 3–12, ISBN 978-3-319-91607-1.
78. D'Onghia, G.; Maiorano, P.; Carlucci, R.; Capezzuto, F.; Carluccio, A.; Tursi, A.; Sion, L. Comparing Deep-Sea Fish Fauna between Coral and Non-Coral "Megahabitats" in the Santa Maria Di Leuca Cold-Water Coral Province (Mediterranean Sea). *PLoS ONE* **2012**, *7*, e44509. [CrossRef]
79. D'Onghia, G.; Capezzuto, F.; Cardone, F.; Carlucci, R.; Carluccio, A.; Chimienti, G.; Corriero, G.; Longo, C.; Maiorano, P.; Mastrototaro, F.; et al. Macro- and Megafauna Recorded in the Submarine Bari Canyon (Southern Adriatic, Mediterranean Sea) Using Different Tools. *Medit. Mar. Sci.* **2015**, *16*, 180. [CrossRef]
80. D'Onghia, G.; Capezzuto, F.; Carluccio, A.; Carlucci, R.; Giove, A.; Mastrototaro, F.; Panza, M.; Sion, L.; Tursi, A.; Maiorano, P. Exploring Composition and Behaviour of Fish Fauna by *in Situ* Observations in the Bari Canyon (Southern Adriatic Sea, Central Mediterranean). *Mar. Ecol.* **2015**, *36*, 541–556. [CrossRef]
81. D'Onghia, G.; Calculli, C.; Capezzuto, F.; Carlucci, R.; Carluccio, A.; Maiorano, P.; Pollice, A.; Ricci, P.; Sion, L.; Tursi, A. New Records of Cold-Water Coral Sites and Fish Fauna Characterization of a Potential Network Existing in the Mediterranean Sea. *Mar. Ecol.* **2016**, *37*, 1398–1422. [CrossRef]

82. D'Onghia, G.; Calculli, C.; Capezzuto, F.; Carlucci, R.; Carluccio, A.; Grehan, A.; Indennidate, A.; Maiorano, P.; Mastrototaro, F.; Pollice, A. Anthropogenic Impact in the Santa Maria Di Leuca Cold-Water Coral Province (Mediterranean Sea): Observations and Conservation Straits. *Deep. Sea Res. Part II Top. Stud. Oceanogr.* **2017**, *145*, 87–101. [CrossRef]
83. D'Onghia, G.; Sion, L.; Capezzuto, F. Cold-Water Coral Habitats Benefit Adjacent Fisheries along the Apulian Margin (Central Mediterranean). *Fish. Res.* **2019**, *213*, 172–179. [CrossRef]
84. Capezzuto, F.; Calculli, C.; Carlucci, R.; Carluccio, A.; Maiorano, P.; Pollice, A.; Sion, L.; Tursi, A.; D'Onghia, G. Revealing the Coral Habitat Effect on Benthopelagic Fauna Diversity in the Santa Maria Di Leuca Cold-water Coral Province Using Different Devices and Bayesian Hierarchical Modelling. *Aquat. Conserv. Mar. Freshw. Ecosyst.* **2019**, *29*, 1608–1622. [CrossRef]
85. Capezzuto, F.; Ancona, F.; Calculli, C.; Sion, L.; Maiorano, P.; D'Onghia, G. Feeding of the Deep-Water Fish *Helicolenus dactylopterus* (Delaroche, 1809) in Different Habitats: From Muddy Bottoms to Cold-Water Coral Habitats. *Deep. Sea Res. Part I Oceanogr. Res. Pap.* **2020**, *159*, 103252. [CrossRef]
86. Spedicato, M.T.; Massutí, E.; Mérigot, B.; Tserpes, G.; Jadaud, A.; Relini, G. The MEDITS Trawl Survey Specifications in an Ecosystem Approach to Fishery Management. *Sci. Mar.* **2019**, *83*, 9. [CrossRef]
87. Capezzuto, F.; Carlucci, R.; Maiorano, P.; Sion, L.; Battista, D.; Giove, A.; Indennidate, A.; Tursi, A.; D'Onghia, G. The Bathyal Benthopelagic Fauna in the North-Western Ionian Sea: Structure, Patterns and Interactions. *Chem. Ecol.* **2010**, *26*, 199–217. [CrossRef]
88. Maiorano, P.; Sion, L.; Carlucci, R.; Capezzuto, F.; Giove, A.; Costantino, G.; Panza, M.; D'Onghia, G.; Tursi, A. The Demersal Faunal Assemblage of the North-Western Ionian Sea (Central Mediterranean): Current Knowledge and Perspectives. *Chem. Ecol.* **2010**, *26*, 219–240. [CrossRef]
89. Maiorano, P.; Sabatella, R.F.; Marzocchi, B.M. (Eds.) *Annuario sullo Stato delle Risorse e sulle Strutture Produttive dei Mari Italiani*; SIBM: Genova, Italy, 2015; 432p.
90. Tursi, A.; Mastrototaro, F.; Matarrese, A.; Maiorano, P.; D'Onghia, G. Biodiversity of the White Coral Reefs in the Ionian Sea (Central Mediterranean). *Chem. Ecol.* **2004**, *20*, 107–116. [CrossRef]
91. Taviani, M.; Remia, A.; Corselli, C.; Freiwald, A.; Malinverno, E.; Mastrototaro, F.; Savini, A.; Tursi, A. First Geo-Marine Survey of Living Cold-Water Lophelia Reefs in the Ionian Sea (Mediterranean Basin). *Facies* **2005**, *50*, 409–417. [CrossRef]
92. Taviani, M.; Angeletti, L.; Beuck, L.; Campiani, E.; Canese, S.; Foglini, F.; Freiwald, A.; Montagna, P.; Trincardi, F. Reprint of 'On and off the Beaten Track: Megafaunal Sessile Life and Adriatic Cascading Processes'. *Mar. Geol.* **2016**, *375*, 146–160. [CrossRef]
93. Morelli, D. La Cartografia Marina: Ricerche Ed Applicazioni Orientate Ai Rischi Geologico-Ambientali in Aree Campione. Ph.D. Thesis, University of Genoa, Genoa, Italy, 2008.
94. Freiwald, A.; Beuck, L.; Rüggeberg, A.; Taviani, M.; Hebbeln, D.; R/V Meteor Cruise M70-1 Participants. The White Coral Community in the Central Mediterranean Sea Revealed by ROV Surveys. *Oceanography* **2009**, *22*, 58–74. [CrossRef]
95. Mastrototaro, F.; D'Onghia, G.; Corriero, G.; Matarrese, A.; Maiorano, P.; Panetta, P.; Gherardi, M.; Longo, C.; Rosso, A.; Sciuto, F.; et al. Biodiversity of the White Coral Bank off Cape Santa Maria Di Leuca (Mediterranean Sea): An Update. *Deep. Sea Res. Part II Top. Stud. Oceanogr.* **2010**, *57*, 412–430. [CrossRef]
96. Mastrototaro, F.; Maiorano, P.; Vertino, A.; Battista, D.; Indennidate, A.; Savini, A.; Tursi, A.; D'Onghia, G. A Facies of Kophobelemnon (C Nidaria, O Ctocorallia) from S Anta M Aria Di L Euca Coral Province (Mediterranean Sea). *Mar. Ecol.* **2013**, *34*, 313–320. [CrossRef]
97. Angeletti, L.; Taviani, M.; Canese, S.; Foglini, F.; Mastrototaro, F.; Argnani, A.; Trincardi, F.; Bakran-Petricioli, T.; Ceregato, A.; Chimienti, G. New Deep-Water Cnidarian Sites in the Southern Adriatic Sea. *Mediterr. Mar. Sci.* **2014**, *15*, 263–273. [CrossRef]
98. Carbonara, P.; Zupa, W.; Follesa, M.C.; Cau, A.; Capezzuto, F.; Chimienti, G.; D'Onghia, G.; Lembo, G.; Pesci, P.; Porcu, C.; et al. Exploring a Deep-Sea Vulnerable Marine Ecosystem: *Isidella elongata* (Esper, 1788) Species Assemblages in the Western and Central Mediterranean. *Deep. Sea Res. Part I Oceanogr. Res. Pap.* **2020**, *166*, 103406. [CrossRef]
99. Pérès, J.-M.; Picard, J. Nouveau Manuel de Bionomie Benthique de La Mer Méditerranée. *Recueil des Travaux de la Station marine d'Endoume* **1964**, *31*, 1–137.
100. Azouz, A. Les Fonds Chalutables de La Région Nord de La Tunisie: 1. Cadre Physique et Biocoénoses Benthiques. *Bull. Inst. Océanogr. Pêche. Salammbô* **1973**, *2*, 473–559.
101. Maynou, F.; Sardà, F.; Tudela, S.; Demestre, M. Management Strategies for Red Shrimp (Aristeus Antennatus) Fisheries in the Catalan Sea (NW Mediterranean) Based on Bioeconomic Simulation Analysis. *Aquat. Living Resour.* **2006**, *19*, 161–171. [CrossRef]
102. Kapiris, K. Feeding Habits of Both Deep-Water Red Shrimps, *Aristaeomorpha foliacea* and *Aristeus antennatus* (Decapoda, Aristeidae) in the Ionian Sea (E. Mediterranean). In *Food Quality*; Kapiris, K., Ed.; InTech: London, UK, 2012; ISBN 978-953-51-0560-2.
103. Maynou, F.; Cartes, J.E. Effects of Trawling on Fish and Invertebrates from Deep-Sea Coral Facies of *Isidella Elongata* in the Western Mediterranean. *J. Mar. Biol. Ass.* **2012**, *92*, 1501–1507. [CrossRef]
104. Cartes, J.E.; LoIacono, C.; Mamouridis, V.; López-Pérez, C.; Rodríguez, P. Geomorphological, Trophic and Human Influences on the Bamboo Coral Isidella Elongata Assemblages in the Deep Mediterranean: To What Extent Does Isidella Form Habitat for Fish and Invertebrates? *Deep. Sea Res. Part I Oceanogr. Res. Pap.* **2013**, *76*, 52–65. [CrossRef]
105. Smith, C.J.; Mytilineou, C.; Papadopoulou, K.N.; Pancucci-Papadopoulou, M.A.; Salomidi, M. ROV Observations on Fish and Megafauna in Deep Coral Areas of the Eastern Ionian. *Rapp. Comm. Int. L'exploration Sci. Mer Méditerranée* **2010**, *39*, 669.

106. Lauria, V.; Garofalo, G.; Fiorentino, F.; Massi, D.; Milisenda, G.; Piraino, S.; Russo, T.; Gristina, M. Species Distribution Models of Two Critically Endangered Deep-Sea Octocorals Reveal Fishing Impacts on Vulnerable Marine Ecosystems in Central Mediterranean Sea. *Sci. Rep.* **2017**, *7*, 8049. [CrossRef] [PubMed]
107. Mastrototaro, F.; Chimienti, G.; Acosta, J.; Blanco, J.; Garcia, S.; Rivera, J.; Aguilar, R. *Isidella Elongata* (Cnidaria: Alcyonacea) Facies in the Western Mediterranean Sea: Visual Surveys and Descriptions of Its Ecological Role. *Eur. Zool. J.* **2017**, *84*, 209–225. [CrossRef]
108. Mytilineou, C.h.; Smith, C.J.; Anastasopoulou, A.; Papadopoulou, K.N.; Christidis, G.; Bekas, P.; Kavadas, S.; Dokos, J. New Cold-Water Coral Occurrences in the Eastern Ionian Sea: Results from Experimental Long Line Fishing. *Deep. Sea Res. Part II Top. Stud. Oceanogr.* **2014**, *99*, 146–157. [CrossRef]
109. Eumofa. The EU Fish Market, 107. 2019. Available online: https://doi.org/10.2771/168390 (accessed on 6 March 2022).
110. Chimienti, G.; De Padova, D.; Mossa, M.; Mastrototaro, F. A Mesophotic Black Coral Forest in the Adriatic Sea. *Sci. Rep.* **2020**, *10*, 8504. [CrossRef]
111. Cau, A.; Follesa, M.C.; Moccia, D.; Bellodi, A.; Mulas, A.; Bo, M.; Canese, S.; Angiolillo, M.; Cannas, R. *Leiopathes glaberrima* Millennial Forest from SW Sardinia as Nursery Ground for the Small Spotted Catshark *Scyliorhinus Canicula*: Catshark Nursery Ground in a Millennial Black Coral Forest. *Aquatic Conserv. Mar. Freshw. Ecosyst.* **2017**, *27*, 731–735. [CrossRef]
112. Hebbeln, D. Report and Preliminary Results of RV POSEIDON Cruise POS 385 "Cold-Water Corals of the Alboran Sea (Western Mediterranean Sea)", Faro-Toulon, 29 May–16 June 2009. *Ber. Fachbereich Geowiss. Univ. Brem.* **2009**, *273*, 1–79.
113. Angeletti, L.; Mecho, A.; Doya, C.; Micallef, A.; Huvenne, V.; Georgiopoulou, A.; Taviani, M. First Report of Live Deep-Water Cnidarian Assemblages from the Malta Escarpment. *Ital. J. Zool.* **2015**, *82*, 291–297. [CrossRef]
114. Thompson, A.; Sanders, J.; Tandstad, M.; Carocci, F.; Fuller, J. Vulnerable Marine Ecosystems: Processes and Practices in the High Seas. In *FAO Fisheries and Aquaculture Technical Paper*; FAO: Roma, Italy, 2016.
115. Freiwald, A.; Fossa, J.H.; Grehan, A.; Koslow, T.; Roberts, J.M. *Cold-Water Coral Reefs: Out of Sight-No Longer out of Mind*; UNEP-WCMC: Cambridge, UK, 2004; ISBN 92-807-2453-3.
116. Roberts, J.M.; Wheeler, A.J.; Freiwald, A. Reefs of the Deep: The Biology and Geology of Cold-Water Coral Ecosystems. *Science* **2006**, *312*, 543–547. [CrossRef] [PubMed]
117. Hebbeln, D.; Van Rooij, D.; Wienberg, C. Good Neighbours Shaped by Vigorous Currents: Cold-Water Coral Mounds and Contourites in the North Atlantic. *Mar. Geol.* **2016**, *378*, 171–185. [CrossRef]
118. Buhl-Mortensen, L.; Vanreusel, A.; Gooday, A.J.; Levin, L.A.; Priede, I.G.; Buhl-Mortensen, P.; Gheerardyn, H.; King, N.J.; Raes, M. Biological Structures as a Source of Habitat Heterogeneity and Biodiversity on the Deep Ocean Margins: Biological Structures and Biodiversity. *Mar. Ecol.* **2010**, *31*, 21–50. [CrossRef]
119. Linley, T.D.; Lavaleye, M.; Maiorano, P.; Bergman, M.; Capezzuto, F.; Cousins, N.J.; D'Onghia, G.; Duineveld, G.; Shields, M.A.; Sion, L. Effects of Cold-water Corals on Fish Diversity and Density (European Continental Margin: Arctic, NE Atlantic and Mediterranean Sea): Data from Three Baited Lander Systems. *Deep. Sea Res. Part II Top. Stud. Oceanogr.* **2017**, *145*, 8–21. [CrossRef]
120. Orsi Relini, L.; Relini, G. The Red Shrimps Fishery in the Ligurian Sea: Mismanagement or Not? *FAO Fish. Rep.* **1985**, *336*, 99–106.
121. Tobar, R.; Sardà, F. Análisis de La Evolución de Las Capturas de Gamba Rosada, *Aristeus antennatus* (Risso, 1816), En Los Últimos Decenios En Cataluña. *Inf. Técn. Inv. Pesq.* **1987**, *142*, 3–20.
122. Sardà, F.; Cartes, J.E. Spatio-Temporal Variations in Megabenthos Abundance in Three Different Habitats of the Catalan Deep-Sea (Western Mediterranean). *Mar. Biol.* **1994**, *120*, 211–219. [CrossRef]
123. Sardà, F.; Company, J.B.; Bahamón, N.; Rotllant, G.; Flexas, M.M.; Sánchez, J.D.; Zúñiga, D.; Coenjaerts, J.; Orellana, D.; Jordà, G.; et al. Relationship between Environment and the Occurrence of the Deep-Water Rose Shrimp *Aristeus antennatus* (Risso, 1816) in the Blanes Submarine Canyon (NW Mediterranean). *Prog. Oceanogr.* **2009**, *82*, 227–238. [CrossRef]
124. Ramirez-Llodra, E.; Company, J.B.; Sardà, F.; Rotllant, G. Megabenthic Diversity Patterns and Community Structure of the Blanes Submarine Canyon and Adjacent Slope in the Northwestern Mediterranean: A Human Overprint?: Megabenthic Diversity in a Mediterranean Canyon and Slope. *Mar. Ecol.* **2010**, *31*, 167–182. [CrossRef]
125. Puig, P.; Company, J.B.; Sardà, F.; Palanques, A. Responses of Deep-Water Shrimp Populations to Intermediate Nepheloid Layer Detachments on the Northwestern Mediterranean Continental Margin. *Deep. Sea Res. Part I Oceanogr. Res. Pap.* **2001**, *48*, 2195–2207. [CrossRef]
126. Fernandez-Arcaya, U.; Rotllant, G.; Ramirez-Llodra, E.; Recasens, L.; Aguzzi, J.; Flexas, M.M.; Sanchez-Vidal, A.; López-Fernández, P.; García, J.A.; Company, J.B. Reproductive Biology and Recruitment of the Deep-Sea Fish Community from the NW Mediterranean Continental Margin. *Prog. Oceanogr.* **2013**, *118*, 222–234. [CrossRef]
127. Orejas, C.; Gori, A.; Iacono, C.L.; Puig, P.; Gili, J.-M.; Dale, M.R. Cold-Water Corals in the Cap de Creus Canyon, Northwestern Mediterranean: Spatial Distribution, Density and Anthropogenic Impact. *Mar. Ecol. Prog. Ser.* **2009**, *397*, 37–51. [CrossRef]
128. Cau, A.; Alvito, A.; Moccia, D.; Canese, S.; Pusceddu, A.; Rita, C.; Angiolillo, M.; Follesa, M.C. Submarine Canyons along the Upper Sardinian Slope (Central Western Mediterranean) as Repositories for Derelict Fishing Gears. *Mar. Pollut. Bull.* **2017**, *123*, 357–364. [CrossRef] [PubMed]
129. Taviani, M.; Angeletti, L.; Canese, S.; Cannas, R.; Cardone, F.; Cau, A.; Cau, A.B.; Follesa, M.C.; Marchese, F.; Montagna, P. The "Sardinian Cold-Water Coral Province" in the Context of the Mediterranean Coral Ecosystems. *Deep. Sea Res. Part II Top. Stud. Oceanogr.* **2017**, *145*, 61–78. [CrossRef]

130. Sabatini, A.; Follesa, M.C.; Locci, I.; Pendugiu, A.A.; Pesci, P.; Cau, A. Assemblages in a Submarine Canyon: Influence of Depth and Time. In *Biodiversity in Enclosed Seas and Artificial Marine Habitats*; Springer: Berlin/Heidelberg, Germany, 2007; pp. 265–271.
131. Yoklavich, M.M.; Greene, H.G.; Cailliet, G.M.; Sullivan, D.E.; Lea, R.N.; Love, M.S. Habitat Associations of Deep-Water Rockfishes in a Submarine Canyon: An Example of a Natural Refuge. *Fish. Bull.* **2000**, *98*, 625–641.
132. Tudela, S.; Simard, F.; Skinner, J.; Guglielmi, P. The Mediterranean Deep-Sea Ecosystems: A Proposal for Their Conservation. In *The Mediterranean Deep-Sea Ecosystems: An Overview of Their Diversity, Structure, Functioning and Anthropogenic Impacts, with a Proposal for Conservation*; IUCN Publications Services: Málaga, Spain; Rome, Italy, 2004; pp. 39–47.
133. Angeletti, L.; D'Onghia, G.; Otero, M.D.M.; Settanni, A.; Spedicato, M.T.; Taviani, M. A Perspective for Best Governance of the Bari Canyon Deep-Sea Ecosystems. *Water* **2021**, *13*, 1646. [CrossRef]
134. Zibrowius, H.; Taviani, M. Remarkable Sessile Fauna Associated with Deep Coral and Other Calcareous Substrates in the Strait of Sicily, Mediterranean Sea. In *Cold-Water Corals and Ecosystems*; Freiwald, A., Roberts, J.M., Eds.; Erlangen Earth Conference Series; Springer: Berlin/Heidelberg, Germany, 2005; pp. 807–819, ISBN 978-3-540-24136-2.
135. Bo, M.; Bertolino, M.; Borghini, M.; Castellano, M.; Covazzi Harriague, A.; Di Camillo, C.G.; Gasparini, G.; Misic, C.; Povero, P.; Pusceddu, A.; et al. Characteristics of the Mesophotic Megabenthic Assemblages of the Vercelli Seamount (North Tyrrhenian Sea). *PLoS ONE* **2011**, *6*, e16357. [CrossRef]
136. Orsi Relini, L.; Relini, G. *Gaidropsarus Granti* from a Ligurian Seamount: A Mediterranean Native Species? *Mar. Ecol.* **2014**, *35*, 35–40. [CrossRef]
137. Sabatini, A.; Follesa, M.C.; Locci, I.; Matta, G.; Palmas, F.; Pendugiu, A.A.; Pesci, P.; Cau, A. Demersal Assemblages in Two Trawl Fishing Lanes Located on the Baronie Seamount (Central Western Mediterranean). *J. Mar. Biol. Ass.* **2011**, *91*, 65–75. [CrossRef]
138. Galil, B.; Zibrowius, H. First Benthos Samples from Eratosthenes Seamount, Eastern Mediterranean. *Senckenbergiana Marit.* **1998**, *28*, 111–121. [CrossRef]
139. Öztürk, B.; Topcu, E.N.; Topaloglu, B. A Preliminary Study On Two Seamounts In The Eastern Mediterranean Sea. *Rapp. Comm. Int. Mer Médit.* **2010**, *39*, 620.
140. Sperone, E.; Parise, G.; Leone, A.; Milazzo, C.; Circosta, V.; Santoro, G.; Paolillo, G.; Micarelli, P.; Tripepi, S. Spatiotemporal Patterns of Distribution of Large Predatory Sharks in Calabria (Central Mediterranean, Southern Italy). *Acta Adriat.* **2012**, *53*, 13–24.
141. De la Torriente Diez, A.; González-Irusta, J.M.; Serrano, A.; Aguilar, R.; Sánchez, F.; Blanco, M.; Punzón, A. Spatial Assessment of Benthic Habitats Vulnerability to Bottom Fishing in a Mediterranean Seamount. *Mar. Policy* **2022**, *135*, 104850. [CrossRef]
142. Trincardi, F.; Foglini, F.; Verdicchio, G.; Asioli, A.; Correggiari, A.; Minisini, D.; Piva, A.; Remia, A.; Ridente, D.; Taviani, M. The Impact of Cascading Currents on the Bari Canyon System, SW-Adriatic Margin (Central Mediterranean). *Mar. Geol.* **2007**, *246*, 208–230. [CrossRef]
143. Budillon, G.; Bue, N.L.; Siena, G.; Spezie, G. Hydrographic Characteristics of Water Masses and Circulation in the Northern Ionian Sea. *Deep. Sea Res. Part II Top. Stud. Oceanogr.* **2010**, *57*, 441–457. [CrossRef]
144. Gacic, M.; Schroeder, K.; Civitarese, G.; Vetrano, A.; Borzelli, G.E. On the Relationship among the Adriatic–Ionian Bimodal Oscillating System (BiOS), the Eastern Mediterranean Salinity Variations and the Western Mediterranean Thermohaline Cell. *BiOS* **2012**, *9*, 2561–2580.
145. D'Onghia, G.; Indennidate, A.; Giove, A.; Savini, A.; Capezzuto, F.; Sion, L.; Vertino, A.; Maiorano, P. Distribution and Behaviour of Deep-Sea Benthopelagic Fauna Observed Using Towed Cameras in the Santa Maria Di Leuca Cold-Water Coral Province. *Mar. Ecol. Prog. Ser.* **2011**, *443*, 95–110. [CrossRef]
146. Sbrana, M.; Zupa, W.; Ligas, A.; Capezzuto, F.; Chatzispyrou, A.; Follesa, M.C.; Gancitano, V.; Guijarro, B.; Isajlovic, I.; Jadaud, A. Spatiotemporal Abundance Pattern of Deep-Water Rose Shrimp, *Parapenaeus Longirostris*, and Norway Lobster, *Nephrops Norvegicus*, in European Mediterranean Waters. *Sci. Mar.* **2019**, *83*, 71–80. [CrossRef]
147. Cartes, J.E.; Maynou, F.; Fanelli, E.; Papiol, V.; Lloris, D. Long-Term Changes in the Composition and Diversity of Deep-Slope Megabenthos and Trophic Webs off Catalonia (Western Mediterranean): Are Trends Related to Climatic Oscillations? *Prog. Oceanogr.* **2009**, *82*, 32–46. [CrossRef]
148. Carpentieri, P.; Colloca, F.; Cardinale, M.; Belluscio, A.; Ardizzone, G.D. Feeding Habits of European Hake (*Merluccius merluccius*) in the Central Mediterranean Sea. *Fish. Bull.* **2005**, *103*, 411–416.
149. Bartolino, V.; Colloca, F.; Sartor, P.; Ardizzone, G. Modelling Recruitment Dynamics of Hake, *Merluccius merluccius*, in the Central Mediterranean in Relation to Key Environmental Variables. *Fish. Res.* **2008**, *92*, 277–288. [CrossRef]
150. Guijarro, B.; Bitetto, I.; D'Onghia, G.; Follesa, M.C.; Kapiris, K.; Mannini, A.; Marković, O.; Micallef, R.; Ragonese, S.; Skarvelis, K. Spatial and Temporal Patterns in the Mediterranean Populations of *Aristaeomorpha foliacea* and *Aristeus antennatus* (Crustacea: Decapoda: Aristeidae) Based on the MEDITS Surveys. *Sci. Mar.* **2019**, *83*, 57–70. [CrossRef]
151. Sion, L.; Zupa, W.; Calculli, C.; Garofalo, G.; Hidalgo, M.; Jadaud, A.; Lefkaditou, E.; Ligas, A.; Peristeraki, P.; Bitetto, I. Spatial Distribution Pattern of European Hake, *Merluccius merluccius* (Pisces: Merlucciidae), in the Mediterranean Sea. *Sci. Mar.* **2019**, *83*, 21–32. [CrossRef]
152. Carlucci, R.; Lembo, G.; Maiorano, P.; Capezzuto, F.; Mariano, C.A.; Sion, L.; Spedicato, M.T.; Ungaro, N.; Tursi, A.; D'Onghia, G. Nursery Areas of Red Mullet (*Mullus Barbatus*), Hake (*Merluccius merluccius*) and Deep-Water Rose Shrimp (*Parapenaeus Longirostris*) in the Eastern-Central Mediterranean Sea. *Estuar. Coast. Shelf Sci.* **2009**, *83*, 529–538. [CrossRef]

153. Murenu, M.; Cau, A.; Colloca, F.; Sartor, P.; Fiorentino, F.; Garofalo, G.; Piccinetti, C.; Manfredi, C.; D'Onghia, G.; Carlucci, R. Mapping the Potential Locations of European Hake (*Merluccius merluccius*) Nurseries in the Italian Waters. In *GIS/Spatial Analyses in Fishery and Aquatic Sciences*; Nishida, T., Kailola, P.J., Hollingworth, C.E., Eds.; FAO Fishery Gis Research Group: Rome, Italy, 2010; Volume 4.
154. Druon, J.-N.; Fiorentino, F.; Murenu, M.; Knittweis, L.; Colloca, F.; Osio, C.; Mérigot, B.; Garofalo, G.; Mannini, A.; Jadaud, A.; et al. Modelling of European Hake Nurseries in the Mediterranean Sea: An Ecological Niche Approach. *Prog. Oceanogr.* **2015**, *130*, 188–204. [CrossRef]
155. Hidalgo, M.; Ligas, A.; Bellido, J.M.; Bitetto, I.; Carbonara, P.; Carlucci, R.; Guijarro, B.; Jadaud, A.; Lembo, G.; Manfredi, C. Size-Dependent Survival of European Hake Juveniles in the Mediterranean Sea. *Sci. Mar.* **2019**, *83*, 207–221. [CrossRef]
156. Spedicato, M.T.; Zupa, W.; Carbonara, P.; Casciaro, L.; Bitetto, I.; Facchini, M.T.; Gaudio, P.; Palmisano, M.; Lembo, G. Lo Stato Delle Risorse Biologiche e Della Pesca Nel Basso Adriatico e Nello Ionio Nord Occidentale. In *Atti del Convegno il Mare Adriatico e le sue Risorse*; Marini, M., Bombace, G., Iacobone, G., Eds.; Carlo Saladino Editore: Palermo, Italy, 2017; pp. 177–208.
157. Carlucci, R.; Bandelj, V.; Ricci, P.; Capezzuto, F.; Sion, L.; Maiorano, P.; Tursi, A.; Solidoro, C.; Libralato, S. Exploring Spatio-Temporal Changes in the Demersal and Benthopelagic Assemblages of the North-Western Ionian Sea (Central Mediterranean Sea). *Mar. Ecol. Prog. Ser.* **2018**, *598*, 1–19. [CrossRef]
158. Carlucci, R.; Capezzuto, F.; Cipriano, G.; D'Onghia, G.; Fanizza, C.; Libralato, S.; Maglietta, R.; Maiorano, P.; Sion, L.; Tursi, A.; et al. Assessment of Cetacean–Fishery Interactions in the Marine Food Web of the Gulf of Taranto (Northern Ionian Sea, Central Mediterranean Sea). *Rev. Fish Biol. Fish.* **2021**, *31*, 135–156. [CrossRef]
159. Carlucci, R.; Manea, E.; Ricci, P.; Cipriano, G.; Fanizza, C.; Maglietta, R.; Gissi, E. Managing Multiple Pressures for Cetaceans' Conservation with an Ecosystem-Based Marine Spatial Planning Approach. *J. Environ. Manag.* **2021**, *287*, 112240. [CrossRef]
160. Cipriano, G.; Carlucci, R.; Bellomo, S.; Santacesaria, F.C.; Fanizza, C.; Ricci, P.; Maglietta, R. Behavioral Pattern of Risso's Dolphin (*Grampus griseus*) in the Gulf of Taranto (Northern Ionian Sea, Central-Eastern Mediterranean Sea). *JMSE* **2022**, *10*, 175. [CrossRef]
161. Foskolos, I.; Koutouzi, N.; Polychronidis, L.; Alexiadou, P.; Frantzis, A. A Taste for Squid: The Diet of Sperm Whales Stranded in Greece, Eastern Mediterranean. *Deep. Sea Res. Part I Oceanogr. Res. Pap.* **2020**, *155*, 103164. [CrossRef]
162. Luna, A.; Sánchez, P.; Chicote, C.; Gazo, M. Cephalopods in the Diet of Risso's Dolphin (*Grampus griseus*) from the Mediterranean Sea: A Review. *Mar. Mammal. Sci.* **2022**, *38*, 725–741. [CrossRef]
163. Ricci, P.; Ingrosso, M.; Cipriano, G.; Fanizza, C.; Maglietta, R.; Renò, V.; Tursi, A.; Carlucci, R. Top-down Cascading Effects Driven by the Odontocetes in the Gulf of Taranto (Northern Ionian Sea, Central Mediterranean Sea). In Proceedings of the IMEKO Metrology for the Sea, Naples, Italy, 5–7 October 2020.
164. Ricci, P.; Manea, E.; Cipriano, G.; Cascione, D.; D'Onghia, G.; Ingrosso, M.; Fanizza, C.; Maiorano, P.; Tursi, A.; Carlucci, R. Addressing Cetacean–Fishery Interactions to Inform a Deep-Sea Ecosystem-Based Management in the Gulf of Taranto (Northern Ionian Sea, Central Mediterranean Sea). *J. Mar. Sci. Eng.* **2021**, *9*, 872. [CrossRef]
165. Roman, J.; Estes, J.A.; Morissette, L.; Smith, C.; Costa, D.; McCarthy, J.; Nation, J.B.; Nicol, S.; Pershing, A.; Smetacek, V. Whales as Marine Ecosystem Engineers. *Front. Ecol. Environ.* **2014**, *12*, 377–385. [CrossRef]
166. Ramirez-Llodra, E.; Tyler, P.A.; Baker, M.C.; Bergstad, O.A.; Clark, M.R.; Escobar, E.; Levin, L.A.; Menot, L.; Rowden, A.A.; Smith, C.R.; et al. Man and the Last Great Wilderness: Human Impact on the Deep Sea. *PLoS ONE* **2011**, *6*, e22588. [CrossRef] [PubMed]
167. Levin, L.A.; Le Bris, N. The Deep Ocean under Climate Change. *Science* **2015**, *350*, 766–768. [CrossRef]
168. Paulus, E. Shedding Light on Deep-Sea Biodiversity—A Highly Vulnerable Habitat in the Face of Anthropogenic Change. *Front. Mar. Sci.* **2021**, *8*, 281. [CrossRef]
169. Díaz, S.; Fargione, J.; Chapin, F.S., III; Tilman, D. Biodiversity Loss Threatens Human Well-Being. *PLoS Biol.* **2006**, *4*, e277. [CrossRef]
170. Worm, B.; Barbier, E.B.; Beaumont, N.; Duffy, J.E.; Folke, C.; Halpern, B.S.; Jackson, J.B.C.; Lotze, H.K.; Micheli, F.; Palumbi, S.R.; et al. Impacts of Biodiversity Loss on Ocean Ecosystem Services. *Science* **2006**, *314*, 787–790. [CrossRef]
171. Cardinale, B.J.; Duffy, J.E.; Gonzalez, A.; Hooper, D.U.; Perrings, C.; Venail, P.; Narwani, A.; Mace, G.M.; Tilman, D.; Wardle, D.A. Biodiversity Loss and Its Impact on Humanity. *Nature* **2012**, *486*, 59–67. [CrossRef]
172. Armstrong, C.W. Cold Water Coral Reef Management from an Ecosystem Service Perspective. *Mar. Policy* **2014**, *50*, 126–134. [CrossRef]
173. Costanza, R.; d'Arge, R.; De Groot, R.; Farber, S.; Grasso, M.; Hannon, B.; Limburg, K.; Naeem, S.; O'Neill, R.V.; Paruelo, J. The Value of the World's Ecosystem Services and Natural Capital. *Nature* **1997**, *387*, 253–260. [CrossRef]
174. Millenniumeco System Assessment. *Ecosystems and Human Well-Being: Wetlands and Water*; World Resources Institute: Washington, DC, USA, 2005; ISBN 1-56973-597-2.
175. Fonds Mondial pour la Nature. *Living Planet Report 2010: Biodiversity, Biocapacity and Development*; World Wide Fund for Nature: Gland, Switzerland, 2010; ISBN 978-2-940443-08-6.
176. Armstrong, C.W.; Grehan, A.J.; Kahui, V.; Mikkelsen, E.; Reithe, S.; van den Hove, S. Bioeconomic Modeling and the Management of Cold-Water Coral Resources. *Oceanography* **2009**, *22*, 86–91. [CrossRef]
177. Foley, N.S. The Ecological and Economic Value of Cold-Water Coral Ecosystems. *Ocean Coast. Manag.* **2010**, *53*, 313–326. [CrossRef]
178. Allen, J.I.; Clarke, K.R. Effects of Demersal Trawling on Ecosystem Functioning in the North Sea: A Modelling Study. *Mar. Ecol. Prog. Ser.* **2007**, *336*, 63–75. [CrossRef]

179. Armstrong, C.W.; Foley, N.S.; Tinch, R.; van den Hove, S. Services from the Deep: Steps towards Valuation of Deep Sea Goods and Services. *Ecosyst. Serv.* **2012**, *2*, 2–13. [CrossRef]
180. Pikitch Habitat Use and Demographic Population Structure of Elasmobranchs at a Caribbean Atoll (Glover's Reef, Belize). *Mar. Ecol. Prog. Ser.* **2005**, *302*, 187–197. [CrossRef]
181. UNEP. Report of the Conference of the Parties to the Convention on Biological Diversity on the Work of Its Ninth Meeting, UNEP/CBD/COP/9/29. In Proceedings of the Conference of the Parties to the Convention on Biological Diversity Ninth Meeting, Bonn, Germany, 19–30 May 2008.
182. Secretariat of the Convention on Biological Diversity. *Global Biodiversity Outlook 4*; Secretariat of the Convention on Biological Diversity: Montréal, QC, Canada, 2014; 155p.
183. UN. Sustainable Fisheries, Including through the 1995 Agreement for the Implementation of the Provisions of the United Nations Convention on the Law of the Sea of 10 December 1982 relating to the Conservation and Management of Straddling Fish Stocks and Highly Migratory Fish Stocks, and Related Instruments; UNGA A/RES/59/25. In Proceedings of the The 59th Session of the United Nations General Assembly, New York, NY, USA, 14 September 2004.
184. UN. Sustainannuable fisheries, including through the 1995 Agreement for the Implementation of the Provisions of the United Nations Convention on the Law of the Sea of 10 December 1982 relating to the Conservation and Management of Straddling Fish. Stocks and Highly Migratory Fish Stocks, and related instruments. UNGA A/RES/61/105. In Proceedings of the The 61th Session of the United Nations General Assembly, New York, NY, USA, 12 September 2006.
185. UN. Sustainable fisheries, including through the 1995 Agreement for the Implementation of the Provisions of the United Nations Convention on the Law of the Sea of 10 December 1982 relating to the Conservation and Management of Straddling Fish Stocks and Highly Migratory Fish Stocks, and related instruments UNGA A/RES/64/72. In Proceedings of the The 64th Session of the United Nations General Assembly, New York, NY, USA, 15 September 2009.
186. Directive, H. Council Directive 92/43/EEC of 21 May 1992 on the Conservation of Natural Habitats and of Wild Fauna and Flora. *Off. J. Eur. Union* **1992**, *206*, 7–50.
187. Marine Spatial Plan Directive 2014/89/EU. Available online: http://www.marsplan.ro/en/common-msp/directive-2014-89-eu.html (accessed on 8 March 2022).
188. Oceana. *Oceana MedNet, MPA Proposal for the Mediterranean Sea. 100 Reasons to Reach 10%*; Oceana: Madrid, Spain, 2011.
189. Micheli, F.; Levin, N.; Giakoumi, S.; Katsanevakis, S.; Abdulla, A.; Coll, M.; Fraschetti, S.; Kark, S.; Koutsoubas, D.; Mackelworth, P.; et al. Setting Priorities for Regional Conservation Planning in the Mediterranean Sea. *PLoS ONE* **2013**, *8*, 17. [CrossRef]
190. Bastari, A.; Pica, D.; Ferretti, F.; Micheli, F.; Cerrano, C. Sea Pens in the Mediterranean Sea: Habitat Suitability and Opportunities for Ecosystem Recovery. *ICES J. Mar. Sci.* **2018**, *75*, 1722–1732. [CrossRef]
191. United Nations. *Transforming Our World: The 2030 Agenda for Sustainable Development*; United Nations: New York, NY, USA, 2015.
192. European Union. Communication From The Commission To The European Parliament, The Council, The European Economic And Social Committee And The Committee Of The Regions Eu. In *Biodiversity Strategy for 2030 Bringing Nature Back into Our Lives*; European Union: Brussels, Belgium, 2020.
193. Taviani, M.; Angeletti, L.; Antolini, B.; Ceregato, A.; Froglia, C.; Lopez Correa, M.; Montagna, P.; Remia, A.; Trincardi, F.; Vertino, A. Geo-Biology of Mediterranean Deep-Water Coral Ecosystems. *Mar. Res. CNR* **2011**, *6*, 705–719.
194. Lembo, G.; Spedicato, M.T. Lo Stato Delle Risorse Demersali Nei Mari Italiani. GSA 18–Adriatico Meridionale. *State Ital. Mar. Fish. Aquac.* **2011**, *1*, 79–87.
195. FAO. *Report of the FAO Workshop on Deep-Sea Fisheries and Vulnerable Marine Ecosystems of the Mediterranean*; Food and Agriculture Organization of the United Nations: Rome, Italy, 2016.
196. European Union. Communication from the Commission to the European Parliament, The Council, The European Economic and Social Committee and the Committee of the Regions. In *Blue Growth Opportunities for Marine and Maritime Sustainable Growth*; COM/2012/0494 final; European Union: Brussels, Belgium, 2012.
197. Danovaro, R.; Fanelli, E.; Aguzzi, J.; Billett, D.; Carugati, L.; Corinaldesi, C.; Dell'Anno, A.; Gjerde, K.; Jamieson, A.J.; Kark, S. Ecological Variables for Developing a Global Deep-Ocean Monitoring and Conservation Strategy. *Nat. Ecol. Evol.* **2020**, *4*, 181–192. [CrossRef] [PubMed]

Article

Composite of Layered Double Hydroxide with Casein and Carboxymethylcellulose as a White Pigment for Food Application

Estee Ngew, Wut Hmone Phue, Ziruo Liu and Saji George *

Department of Food Science and Agricultural Chemistry, McGill University, Sainte-Anne-de-Bellevue, QC H9X 3V9, Canada; estee.ngew@mail.mcgill.ca (E.N.); wut.phue@mail.mcgill.ca (W.H.P.); ziruo.liu@mail.mcgill.ca (Z.L.)

* Correspondence: saji.george@mcgill.ca

Abstract: Titanium dioxide (TiO_2) is commonly used in food, cosmetic, and pharmaceutical industries as a white pigment due to its extraordinary light scattering properties and high refractive index. However, as evidenced from recent reports, there are overriding concerns about the safety of nanoparticles of TiO_2. As an alternative to TiO_2, Mg-Al layered double hydroxide (LDH) and their composite containing casein and carboxymethyl cellulose (CMC) were synthesized using wet chemistry and compared with currently used materials (food grade TiO_2 (E171), rice starch, and silicon dioxide (E551)) for its potential application as a white pigment. These particles were characterized for their size and shape (Transmission Electron Microscopy), crystallographic structure (X-Ray Diffraction), agglomeration behavior and surface charge (Dynamic Light Scattering), surface chemistry (Fourier Transform Infrared Spectroscopy), transmittance (UV–VIS spectroscopy), masking power, and cytotoxicity. Our results showed the formation of typical layered double hydroxide with flower-like morphology which was restructured into pseudo-spheres after casein intercalation. Transmittance measurement showed that LDH composites had better performance than pristine LDH, and the aqueous suspension was heat and pH resistant. While its masking power was not on a par with E171, the composite of LDH was superior to current alternatives such as rice starch and E551. Sustainability score obtained by MATLAB® based comparison for price, safety, and performance showed that LDH composite was better than any of the compared materials, highlighting its potential as a white pigment for applications in food.

Keywords: white pigment; E171 alternative; layered double hydroxides; casein; carboxymethyl cellulose; sustainability index

Citation: Ngew, E.; Phue, W.H.; Liu, Z.; George, S. Composite of Layered Double Hydroxide with Casein and Carboxymethylcellulose as a White Pigment for Food Application. *Foods* **2022**, *11*, 1120. https://doi.org/10.3390/foods11081120

Academic Editor: Beatriz Gandul-Rojas

Received: 17 February 2022
Accepted: 8 April 2022
Published: 13 April 2022

1. Introduction

E171 (food grade titanium dioxide (TiO_2)) is one of the most widely used food additives containing nanoparticles [1]. Pigment grade TiO_2 with its primary size ranging from 200–350 nm shows a high refractive index (2.6–2.9), negligible absorption in visible range of light spectrum, high scattering, and excellent masking power [2]. In addition, it is stable across a wide range of measurements of pH, temperature, and humidity, and it does not react with matrix components, making it a desired material for food and pharmaceutical products. E171 is copiously incorporated as a white pigment and a masking agent in over 900 commonly consumed food products such as dairy products and analogues, edible ices, confectionaries, surimi and similar products, food supplements, and seasonings and sauces [3,4].

TiO_2 is generally recognized as safe (GRAS) according to the U.S. Food and Drug Administration (FDA) and is permitted to be used in food up to 1% without being declared on labels. Recent studies point towards potential adverse health outcomes from oral administration of TiO_2 nanoparticles [2,5]. E171 particles are not designed to be nanoparticles, but pristine and recovered E171 particles from several food products showed a substantial proportion of nanoparticles (below 100 nm) [4,6]. The presence of nanoparticles in E171 has triggered concerns regarding the undesirable impacts of TiO_2 nanoparticles on human health. In addition, recent publications from our group have reported the negative impacts of TiO_2 nanoparticles on the functional integrity of intestinal epithelial cells and potential allergenicity of food allergens [7,8]. Pinget et al. also found that food grade TiO_2 could impair the homeostasis in gut microbiota-host interactions [3]. The European food safety authority (EFSA) recently deemed that E171 is "not safe" as a food additive after evaluating the outcomes from recent health risk assessments [9]. Consequently, the European Union moved to ban the application of TiO_2 as a food additive under Regulation (EC) No. 2022/63 in the Official Journal of the European Union, with a transition period until 7 August 2022. This and the growing consumer demand for natural and organic foods have resulted in food manufacturers using white pigment alternatives like rice starch (RS), silicon dioxide (E551), and calcium phosphate.

Layered double hydroxides (LDH) have recently fascinated researchers for their wide range of applications in various fields due to their biocompatibility, cost and resource effective methods of synthesis, and suitability for modification [10–12]. LDH is composed of positively charged two-dimensional (2D) metal hydroxide layers which stack alternatively with interlayers of anions, forming a three-dimensional (3D) "lasagna" structure. Structurally, the 2D "lasagna sheets" are composed of divalent cations octahedrally surrounded by OH–ions where the octahedra share edges forming an infinite 2D layer. The positively charged brucite-like (M^{2+} $(OH)_2$) lasagna sheets are separated by an interlayer region containing charge compensating anions and solvation molecules, forming a 3D structure held together via electrostatic interactions and hydrogen bonds [13]. Anionic organic species like peptides, amino acids, and proteins could be intercalated in between the (M^{2+} $(OH)_2$) layers through anion exchange chemistry [14]. LDH-composites thus generated have shown promising applications in medicine and energy conversion as well as storage and environmental remediation [15].

In this article, we report the generation of composite materials where Mg-Al LDH was intercalated with casein through anion exchange, and the surface was further modified with carboxymethylcellulose (CMC). LDH and its composites were characterized and compared to currently used white pigments for their physicochemical properties, masking power and safety. We report the superior performance, cost, and safety features of LDH composite in comparison to generally used E171 alternatives for its potential applications in products intended for human consumption.

2. Materials and Methodologies

Food additive silicon dioxide particles E551 (SIPERNAT 22) were obtained from Evonik Corporation (Parsippany, NJ, USA) and food additive titanium dioxide particles (E171, cat # 13463-67-7) were obtained from Minerals-Water, UK. All particles were used as received.

Commercial bovine milk casein (cat # 5890), magnesium oxide (MgO), aluminum oxide (Al_2O_3), sodium hydroxide (NaOH), rice starch (RS), and carboxymethyl cellulose (CMC) with 0.55–1.0 degree of substitution (DS) were purchased from Sigma Chemical Co. (St-Louis, MO, USA).

Stock solutions of particles (E171, E551) and rice starch were prepared by dispersing 10 mg/mL in deionized (DI) water obtained from a Milli-Q water system (Millipore Sigma, Milford, MA, USA).

2.1. Synthesis of Mg-Al LDH

Mg-based LDH was chosen because it is white color [16]; and Al^{3+} was chosen as the "guest" metal cation due to its ability to stabilize the α- form of hydroxides during LDH formation [17]. LDH was synthesized using a hydrothermal method as detailed elsewhere [18,19]. We chose the hydrothermal method as it improves the crystallinity of LDH which is desirable for the scattering of light within the LDH structure [20].

Accordingly, we mixed MgO: Al_2O_3 in a molar ratio of 2 in 50 mL of deionized water using ultrasonication for 30 m in a clean glass bottle. A Mg/Al ratio of 2 was chosen to form a clear hydrotalcite phase and higher basal spacing which is preferred for the anionic exchange reaction in the following steps [21,22]. One molar NaOH was added dropwise into the bottle under vigorous stirring until the pH of the solution reached ~10. The obtained mixture was stirred at 110 °C for 12 h in the tightly closed bottle, followed by ageing in the oven at 110 °C for 10 days for a dissociation-deposition-diffusion mechanism to mediate the formation of LDH [19,23]. The synthesized LDH was then washed with DI water twice at 12,500 rpm and the precipitate was freeze-dried overnight to obtain dry powder of LDH.

2.2. Synthesis of Mg-Al-Casein LDH (CLDH)

CLDH was synthesized by rehydration of LDH in the presence of casein. For this, 1.5 g LDH and 2.5 g casein were added to 100 mL DI water under vigorous stirring, followed by 30 min bath sonication to fully disperse the particles. NaOH (1 M) was added dropwise to the suspension under vigorous stirring. The suspension (pH~10) was further stirred in a tightly sealed bottle for 10 h. At this relatively high pH way above the isoelectric point of casein (pI = 4.6), casein was intercalated into the LDH interlayer region between the $[Mg_2Al\,(OH)_6]^+$ "lasagna sheets". Subsequently, the white suspension was collected by centrifugation at 12,500 rpm and washed with DI water twice to remove excess casein. The white precipitate from final wash was then freeze dried overnight to obtain CLDH.

2.3. Synthesis of Carboxymethyl Cellulose (CMC) Modified CLDH (CCLDH)

CMC (1.0 g) was dissolved in DI water (500 mL) with vigorous stirring at room temperature for 5 h for it to be fully dispersed. CLDH suspension in water (8 mg/mL) was added dropwise into the CMC solution such that the mass ratio of CMC to CLDH is 9 into a glass bottle. After 6 h, the CCLDH suspension was washed with DI water thrice, and was dried overnight in a freeze dryer.

2.4. Materials Characterization

X-Ray Diffraction (XRD) analysis was performed to observe and compare crystal structure of LDH, CLDH, and CCLDH composite and other particles used in this study. For this, the powdered particles were placed on the XRD specimen holder and pressed with a glass plate to fill the holder. The XRD patterns were recorded using a diffractometer (Bruker D8 Discover diffractometer with VANTEC–2000 detector system and Cu source) and analyzed with a diffraction angle range between 0 and 50° (2 theta). The voltage, current and pass time used were 40 Kv, 40 mA, and 1 s, respectively.

Attenuated total reflectance Fourier-transform infrared spectroscopy (ATR-FTIR, Bruker, Billerica, MA, USA) was used to identify the functional groups of the particles. For this, the dried powder samples were placed on the ATR probe and pressed. The FTIR spectra were taken in the range of 400–4000 cm^{-1}, with a resolution of 4 cm^{-1} and 24 scans. Spectra were deconvoluted and analyzed using OMNIC software.

Transmission electron microscope (TEM) images were obtained using transmission electron microscope (Field Electron and Ion Company (FEI), Hillsboro, OR, USA) operating at 120 kV. For this, 1 mg/mL of particles (LDH, CLDH, CCLDH) were dispersed in ethanol and bath sonicated for 15 min. The suspension was further diluted in ethanol to 100 μg/mL. Five μL of samples were dropped on the TEM grid, air dried for 30 min, and imaged. Size distribution analysis for E171 was performed on at least 20 particles identified on the grid using ImageJ software (NIH, Bethesda, MD, USA).

The average hydrodynamic diameter, polydispersity index (PdI) and zeta potential of the particles were determined by dynamic light scattering (DLS) on a Zetasizer nano-ZS (Malvern Instruments, Malvern, UK) at 25 °C. The particles were dispersed and diluted to

50 μg/mL in deionized water and were measured twice. Size results are given as intensity distribution by the mean diameter with its standard deviation.

2.5. Transmittance and Opacity Measurements

The UV-Visible absorbance spectra could be used to associate with opacity [24]. The total reflectance spectra of the synthesized particles were measured and compared with E171, E551, and RS. For this, equal quantities of dry powders of particles were pressed onto the wall of the quartz cuvette of a UV–VIS diffuse reflectance spectrophotometer (Lambda 750, PerkinElmer Life and Analytical Sciences, Shelton, CT, USA) operated by UV WinLab™ software (Version 4.0, PerkinElmer Life and Analytical Sciences, Shelton, CT, USA), and % reflectance was measured in the range of 200–800 nm with 1 nm interval.

Opacity is defined as the proficiency of the pigment to mask the features of the underlying substrate [25]. Opacities of the synthesized particles were compared with that of E171 by measuring % transmittance. For transmittance measurement, 100 μL of particle suspensions (1 mg/mL) were dispensed into wells of 96 well plate and % transmittance at 550 nm was measured using a microplate reader (Spectra Max i3x, Molecular Devices, San Jose, CA, USA). Masking power comparisons between E171 and synthesized particles were conducted visually with optimized concentrations of hydroxy propyl methylcellulose (HPMC) (85% *w*/*v*) and plasticizer (Glycerol 10% *w/v*) for preparation of a film layer. Firstly, dry particles of E171, E551, and RS were suspended (1 mg/mL) in DI water, while the synthesized particles in dry powder form were suspended (1 mg/mL) at pH ~10 adjusted solution (with 0.1 M NaOH). These suspensions were individually mixed with slurry containing 85% hydroxy propyl methylcellulose (HPMC), 10% glycerol, and 5% solution. The resulting thick pastes (1 mL) of particles were spread evenly on a glass plate using a paint brush and left to dry overnight at room temperature. The glass plate was imaged against a worded background for visual comparison of masking power, recorded using digital camera.

2.6. Characterization of Stability of Particle Suspension to pH and Temperature

Stabilities of aqueous suspensions of LDH, CLDH, CCLDH, and E171, E551 (1 mg/mL) were evaluated as a function of time and under different pH and temperature conditions. For this, the particle suspensions were probe sonicated for three intervals of 15 s to evenly disperse the particles in DI water.

2.6.1. Effect of pH

Stability of particle suspensions as a function of pH was performed by adjusting the pH of the suspension using 0.1 M NaOH and 1 M HNO_3 to obtain pH levels of 3 and 9. These two pH levels were chosen as they represent the acidic and alkaline pH typically encountered in food processing. The suspension of particles with pH adjusted were left undisturbed for 30 min at room temperature, and the % transmittance (550 nm) was measured to assess the extent of particle sedimentation. The experiment was repeated in triplicates, and the mean value was plotted as % transmittance vs. pH.

2.6.2. Effect of Temperature

Stability of particle suspension as a function of temperature was determined by heating 1 mg/mL samples in Eppendorf tubes in a thermostat water bath to 25, 50, 75, and 100 °C. After holding samples at the stipulated temperatures for 5 min, suspensions were cooled in ice bath to room temperature, and 100 μL of each suspension was aliquoted into a well of 96 wells plate. The % transmittance at 550 nm was determined using a microplate reader. The results of the three replicates were averaged and a graph of % transmittance vs. temperature was plotted with mean values and standard deviations.

2.7. Casein Antigenicity Study

One of the potential challenges for the application of the CCLDH composite would be the allergenicity of casein. Therefore, we compared the antigenicity of casein before and after conjugating with LDH using enzyme-linked immunosorbent assay (ELISA) as reported [8]. The ELISA plate was prepared by adding 100 µL of antigen solution (casein, CLDH and CCLDH (100 ng/mL) in coating buffer (0.1 M Na_2HPO_4 in DI water- pH 9.5)) to individual wells of a 96-well plate (Nunc Maxisorp 96-well ELISA plate). The plate was kept overnight at 4 °C for adsorption of antigen (casein, CLDH or CCLDH) to well surface. The uncoated antigen was discarded, and the residual free binding sites were blocked with blocking buffer (2% BSA in 1× TBS containing 50 mM Tris-HCl and 150 mM NaCl) for 1 h at 37 °C. After discarding the blocking buffer, primary antibody (1: 2000) (anti-casein rabbit antibody- cat # ab166596) was added in the blocking buffer and incubated for 1 h at room temperature. Excess of primary antibody (remaining in solution) was discarded, and the wells were washed four times with washing buffer (0.05% tween−20 in 1× TBS). The wells were then incubated with secondary antibody (1: 10,000) (anti-rabbit antibody- cat # 6721) in blocking buffer for 1 h at RT. Subsequently, the excess unbound secondary antibody was discarded, and the wells were washed four times with washing buffer. These wells were further added with 100 µL of substrate solution (5 mM of tetramethylbenzidine-TMB) and incubated for 3 min at RT for the enzymatic color development, followed by the addition of stop solution (100 µL of 2 M sulphuric acid) to terminate the reaction. Absorbance at a wavelength of 450 nm was determined by using a microplate reader. The amount of primary antibody binding onto antigen (casein) was determined using the standard curve, and the antigenicity is expressed as casein antigen response (ng/mL). The result presented is an average of values from triplicate experiments.

2.8. Cytotoxicity Study

Particle dispersion in serum-free cell culture media: Particle stock solution (1 mg/mL) was prepared by dispersing the particle powder in deionized water by probe sonication for 30 s of five intervals. The sonicated particle suspensions in DI water were further diluted in serum-free Dulbecco's modified eagle medium (DMEM) medium.

Cell culture and co-incubation with particles: Human epithelial Caco-2 cells between passages 20–40 were cultured in DMEM (Invitrogen, Grand Island, NY, USA) supplemented with 10% (*v*/*v*) fetal bovine serum (FBS; Sigma Aldrich, St. Louis, MO, USA), and 1% antibiotics (penicillin/streptomycin; Invitrogen) 10 cm^2 cell culture plates. These plates were incubated at 37 °C in a 5% CO_2 incubator to reach 70–80% confluency. The cells were trypsinized, washed and seeded at 10,000 cells in 100 µL per well in a 96-well plate. Seeded cells were cultured overnight at 37 °C in a 5% CO_2 incubator for complete attachment before particles exposure. The stock solution of particles dispersed in DI water was further diluted in serum-free DMEM medium, and particles exposures were carried out at concentrations of 25, 12.5, 6.25, 3.12, 1.56, and 0 µg/mL for 24 h.

Effect of particles on cell viability: Cytotoxicity of the particles on Caco-2 cells were assessed by the Alamar Blue (resazurin) assay after exposure to the desired concentration of the particles. For this, the exposure media was removed from the wells, and the cells were incubated with 100 µL resazurin solution (50 ppm) and incubated at 37 °C in 5% CO_2 for 4 h. Absorbance at 570 nm was recorded using a microplate reader (Spectra Max M2, Molecular Devices, Sunnyvale, CA, USA). The readings were exported to an Excel file for comparison and statistical analysis.

Cytotoxicity was calculated as:

$$\text{Cell viability } (\%) = \frac{\text{Abs test}}{\text{Abs control}} \times 100 \quad (1)$$

2.9. Application of Alternatives Assessment Framework

Developing sustainable alternatives to existing solution demands evaluation not only for performance but also for its cost and safety. The alternatives assessment framework allows us to identify sustainable alternatives to existing chemicals that needs replacement [26]. We conducted a preliminary alternative assessment of CCLDH for its potential application as a food pigment. The assessment conforms to the US National Academy's Framework to Guide Selection of Chemical Alternatives. Accordingly, the Efficacy (% reflectance), Cost (USD/g) (Table S1), and Safety (NOAEL) (Table S2) were used for a multiparametric analysis using MATLAB, as detailed in Supplementary Materials.

3. Results

3.1. Synthesis and Characterization of Particles

Figure 1a–f present the TEM images of particles tested in this study. In general, these particles had spherical or pseudospherical morphologies. Pristine LDH particles showed the characteristic flower-like morphology with hexagonal shape of the flaky sheets of LDH crystallites (Figure 1a). Significant changes in morphological features were observed when pristine LDH was modified by intercalation of casein. The presence of casein in the CLDH composite gave rise to the ordered inorganic lamellae represented by the darker lines (Figure 1b), which was similar to the Ca-Al-casein LDH reported earlier [27]. The schematic diagram (Figure 1g) shows the possible mechanism of casein intercalation through anion exchange when the negatively charged groups (–COO^-) of casein gets electrostatically adsorbed between the positively charged LDH layers. The TEM micrograph of CCLDH (Figure 1c) showed mesoporous plates and rod-like cores of CLDH rendered by ordered CMC. Figure 1h demonstrates the scheme of CMC layered onto the CLDH particles. RS showed amorphous morphology with larger discrete particles with a size of ~5 µm (Figure 1d). E171 showed more consistent shapes with an estimated particle size of 100–120 nm (Figure 1e). Size distribution analysis (Figure S1) showed that ~20% of particles in E171 had size less than 100 nm (to be categorized as nanoparticles). Similar to previous observation from our group, E551 (food grade silica) were heavily agglomerated where the size of individual agglomerate was ~30 nm [28].

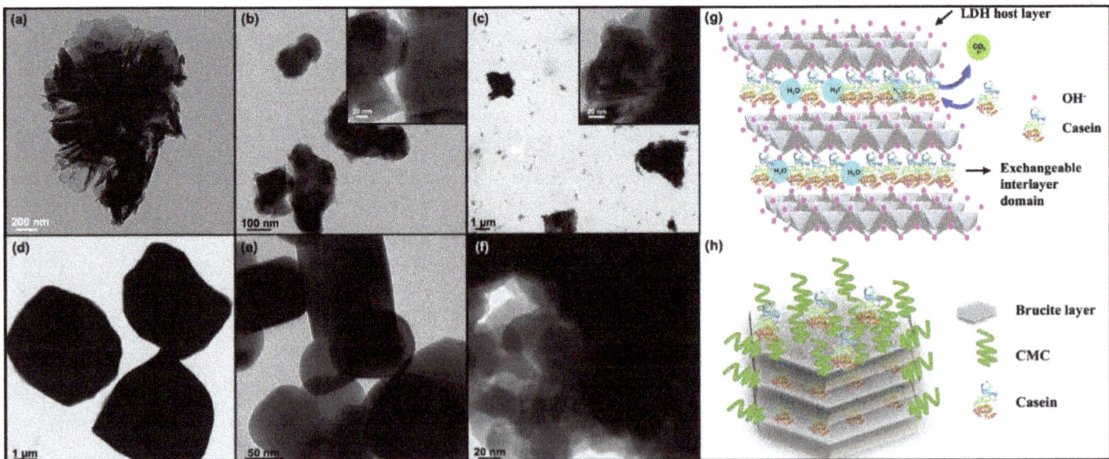

Figure 1. TEM micrographs of layered double hydroxide (LDH) powders: (**a**) LDH, (**b**) CLDH, (**c**) CCLDH, (**d**) Rice Starch, (**e**) E171, (**f**) E551, (**g**) scheme showing anionic exchange facilitating the intercalation of casein molecules between the brucite layers, (**h**) schematic of the CCLDH composite.

These particles were further characterized for their agglomeration behavior, surface charge, surface chemistry, and crystallinity. Based on the DLS results (Figure 2a), polydispersity of CLDH decreased significantly from pristine LDH but showed a significant increase after CMC adsorption. Since the LDH samples were poly-dispersed, it was hard to evaluate the mean size based on the TEM images; thus, the size distributions of the particles were measured in solution by DLS (Figure S2). The hydrodynamic diameter (Figure 2a) of pristine LDH was larger (>1 µm) in comparison to CLDH (<500 nm). The average diameter by volume (Figure 2a) of CCLDH had also increased compared to CLDH. The zeta potential (Figure 2b) of pristine LDH was +10.24 mV while that of casein intercalated (CLDH) was +6.03 mV. CCLDH showed a net negative zeta-potential of −11.96 mV. Zeta potential values of CLDH were close to those of pristine LDH than that of casein, suggesting that the LDH accommodated the casein molecules between the LDH brucite layers. CCLDH however showed a negative zeta potential value because of the surface adsorption of CMC onto CLDH.

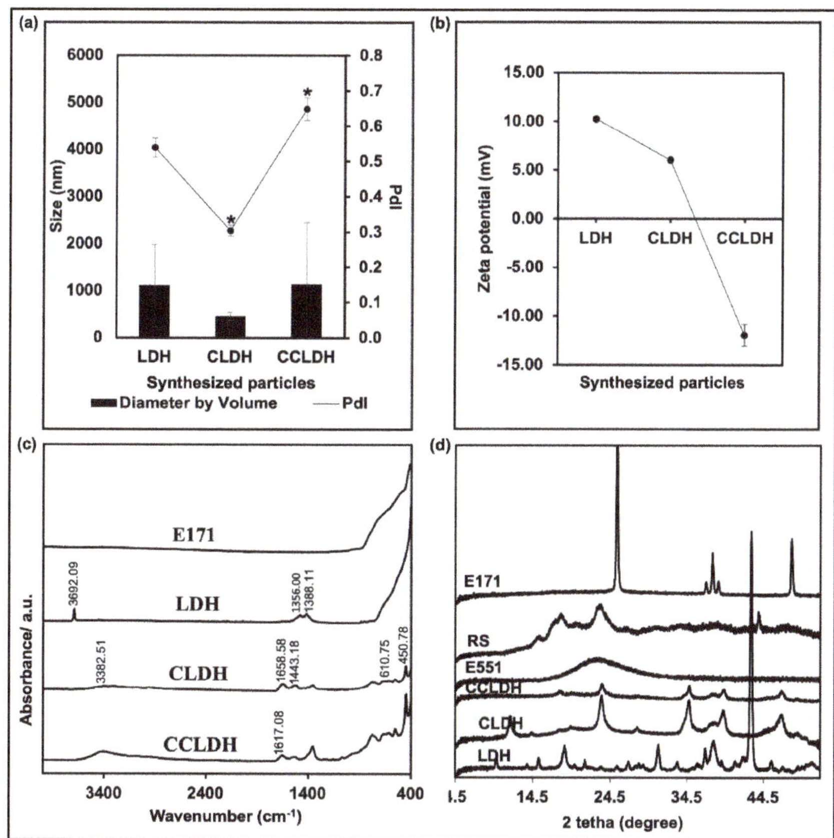

Figure 2. (**a**) Hydrodynamic diameter (bar graph) and polydispersity index (line graph) of particles (50 µg/mL) suspended in water. * Indicates statistical significance (*t*-test) in polydispersity (PdI) of CLDH in comparison to pristine LDH and CCLDH compared to CLDH, $p \leq 0.05$, $N = 2$; (**b**) Zeta potential of LDH particles and its composites; (**c**) FTIR spectra of the particles; particles were placed directly on the ATR probe prior recording the spectra for a wavenumber range of 4000–400 cm^{-1} using ATR–FTIR; (**d**) XRD spectra of the tested particles. The spectra of dry powder of samples were captured using Bruker D8 Discovery X-ray Diffractometer.

ATR-FTIR was performed to assess the surface groups present in the LDH composites and the commercially used white food pigments, and the result is summarized in Figure 2c.

The peak of 3692.09 cm^{-1} on the LDH graph features the vibration of −OH groups present in the LDH inorganic Mg $(OH)_2$ layer [29]. The sharp peak at 1356.00 cm^{-1} indicates the adsorption of CO_2 contamination from air during the collection process which is difficult to be excluded during LDH synthesis [30]. This contamination occupies some sites, which may lead to incomplete anionic exchange for casein intercalation [30].

The strong peak at 1388.11 cm^{-1} [due to $\nu(CO_3{}^{2-})$] and bands below 1000 cm^{-1} (due to M–O vibrations and M-O-H bending of LDH) are all characteristics of Mg-Al LDH [28,31–33]. The peak at 450.78 cm^{-1} in CLDH is a unique characteristic of Mg-Al LDH materials at 2:1 ratio [31]. This strong peak and a broad peak at 610.75 cm^{-1} are attributed to the lattice vibrations of Mg_2Al-OH layer [34]. The band at 1443.18 cm^{-1} from CLDH is attributed to the stretching vibration of C = O symmetric stretching, verifying the presence of casein. The FT-IR spectrum of CCLDH is the combination of both the CMC and CLDH spectra, which is the good reason for the successful preparation of CCLDH. The casein peak of 1658.58 cm^{-1} had also broadened to 1617.08 cm^{-1}.

The phase purity and crystallinity of the CLDH and CCLDH were analyzed and compared to the pristine LDH, alongside other industrial used white pigments by measuring XRD peaks (Figure 2d). In general, the LDH particles possessed defined basal diffraction peaks, suggesting their crystallinity. The varied phase composition of the synthesized LDH-composites indicated the presence of a variety of hydroxides and oxide hydroxides. The diffraction peaks at 15.3° for a rhombical symmetry is characteristic of hydrothermally synthesized LDH [19]. Strong diffraction peaks of the LDH due to a long aging time are in good agreement with those reported for Mg-Al-OH LDH [19]. The larger and more intense peak of LDH is attributed to its larger crystallite size compared to the smaller ones (less intense diffraction peaks), suggesting a correlation of size-crystallinity in LDHs [35]. Notably, the increased intensity of the rhombohedral symmetry peak at 23.6° in the composite material suggested successful intercalation of casein into the crystal structure. This finding concurs with the FTIR results as well as with the results from TEM images. LDH peaks disappeared in CCLDH suggesting the rendering of the negatively charged CMC over the positively charged surface of LDH [36]. All in all, physicochemical characterization of parental LDH and variants synthesized demonstrated a typical floral morphology of primary particle size from 100 to 1000 nm, well dispersed in aqueous solution where the original positive surface charge changed to negative when CLDH was surface functionalized with CMC, and the crystalline characteristics of the original LDH were reduced when it was intercalated with casein and subsequently functionalized with CMC. E551 particles were amorphous as evidenced by the broad peak in XRD spectra, while E171 were anatase [7]. RS, however, showed both crystalline and amorphous XRD patterns [36].

3.2. Reflectance, Stability of Particle Suspensions, and Masking Power

Among the synthesized particles, CCLDH showed excellent reflectance at the visible range which was second only to E171 (Figure 3a). This result complies with the masking power test whereby CCLDH had the strongest masking property in comparison with other LDH composites (Figure 4a). As shown in Figure 3a, the reflectance increased from 90% to 100% when the CLDH was engineered with CMC.

Since food additives are commonly exposed to different temperatures and pH during processing or preparation, we investigated if the aqueous suspensions of particles tested were stable at elevated temperatures and in pH range [25,37]. We observed that suspensions of CLDH and CCLDH were more thermally stable compared to pristine LDH, possibly due to the free organic anion after casein intercalation. Further, adsorption of CMC onto the LDH surface could have prevented aggregation of LDH because of static and stearic hindrance. Both RS and E551 showed increased % transmittance as the temperature increased, which suggested that the opacity of currently used white pigment alternatives is not stable at elevated temperatures. In addition, the particle suspension seemed stable at alkaline and neutral pH, but an increase in % transmittance of the LDH composites was observed at pH 3 as LDH particles started to aggregate. Notably, RS was not stable at an alkaline pH as evidenced by an increased % transmittance. E551, however, was stable at an alkaline pH similar to that of the LDH composites.

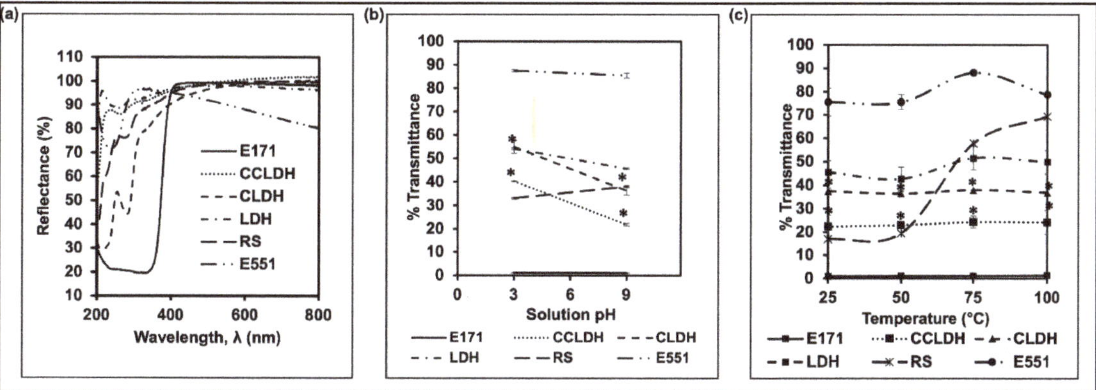

Figure 3. (**a**) Reflectance of powdered particles (both synthesized and commercially used white pigment) were measured in the range of 200–800 nm with 1 nm interval, demonstrating that all the synthesized LDH particles had a good reflectance in the visible range; (**b**,**c**) % transmittance (550 nm) at varied temperature and pH, respectively, suggested relatively high stability of LDHs at high temperatures, but not in acidic pH. * Indicate statistical significance (t-test) in % transmittance of LDH composites in comparison to pristine LDH, $p \leq 0.05$, $N = 3$.

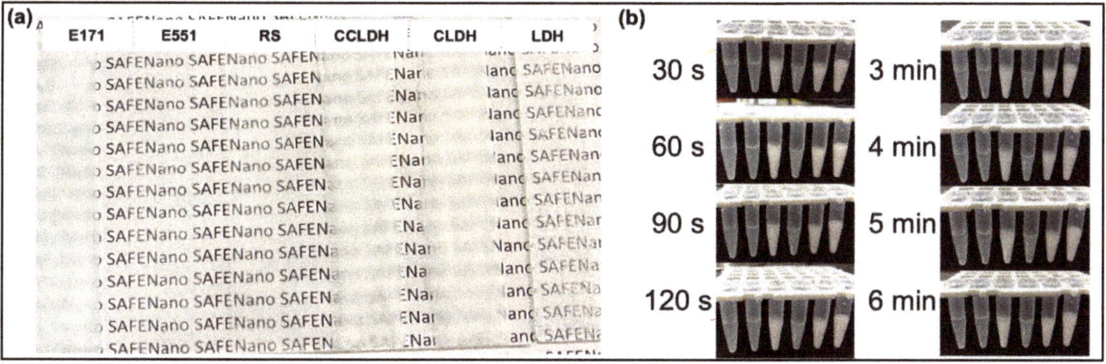

Figure 4. (**a**) Masking ability of the synthesized LDH and its composites was compared with that of commercially used E171 and its alternatives by suspending them into a mixture to form a thick paste to be painted on a glass plate and visually compared against a worded background. From left: E171, E551, RS, CCLDH, CLDH, and LDH. The LDH composites showed better masking. (**b**) Suspension stability of particles (from left: LDH, CLDH (1 mg/mL), CLDH (10 mg/mL), CCLDH (1 mg/mL), CCLDH CLDH (10 mg/mL), and E171) dispersed in DI water were visually compared from time-lapse images taken at 30 s intervals up to 180 s, followed by 60 s intervals from 3 min to 6 min.

The masking power of LDH composites was compared with E171 and other TiO_2 alternatives by painting slurries containing particles on to a clean glass sheet and visually compared over a worded background. Even though the masking power was not as good as E171 (Figure 4a) LDH composites were superior to current alternatives used in the industry. Notably, pure LDH was not suspended well in the slurry and showed granular appearance while composites showed smooth smearing (Figure 4a). The aim of the research was to develop a white and opaque food pigment. Therefore, balancing the pigment structure and its stability in aqueous suspension is crucial. As seen in the time lapse images of particle suspension, casein intercalated LDH (CLDH) and modified with CMC (CCLDH) showed a more stable particle suspension.

3.3. Antigenicity and Cytotoxicity Assessment of LDH Composites

Since LDH particles are efficient delivery carriers of proteins, we also explored the possibility of using LDH as a modulator to prevent the binding of immunoglobulin E (IgE)-mediated hypersensitivity to casein which is often associated with milk allergy symptoms. As shown in Figure 5a, the antigen response of caseins intercalated in the CLDH reduced significantly which confirms its ability to suppress casein-specific IgE binding capacity, therefore decreasing the casein antigenicity. However the same response was not maintained in CCLDH, although the effect was still significantly lesser than pure casein.

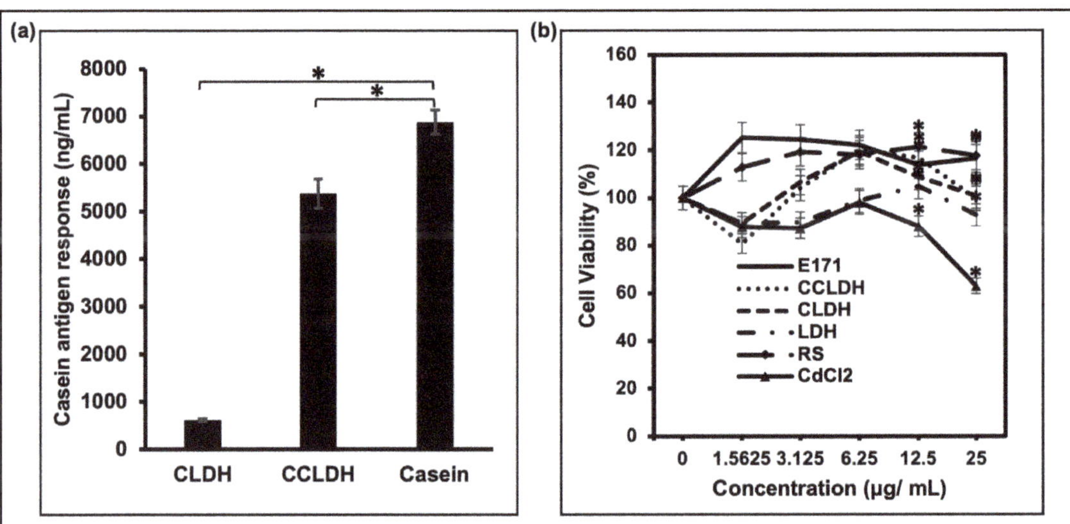

Figure 5. (**a**) IgE binding capacity was determined by ELISA, and its response was expressed as the casein antigen response (ng/mL) of casein samples under different LDH compounds. * Indicates statistically significant (*t*-test) difference in casein antigenic response of LDH composites with casein in comparison to raw untreated casein molecules, $p \leq 0.05$, $N = 3$. (**b**) Particle concentration dependent Caco-2 cell viability changes post exposure to particles tested. The result suggested low/no cytotoxicity of LDH particles. * Indicates the significant differeces of the tested particles from the control ($CdCl_2$) as verified by statistical *t*-test ($p \leq 0.05$, $N = 3$).

The cytotoxicity response was evaluated based on the resazurin assay after the Caco-2 cells were treated with different concentations (0–25 μg/mL) of the synthesized pigments, its alternatives and E171 with cadmium chloride ($CdCl_2$) as the positive control. The in vitro data clearly depicited that all of the synthesized LDH composites did not elicit any obvious cytotoxicity to Caco-2 with a viability above 80%.

3.4. Comparison of Particles for Efficacy, Cost, and Safety

To demonstrate the applicability of the LDH composite as white pigments, we compared the tested materials in their performance, cost, and safety using MATLAB® analysis (Figure 6). The reliability of E171 was scrutinized against its current alternatives and CCLDH based on three factors: efficacy, cost, and safety. Each factor was scored from 0 to 3 as detailed in Supplementary Materials (Table S3). The scores were used to develop a three-dimensional matrix using MATLAB® (Figure 6). As seen in the heatmap, CCLDH was better in comparison to other white pigments in this multi-parametric comparison. E551 with its poor performance and cost had the lowest score, followed by E171 and rice starch.

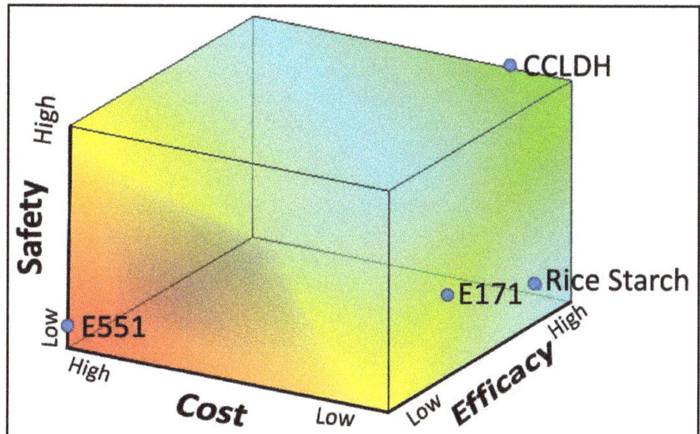

Figure 6. A three-dimensional matrix was developed by MATLAB® for comparing the efficacy (% reflectance), cost (USD/g), and safety (NOAEL) of CCLDH to TiO_2 alternatives and E171. The application desirability of the four particles in food are expressed in the phase diagram developed by MATLAB®, with red representing likely low success, blue likely high success, and green having considerable success as white food pigments. Position of each particle in the matrix was determined from individual scores for the three parameters assessed.

4. Discussion

TiO_2 is used copiously in several food, cosmetic, and pharmaceutical products as a white pigment. Recent health risk assessment studies, however, have indicated potential toxicity of dietary TiO_2. For instance, oral administration of TiO_2 nanoparticles was reported to cause epithelial hyperplasia and preneoplastic lesions in rodent models [5,38,39]. The obscurity on the health risk associated with TiO_2 has incited the European Commission to request the European Food Safety Authority (EFSA) to reassess the safety intake of E171 (food grade TiO_2) as a food additive. Thus, there is a surging demand for TiO_2 (E171) alternatives in the food industry. Finding an alternative to TiO_2 which has been extensively and widely applied across sectors is challenging from the perspective of meeting the constraints of performance, cost, and safety. Here, we report the synthesis and characterization of an LDH-based composite material as a potential alternative to E171 as a white food pigment.

LDH is a class of anionic clay particles with the characteristic hexagonal shape of flake-like sheets as revealed in the TEM image (Figure 1a). The flaky sheets are characteristic of the brucite-like layers, while irregular hexagonal shapes are due to more OH groups being exposed to the aqueous phase during ageing [40]. Mg-Al LDH was chosen as the constituent material due to its versatile properties such as high chemical stability, controllable particle size, and most importantly, it can be white in color [16]. While Balcomb et al., 2015, reported Zn-Al to be white, our results (data not included) showed that it was not as white as Mg-Al LDH after casein intercalation. Moreover, presence of Zn could compromise the safety of the material [41]. Casein, one of the major milk proteins when in micelle form scatters light and imparts opacity, contributing to the characteristic white color of milk. Therefore, we modified the pristine LDH by intercalating it with casein. Casein, which becomes anionic above its isoelectric point, is intercalated between the highly charged LDH layers through anion exchange (see scheme in Figure 1g). The homogeneous distribution and stabilization of the organic casein at the molecular level are driven by an integrated hydrogen-bonding network and Van der Waals interaction, as well as the host-guest interactions [42]. In particular, the negatively charged groups of casein (-COO^-) would electrostatically interact with the LDH host layers, while its positively charged -NH^{3+} groups would repel, exposing the aqueous phase [43]. Comparing Figure 1a with b and inferring from the results of DLS, FTIR, and XRD, it is evident that the LDH lamellar structure acted as lasagna sheets having casein as the lasagna filling. The smaller agglomeration size of CLDH as compared to LDH (DLS data Figure 2a) and XRD data (Figure 2d) suggested that casein intercalation facilitated the dispersion of CLDH by changing the crystallographic structure of LDH. While the resulting CLDH showed improved suspension stability as depicted in Figure 4b, it was not on a par with E171. However, surface modification of CLDH with CMC prolonged the suspension stability of composite material suggesting the role of CMC in improving the electrostatic and steric hindrance-based repulsion between particles.

Light scattering at the boundaries within the 3D crystals of LDH is thought to play an important role in the white color appearance of the LDH. CLDH, however, has several types of interactions including stacking and charge interactions from sandwiching of casein molecules between the brucite layer, imitating a micelle-like structure. Surprisingly, CLDH exhibited a strong anti-reflectance behavior in the near UV region. This could be due to the incomplete anionic-exchange and increased amount of casein on the LDH surface, leading to a lower opacity. As described in the introduction, the "filling" of the interlayer lasagna sheets is mainly composed of water molecules and anions. When casein intercalation takes place, the former anion would be desorbed slowly from the LDH lattice. As the interlayer gets fully occupied, the excess casein could be exposed at the particle-water (aqueous phase) interphase [44]. Surface modification of CLDH with CMC (CCLDH) increased the reflectance possibly due to the ordered CMC acting as milk fat replacers entrapping the CLDH aggregates, forming a micelle-like structure [45]. The negative zeta potential of CCLDH could be due to the adsorption of CMC to CLDH through H-bonding between the hydroxide of LDH brucite layers and amide groups of the CMC molecules. In addition, the LDH composites suspensions showed high stability at elevated temperatures and alkaline pH after intercalation of casein due to the host-guest interaction involving hydrogen bonds as demonstrated by FTIR analysis. Addition of CMC also effectively protected the LDH particle suspension from heat-induced aggregation possibly because of steric stabilization mediated by multilayer structures [46]. Notably, LDH and its composites showed a general tendency of decreasing suspension stability as a function of decreasing pH. The basal surface is positively charged due to the isomorphous substitution of Mg by Al while the charges from the edges arise from the pH-dependent -OH groups. Therefore, a topotactic reaction may take place causing layer erosion at acidic pH [47]. This results in a decrease of electrostatic repulsions leading to the collapse or aggregation of LDH. In addition, the intercalated casein could be released as the LDH layers get delaminated at low pH [48].

Interestingly, intercalating the pristine LDH with casein improved its stability when dispersed in water, but this finding also evokes recognition of the LDH composite by the immune system. Based on the TEM micrographs in Figure 1a, the mixed (horizontal and vertical) platelet morphology allows the composite to form porous structure, acting as a protective layer encapsulating the casein molecule, thus attenuating the allergenicity of casein to a great extent [15]. Loading of casein into the interlayer galleries of the LDH octahedral sheets significantly reduced the antigenic IgE response to casein, therefore inhibiting the immunoreactivity. In comparison to CLDH, the antigenicity of CCLDH was not that apparent in comparison to casein. More studies are required to rule out the possibility of non-specific interaction between IgE and CCLDH containing CMC on the surface.

The predisposing factor of the successful application of an E171 alternative would be its efficacy in terms of intended function (expressed as % reflectance). However, industrial adaptation of the alternative food pigment should also consider other factors like scalability, reproducibility, cost, and safety. CCLDH, an anionic clay composite, is indeed an ideal white pigment alternative as shown in the alternative assessment analysis when these factors were considered (Figure 6). While this comparison was made on general grounds, more studies are warranted for determining the suitability of LDH composites in specific food applications.

5. Conclusions

Our findings contribute to the potential application of LDH composites as an alternative white food pigment. Mg-Al LDH was synthesized using the hydrothermal method. Masking power and suspension stability of the pristine LDH were substantially improved when casein was intercalated, followed by surface coating with CMC to form CCLDH. Casein antigenicity was also found to be suppressed when intercalated between the brucite layers. The ever-increasing demand for TiO_2 alternatives as white pigment for food and pharmaceutical applications occurs under the constrains of today's consumer choices on products that are not only delicious and visually appealing but also safe and better for human and environmental health. In this regard, results from the current study suggest promising applications of LDH composite as a white pigment for food, pharmaceutical, and cosmetic products. Nonetheless, the compatibility of LDH composites with common food ingredients and elaborate safety assessment using animal models remain to be explored.

Supplementary Materials: The following supporting information can be downloaded at: https://www.mdpi.com/article/10.3390/foods11081120/s1, Figure S1: TEM image of E171 used in the study alongside SEM images from our previous studies as a representative sample; Figure S2: Size distribution analysis of the synthesized particles; Table S1: Cost analysis of CCLDH synthesized in this study; Table S2: No Observed Adverse Effect Level (NOAEL) of tested materials in this study; Table S3: Raw data for MATLAB® generation derived from experimental results based on the scoring system. Refs. [49–52] are cited in the Supplementary Materials.

Author Contributions: Conceptualization, S.G.; methodology, E.N.; investigation, E.N., Z.L., W.H.P., S.G.; data curation, E.N., S.G.; writing—original draft preparation, E.N.; writing—review and editing, W.H.P., Z.L., S.G.; supervision, S.G. project administration, S.G.; funding acquisition, S.G. All authors have read and agreed to the published version of the manuscript.

Funding: This research was funded by Canada Research Chair/George, grant number 248475.

Institutional Review Board Statement: Not applicable.

Informed Consent Statement: Not applicable.

Data Availability Statement: Not applicable.

Acknowledgments: We acknowledge funding from CRC/George/X-coded/248475 for supporting this research work.

Conflicts of Interest: The authors declare that there is no conflict of interest regarding the publication of this article.

References

1. George, S.; Kaptan, G.; Lee, J.; Frewer, L. Awareness on adverse effects of nanotechnology increases negative perception among public: Survey study from Singapore. *J. Nanopart. Res.* **2014**, *16*, 2751. [CrossRef]
2. Sha, B.; Gao, W.; Wang, S.; Li, W.; Liang, X.; Xu, F.; Lu, T.J. Nano-titanium dioxide induced cardiac injury in rat under oxidative stress. *Food Chem. Toxicol.* **2013**, *58*, 280–288. [CrossRef] [PubMed]
3. Pinget, G.; Tan, J.; Janac, B.; Kaakoush, N.O.; Angelatos, A.S.; O'Sullivan, J.; Koay, Y.C.; Sierro, F.; Davis, J.; Divakarla, S.K.; et al. Impact of the Food Additive Titanium Dioxide (E171) on Gut Microbiota-Host Interaction. *Front Nutr.* **2019**, *6*, 57. [CrossRef]
4. Weir, A.; Westerhoff, P.; Fabricius, L.; Hristovski, K.; von Goetz, N. Titanium dioxide nanoparticles in food and personal care products. *Environ. Sci. Technol.* **2012**, *46*, 2242–2250. [CrossRef] [PubMed]
5. Bettini, S.; Boutet-Robinet, E.; Cartier, C.; Coméra, C.; Gaultier, E.; Dupuy, J.; Naud, N.; Taché, S.; Grysan, P.; Reguer, S.; et al. Food-grade TiO_2 impairs intestinal and systemic immune homeostasis, initiates preneoplastic lesions and promotes aberrant crypt development in the rat colon. *Sci. Rep.* **2017**, *7*, 40373. [CrossRef] [PubMed]
6. Geiss, O.; Bianchi, I.; Senaldi, C.; Bucher, G.; Verleysen, E.; Waegeneers, N.; Brassinne, F.; Mast, J.; Loeschner, K.; Vidmar, J.; et al. Particle size analysis of pristine food-grade titanium dioxide and E 171 in confectionery products: Interlaboratory testing of a single-particle inductively coupled plasma mass spectrometry screening method and confirmation with transmission electron microscopy. *Food Control* **2021**, *120*, 107550. [PubMed]

7. Xu, K.; Basu, N.; George, S. Dietary nanoparticles compromise epithelial integrity and enhance translocation and antigenicity of milk proteins: An in vitro investigation. *NanoImpact* **2021**, *24*, 100369. [CrossRef]
8. Phue, W.H.; Xu, K.; George, S. Inorganic food additive nanomaterials alter the allergenicity of milk proteins. *Food Chem. Toxicol.* **2022**, *162*, 112874. [CrossRef]
9. EFSA Panel on Food Additives and Flavourings (FAF); Younes, M.; Aquilina, G.; Castle, L.; Engel, K.-H.; Fowler, P.; Fernandez, M.J.F.; Fürst, P.; Gundert-Remy, U.; Gürtler, R.; et al. Safety assessment of titanium dioxide (E171) as a food additive. *EFSA J.* **2021**, *19*, e06585.
10. Zhao, Y.; Jia, X.; Waterhouse, G.I.N.; Wu, L.-Z.; Tung, C.-H.; O'Hare, D.; Zhang, T. Layered Double Hydroxide Nanostructured Photocatalysts for Renewable Energy Production. *Adv. Energy Mater.* **2016**, *6*, 1501974. [CrossRef]
11. Machado, J.P.E.; de Freitas, R.A.; Wypych, F. Layered clay minerals, synthetic layered double hydroxides and hydroxide salts applied as pickering emulsifiers. *Appl. Clay Sci.* **2019**, *169*, 10–20. [CrossRef]
12. Guillot, S.; Bergaya, F.; de Azevedo, C.; Warmont, F.; Tranchant, J.-F. Internally structured pickering emulsions stabilized by clay mineral particles. *J. Colloid Interface Sci.* **2009**, *333*, 563–569. [CrossRef] [PubMed]
13. Wang, Q.; O'Hare, D. Recent Advances in the Synthesis and Application of Layered Double Hydroxide (LDH) Nanosheets. *Chem. Rev.* **2012**, *112*, 4124–4155. [CrossRef]
14. Reinholdt, M.X.; Kirkpatrick, R.J. Experimental Investigations of Amino Acid-Layered Double Hydroxide Complexes: Glutamate−Hydrotalcite. *Chem. Mater.* **2006**, *18*, 2567–2576. [CrossRef]
15. Gu, Z.; Atherton, J.J.; Xu, Z.P. Hierarchical layered double hydroxide nanocomposites: Structure, synthesis and applications. *Chem. Commun.* **2015**, *51*, 3024–3036. [CrossRef] [PubMed]
16. Balcomb, B.; Singh, M.; Singh, S. Synthesis and characterization of layered double hydroxides and their potential as nonviral gene delivery vehicles. *ChemistryOpen* **2015**, *4*, 137–145. [CrossRef]
17. Kovalenko, V.; Kotok, V.; Yeroshkina, A.; Zaichuk, A. Synthesis and characterisation of dye-intercalated nickel-aluminium layered double hydroxide as a cosmetic pigment. *East. J. Enterp. Technol.* **2017**, *89*, 27–33.
18. Costantino, U.; Marmottini, F.; Nocchetti, M.; Vivani, R. New Synthetic Routes to Hydrotalcite-Like Compounds−Characterisation and Properties of the Obtained Materials. *Eur. J. Inorg. Chem.* **1998**, *1998*, 1439–1446. [CrossRef]
19. Xu, Z.P.; Lu, G.Q. Hydrothermal Synthesis of Layered Double Hydroxides (LDHs) from Mixed MgO and Al2O3: LDH Formation Mechanism. *Chem. Mater.* **2005**, *17*, 1055–1062. [CrossRef]
20. Wijitwongwan, R.; Intasa-ard, S.; Ogawa, M. Preparation of Layered Double Hydroxides toward Precisely Designed Hierarchical Organization. *ChemEngineering* **2019**, *3*, 68. [CrossRef]
21. Lee, J.-Y.; Gwak, G.-H.; Kim, H.-M.; Kim, T.-i.; Lee, G.J.; Oh, J.-M. Synthesis of hydrotalcite type layered double hydroxide with various Mg/Al ratio and surface charge under controlled reaction condition. *Appl. Clay Sci.* **2016**, *134*, 44–49. [CrossRef]
22. Olfs, H.W.; Torres-Dorante, L.O.; Eckelt, R.; Kosslick, H. Comparison of different synthesis routes for Mg–Al layered double hydroxides (LDH): Characterization of the structural phases and anion exchange properties. *Appl. Clay Sci.* **2009**, *43*, 459–464. [CrossRef]
23. Boclair, J.W.; Braterman, P.S. Layered double hydroxide stability. 1. Relative stabilities of layered double hydroxides and their simple counterparts. *Chem. Mater.* **1999**, *11*, 298–302. [CrossRef] [PubMed]
24. Li, J.; Qin, Y.; Jin, C.; Li, Y.; Shi, D.; Schmidt-Mende, L.; Gan, L.; Yang, J. Highly ordered monolayer/bilayer TiO2 hollow sphere films with widely tunable visible-light reflection and absorption bands. *Nanoscale* **2013**, *5*, 5009–5016. [CrossRef] [PubMed]
25. Sharafudeen, R. A spectroscopic method for quick evaluation of tint strength and tint tone of titania (rutile) pigment and factors affecting them. *Color Res. Appl.* **2019**, *44*, 44–49. [CrossRef]
26. Bechu, A.; Ghoshal, S.; Moores, A.; Basu, N. Are Substitutes to Cd-Based Quantum Dots in Displays More Sustainable, Effective, and Cost Competitive? An Alternatives Assessment Approach. *ACS Sustain. Chem. Eng.* **2022**, *10*, 2294–2307. [CrossRef]
27. Yu, B.; Bian, H.; Plank, J. Self-assembly and characterization of Ca–Al–LDH nanohybrids containing casein proteins as guest anions. *J. Phys. Chem. Solids* **2010**, *71*, 468–472. [CrossRef]
28. Phue, W.H.; Liu, M.; Xu, K.; Srinivasan, D.; Ismail, A.; George, S. A Comparative Analysis of Different Grades of Silica Particles and Temperature Variants of Food-Grade Silica Nanoparticles for Their Physicochemical Properties and Effect on Trypsin. *J. Agric. Food Chem.* **2019**, *67*, 12264–12272. [CrossRef]
29. Ansari, A.; Ali, A.; Asif, M.; Uzzaman, S. Microwave-assisted MgO NP catalyzed one-pot multicomponent synthesis of polysubstituted steroidal pyridines. *New J. Chem.* **2017**, *1*, 42. [CrossRef]
30. Xu, Z.P.; Kurniawan, N.D.; Bartlett, P.F.; Lu, G.Q. Enhancement of Relaxivity Rates of Gd-DTPA Complexes by Intercalation into Layered Double Hydroxide Nanoparticles. *Chem. Eur. J.* **2007**, *13*, 2824–2830. [CrossRef]
31. Hernandez-Moreno, M.J.; Ulibarri, M.A.; Rendon, J.L.; Serna, C.J. IR characteristics of hydrotalcite-like compounds. *Phys. Chem. Miner.* **1985**, *12*, 34–38.
32. Cavani, F.; Trifirò, F.; Vaccari, A. Hydrotalcite-type anionic clays: Preparation, properties and applications. *Catal. Today* **1991**, *11*, 173–301. [CrossRef]
33. Kim, H.-J.; Lee, G.J.; Choi, A.-J.; Kim, T.-H.; Kim, T.-i.; Oh, J.-M. Layered Double Hydroxide Nanomaterials Encapsulating Angelica gigas Nakai Extract for Potential Anticancer Nanomedicine. *Front. Pharmacol.* **2018**, *9*, 723. [CrossRef] [PubMed]
34. Trujillano, R.; González-García, I.; Morato, A.; Rives, V. Controlling the Synthesis Conditions for Tuning the Properties of Hydrotalcite-like Materials at the Nano Scale. *ChemEngineering* **2018**, *2*, 31. [CrossRef]

35. Behzadi Nia, S.; Pooresmaeil, M.; Namazi, H. Carboxymethylcellulose/layered double hydroxides bio-nanocomposite hydrogel: A controlled amoxicillin nanocarrier for colonic bacterial infections treatment. *Int. J. Biol. Macromol.* **2020**, *155*, 1401–1409. [CrossRef]
36. Musa, A.; Imam, M.; Ismail, M. Physicochemical properties of germinated brown rice (*Oryza sativa* L.) starch. *Afr. J. Biotechnol.* **2011**, *10*, 6281–6291.
37. Farnood, R. Optical properties of paper: Theory and practice. *Pulp Pap. Fud. Res. Soc. Bury* **2009**, 273. [CrossRef]
38. Urrutia-Ortega, I.M.; Garduño-Balderas, L.G.; Delgado-Buenrostro, N.L.; Freyre-Fonseca, V.; Flores-Flores, J.O.; González-Robles, A.; Pedraza-Chaverri, J.; Hernández-Pando, R.; Rodríguez-Sosa, M.; León-Cabrera, S.; et al. Food-grade titanium dioxide exposure exacerbates tumor formation in colitis associated cancer model. *Food Chem. Toxicol.* **2016**, *93*, 20–31. [CrossRef]
39. Proquin, H.; Rodríguez-Ibarra, C.; Moonen, C.G.J.; Ortega, I.M.U.; Briedé, J.J.; de Kok, T.M.; van Loveren, H.; Chirino, Y.I. Titanium dioxide food additive (E171) induces ROS formation and genotoxicity: Contribution of micro and nano-sized fractions. *Mutagenesis* **2016**, *32*, 139–149. [CrossRef]
40. Xu, Z.P.; Braterman, P.S. Synthesis, structure and morphology of organic layered double hydroxide (LDH) hybrids: Comparison between aliphatic anions and their oxygenated analogs. *Appl. Clay Sci.* **2010**, *48*, 235–242. [CrossRef]
41. George, S.; Pokhrel, S.; Xia, T.; Gilbert, B.; Ji, Z.; Schowalter, M.; Rosenauer, A.; Damoiseaux, R.; Bradley, K.A.; Mädler, L.; et al. Use of a Rapid Cytotoxicity Screening Approach to Engineer a Safer Zinc Oxide Nanoparticle through Iron Doping. *ACS Nano* **2010**, *4*, 15–29. [CrossRef] [PubMed]
42. Zhang, H.; Zou, K.; Guo, S.; Duan, X. Nanostructural drug-inorganic clay composites: Structure, thermal property and in vitro release of captopril-intercalated Mg–Al-layered double hydroxides. *J. Solid State Chem.* **2006**, *179*, 1792–1801. [CrossRef]
43. Zhang, L.-X.; Hu, J.; Jia, Y.-B.; Liu, R.-T.; Cai, T.; Xu, Z.P. Two-Dimensional Layered Double Hydroxide Nanoadjuvant: Recent Progress and Future Direction. *Nanoscale* **2021**, *13*, 7533–7549. [CrossRef] [PubMed]
44. Barahuie, F.; Hussein, M.Z.; Fakurazi, S.; Zainal, Z. Development of drug delivery systems based on layered hydroxides for nanomedicine. *Int. J. Mol. Sci.* **2014**, *15*, 7750–7786. [CrossRef]
45. Arancibia, C.; Costell, E.; Bayarri, S. Fat replacers in low-fat carboxymethyl cellulose dairy beverages: Color, rheology, and consumer perception. *J. Dairy Sci.* **2011**, *94*, 2245–2258. [CrossRef]
46. Liu, L.; Zhao, Q.; Liu, T.; Kong, J.; Long, Z.; Zhao, M. Sodium caseinate/carboxymethylcellulose interactions at oil–water interface: Relationship to emulsion stability. *Food Chem.* **2012**, *132*, 1822–1829. [CrossRef]
47. Alcântara, A.C.S.; Darder, M.; Aranda, P.; Ruiz-Hitzky, E. Zein-layered hydroxide biohybrids: Strategies of synthesis and characterization. *Materials* **2020**, *13*, 825. [CrossRef]
48. Rojas, R.; Palena, M.C.; Jimenez-Kairuz, A.F.; Manzo, R.H.; Giacomelli, C.E. Modeling drug release from a layered double hydroxide–ibuprofen complex. *Appl. Clay Sci.* **2012**, *62–63*, 15–20. [CrossRef]
49. SIAM. Organoclays Category. In *SIDS Initial Assessment Profile*; The Institute of Certified Cost & Management Accountants: DE, USA, 2007.
50. EFSA Panel on Food Additives and Nutrient Sources added to Food (ANS); Mortensen, A.; Aguilar, F.; Crebelli, R.; Di Domenico, A.; Dusemund, B.; Frutos, M.J.; Galtier, P.; Gott, D.; Gundert-Remy, U.; et al. Re-evaluation of oxidised starch (E 1404), monostarch phosphate (E 1410), distarch phosphate (E 1412), phosphated distarch phosphate (E 1413), acetylated distarch phosphate (E 1414), acetylated starch (E 1420), acetylated distarch adipate (E 1422), hydroxypropyl starch (E 1440), hydroxypropyl distarch phosphate (E 1442), starch sodium octenyl succinate (E 1450), acetylated oxidised starch (E 1451) and starch aluminium octenyl succinate (E 1452) as food additives. *EFSA J.* **2017**, *15*, e04911.
51. EFSA Panel on Food Additives and Nutrient Sources added to Food (ANS); Younes, M.; Aggett, P.; Aguilar, F.; Crebelli, R.; Dusemund, B.; Filipič, M.; Frutos, M.J.; Galtier, P.; Gott, D.; et al. Re-evaluation of silicon dioxide (E 551) as a food additive. *EFSA J.* **2018**, *16*, e05088.
52. EFSA Panel on Food Additives and Nutrient Sources added to Food (ANS). Re-evaluation of titanium dioxide (E 171) as a food additive. *EFSA J.* **2016**, *14*, e04545. [CrossRef]

 *foods*

MDPI

Article

A Search Engine Concept to Improve Food Traceability and Transparency: Preliminary Results

Caterina Palocci [1,*], Karl Presser [2], Agnieszka Kabza [2], Emilia Pucci [3] and Claudia Zoani [3]

[1] Department of Enterprise Engineering, University of Rome Tor Vergata, 00133 Rome, Italy
[2] Premotec GmbH, 8400 Winterthur, Switzerland; karl.presser@premotec.ch (K.P.); agnieszka.kabza@premotec.ch (A.K.)
[3] Italian National Agency for New Technologies, Energy and Sustainable Economic Development (ENEA), Department for Sustainability, Biotechnology and Agroindustry Division (SSPT-BIOAG), Casaccia Research Centre, 00123 Rome, Italy; emilia.pucci@enea.it (E.P.); claudia.zoani@enea.it (C.Z.)
* Correspondence: caterina.palocci@uniroma2.it

Abstract: In recent years, the digital revolution has involved the agrifood sector. However, the use of the most recent technologies is still limited due to poor data management. The integration, organisation and optimised use of smart data provides the basis for intelligent systems, services, solutions and applications for food chain management. With the purpose of integrating data on food quality, safety, traceability, transparency and authenticity, an EOSC-compatible (European Open Science Cloud) traceability search engine concept for data standardisation, interoperability, knowledge extraction, and data reuse, was developed within the framework of the FNS-Cloud project (GA No. 863059). For the developed model, three specific food supply chains were examined (olive oil, milk, and fishery products) in order to collect, integrate, organise and make available data relating to each step of each chain. For every step of each chain, parameters of interest and parameters of influence—related to nutritional quality, food safety, transparency and authenticity—were identified together with their monitoring systems. The developed model can be very useful for all actors involved in the food supply chain, both to have a quick graphical visualisation of the entire supply chain and for searching, finding and re-using available food data and information.

Keywords: smart data; search engine concept; search engine visualisation; interoperability; food supply chain; food safety; nutritional quality; traceability; authenticity; food transparency

Citation: Palocci, C.; Presser, K.; Kabza, A.; Pucci, E.; Zoani, C. A Search Engine Concept to Improve Food Traceability and Transparency: Preliminary Results. *Foods* **2022**, *11*, 989. https://doi.org/10.3390/foods11070989

Academic Editors: Theodoros Varzakas and Maria Lisa Clodoveo

Received: 29 January 2022
Accepted: 26 March 2022
Published: 29 March 2022

1. Introduction

Recently, the food industry has become invested by technological innovations: the route toward the fourth industrial revolution (Industry 4.0) has started [1].

The spread of digitisation and interconnection led to a growth in the quantity of data worldwide and in 1997 the term 'big data' was created [2]. There are different definitions of big data in literature because the concept contains several facets. A common understanding is that data is considered as big data if its variety, volume, velocity, veracity, variability, complexity, or value is too hard or impossible to handle with existing data management, data storage or data analysis methods [3,4]. Therefore, new methods are needed to manage, store and analyse big data; this includes new statistical thinking and methods, allowing data itself to identify pertinent variables and patterns that shape the observed outcome [5].

Data are used by companies in several fields: marketing and sales, finance and control, information systems, purchasing, manufacturing, and supply chain [6]. In the agrifood sector, the use of big data is still new [5]. Until now, it has been used to optimise production and to ensure quality and safety [7]; data itself is not useful if it is not used to create value. Today, there are literature debates on how "big" data can become "smart" by transforming the huge quantity of data into strategic knowledge for decision-makers [8].

The rise of the European Open Science Cloud (EOSC) [9] highlights the need for open data sharing, integration, and interoperability, following the FAIR approach [10]. The FAIR principles define that data should be findable, accessible, interoperable, and re-usable. The ambition of EOSC is to develop a web of FAIR data and services for the research community and beyond. The FAIR principles are therefore supporting that data is publicly available for research, industry, consumers, and governments as much as possible. The EOSC turned into an association, and it is collaborating with the European Commission to reach this goal.

The Digital Economy and Society Index (DESI) summarises the digital performances of EU countries in different aspects, such as digital skills, online activity or digital public services, and it tracks changes over time. An analysis of the 2021 index shows a varied picture between member states [11]. This is one reason why digitalisation represents a strategic priority on the political agenda of EU institutions. The use of new technologies gives opportunities to cope with challenges related to the environment, food safety, inclusion, sustainability and transparency of agrifood systems at the national, regional and international level. Digitalisation allows for tracing and tracking across and through the various processes in the chain, enabling greater control over the products and the chain itself, thus promoting consumer confidence in the production system. Information and communication technologies (ICT) support and improve the efficiency of agrifood marketing, product quality and quality maintenance [12].

The main technologies explored in the agrifood sector are smart sensing and monitoring systems (e.g., Internet of Things—IoT, with sensors) using Edge or 5G mobile networks, artificial intelligence (AI), app-based services or blockchain technology (BT) [13,14]. Blockchain was born in 2008 with the cryptocurrency called "Bitcoin" [15], and in the last years it has started to be used by the food industry [16,17]. Its proprieties of decentralisation, transparency and immutability [18] are just some of the strengths that increase research on the topic and its use worldwide [19]. Blockchain was first applied to trace food, but it can also be used, for example, to better monitor food safety, prevent food fraud, reduce waste and give transparency to the consumer [16]. The use of blockchain and big data are proposed to improve agrifood traceability [17,20,21] and to let companies evolve in smart, data driven systems. Transparency and data-sharing between national governments, agencies and industries are the key to better work on risk management, detecting and preventing fraudulent practices and taking actions to inform consumers.

Nowadays, the agrifood sector faces multiple challenges such as population growth, climate change, greenhouse gas emissions, loss of biodiversity, the threat to food security from over-fishing, soil erosion and water shortages. In addition, globalisation in the food trade has led to complexity and fragmentation in the agrifood sector: distances have become bigger and the requirement to keep safety along the supply chain is fundamental to preventing diseases [22]. These challenges currently require a broad vision that should consider all the following interconnected aspects related to food (Figure 1): quality (intended as nutritional and sensorial quality), safety, authenticity, integrity, traceability, transparency and sustainability.

The European Commission aims to assure a high level of food safety and animal and plant health within the EU, through coherent Farm to Fork measures and adequate monitoring, while ensuring an effective internal market with the implementation of a worldwide integrated food safety policy. Of increasing importance is the *One Health* approach, considering the close connection between human health, animal health, and environmental health; therefore food, animal feed, animal and human health, environmental contamination, and environmental impact are closely linked. In order to achieve Goal 2 of the United Nations Sustainable Development Goals (SDGs), another aspect to be considered in every plan and process is sustainability. In the EU, around 88 million tonnes of food waste is generated every year. Primary production, processing and wholesale and retail altogether contribute to the 35% of the total [23].

Figure 1. Relationship between food transparency, food integrity, food traceability, food quality, food safety, food authenticity and sustainability.

The sharing of knowledge and information between players allows for the prevention of food losses and surpluses (reducing food waste) and the education of consumers on the conscious use of food.

Safety, quality and authenticity should be guaranteed by sustainability principles. These three concepts should always be taken into consideration to prevent food adulteration and contamination (WHO estimates each year 600 million foodborne illnesses and 420,000 deaths worldwide [24]) to maintain quality during primary production, processing, logistics, retail and post-retail, and to guarantee the nature, identity, claims, and origins of foods. Traceability acquired relevance after a series of food scandals and safety incidents [25]. As a result, today, storing information for tracing in food supply chains has become mandatory worldwide [26], while integrity is a multidisciplinary issue covering all aspects of the food chain from producers to consumers [27].

Existing food nutrition security data, knowledge, and tools for health and agrifood sciences—although widespread—are fragmented, and access is unevenly distributed for users. This means data are not readily FAIR.

This data fragmentation is not only a reality for food security, but also for the whole domain of scientific food data. Nowadays, we are living in a globalised food trade where the competition for innovation and reputation between different academic, governmental, private and new technologies is increasing the amount of data that is being produced. In such a scenario, datasets are widely spread and fragmented, and it is more and more challenging to keep an overview of the existing data that can be re-used for new research and investigations.

This work aims to overcome these challenges by providing a concept for a search engine that will give an overview of the existing datasets covering different food areas and improve food traceability and transparency. The concept was developed in the FNS-Cloud project, and it is not limited to food data but is more generally applicable. Therefore, only the concept is presented while implementation is depending on the data domain. The concept supports the FAIR principles and therefore the idea of open science; it uses some visualisation and it allows for the identification of datasets with information on content, interoperability, quality and accessibility along the food lifecycle. The concept therefore allows for access to and the discovery of data and metadata integration and analysis based on research queries about nutritional quality, food safety, authenticity and transparency to support all actors of the food system: primary producers, processors, distributors, retailers, consumers, researches, inspection and control agencies, certification bodies, authorities and policy makers (Figure 2).

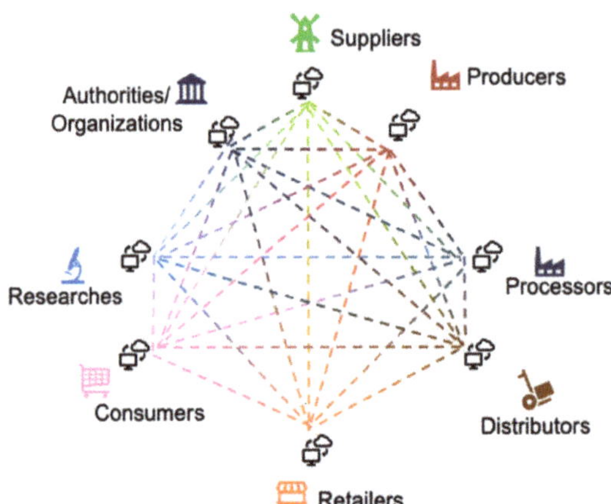

Figure 2. Interoperability network through food system [28].

Section 2 describes the methodology and how three food supply chains were investigated to identify the characteristics for a search engine concept. Section 3 presents the resulting search engine concept, and Section 4 draws conclusions about the concept.

2. Methodology

The use of data as strategic knowledge in the agrifood sector is a resource for multiple stakeholders. The traceability search engine concept was designed using three specific food supply chains as a model to make sure that the concept is more generally valid considering the differences—sometimes very relevant—that occur among the different production chains. The olive oil, milk and fishery products food chains were selected as a model based on the following criteria:

1. To consider products of both vegetable (olive oil) and animal origin (milk and fishery products);
2. To consider products of interest (e.g., economically important) in various European geographical areas (the Mediterranean area for olive oil, central Europe for milk, northern and Mediterranean area for fishery products);
3. To consider products of different supply chains that allow for covering multiple issues and research questions such as nutrition claims (olive oil), the definition of geographical (milk and fishery products) and botanical (olive oil) origin and production sustainability (fishery products);
4. The possibility to extend the case studies to further products obtained by their processing (e.g., dairy products).

For the olive oil, processes that lead to edible virgin olive oils (extra virgin olive oil and virgin olive oil), as defined in Reg. No 1308/2013 (cons. 2020) [29], were included. For the milk, cow milk was specifically considered without including dairy products. For fishery products, all seawater or freshwater animals (except for live bivalve molluscs, live echinoderms, live tunicates and live marine gastropods, and all mammals, reptiles and frogs) whether wild or farmed were considered including all edible forms, parts and products of such animals, as defined in Reg. No n.853/2004 (cons. 2021) [30]. To cover different types of fishery products three different specific supply chains were examined as examples of three food product categories:

1. Sole supply chain—as representative of a marine wild-caught, medium size, and lean fish;
2. Salmon supply chain—as representative of the aquaculture line, large size, and fatty fish;

3. Anchovies supply chain—as representative of a marine wild-caught, small size, medium fat fish.

First of all, each step of the food chain, with all possible routes, was mapped from primary production to human intake.

The mapping was reported in a flowchart for several purposes: to enable an easier comprehension, to support data analysis step by step, and to support visualisation.

Then, a table was created in which each step was described in detail and the inputs and outputs of the steps were identified. The input and the output of a step describes the status of incoming and outgoing food in order to identify them uniquely.

When applicable, terms were reported with their official definitions (e.g., for FAO, WHO, European Regulations, or EFSA documents).

Finally, every step was examined to understand its effect on nutritional quality, safety, authenticity and transparency, and to obtain a catalogue of significant parameters.

For nutritional quality, intrinsic attributes of food such as chemical composition, physical structure, biochemical changes, nutritional value and nutraceuticals (i.e., the capacity, due to the chemical components, to bring benefit to human health) were considered, as well as shelf-life and the way packaging interacts with the food. For food safety, microbial and chemical contamination were considered (hazards from pathogens, microbial spoilage, presence of mycotoxins, heavy metals, pesticides, etc.). The conditions and practices that influence the nutritional quality of intrinsic attributes of food, and that influence food safety (leading to contamination and foodborne illness) were examined. Authenticity and transparency were considered concerning chemical or genetic markers and profiles that allow for the demonstration of geographical, botanical, or zoologic origin, or for the identification of frauds.

For each criterion (nutritional quality, safety, authenticity and transparency) and each step the following parameters were identified:

- Parameters of interest
- Parameters of influence

The parameters of interest (data) are considered as analytes (e.g., a chemical, physical or microbiological substance or component) or as nominal properties/characteristics (e.g., profile or taste) that may be subjected to change depending on the conditions in the step under investigation. The parameters of interest were reported for each step, or in some cases for multiple pooled steps, with the specification of the matrix (e.g., semi-finished products). Some of them were compulsorily provided by the manufacturers, while others were measured only voluntarily.

The parameters of influence (metadata) are considered as conditions that can have an effect on or modify the levels of the parameters of interest. They can refer to the matrix of the corresponding stage, or to other aspects (e.g., the environmental, processes, conditions). For example, the physical–chemical characteristics of soil can influence the bioavailability of toxic and potentially toxic elements and therefore their content in the olives; pedoclimatic conditions such as temperature, rainfall and distance from the sea can affect the isotope ratios and nitrogen level and sun exposure can affect the content of polyphenols.

Finally, for each parameter (or class of parameters) and each step, monitoring systems (e.g., indicators/measurement devices) of the parameters of interest and influence were mapped by differentiating between those for offline detection, such as analytic laboratory methodologies, and those permitting in situ and in-line monitoring, such as IoT sensors.

Based on this examination, a concept for a traceability search engine was created that allowed searches for the different aspects described above. Such a concept is presented in the results section.

The above-described methodology has been applied for all of the three supply chains. In order to provide a practical example, what has been elaborated for the olive oil supply chain is presented in Figure 3 (flowchart), Table 1 (inputs and outputs of the supply chain's steps) and Table 2 (catalogue of parameters of interest and parameters of influence).

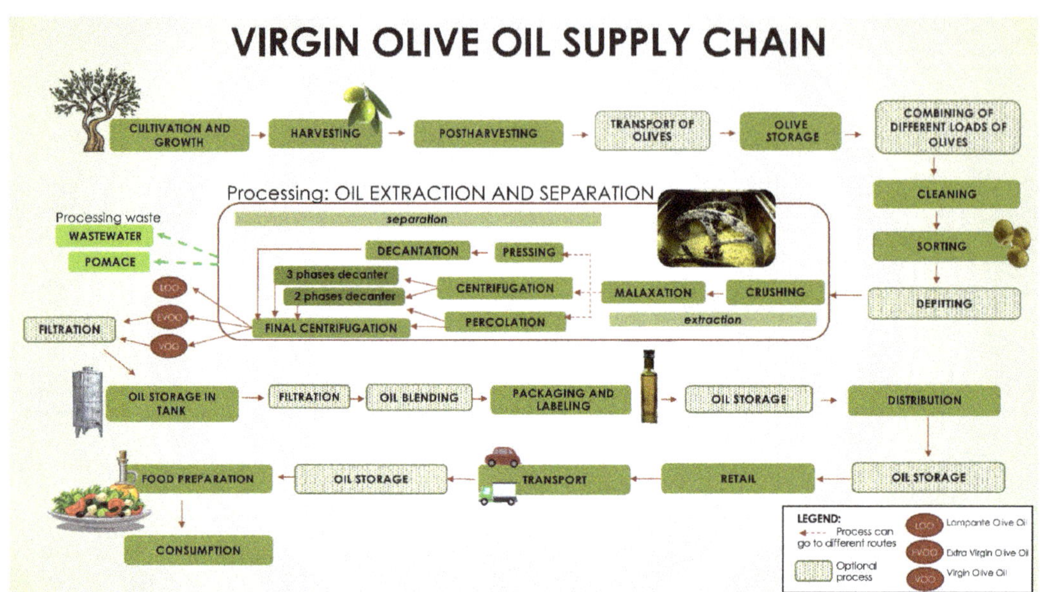

Figure 3. Flowchart of the edible virgin olive oils supply chain.

Table 1. Extract of the table describing each step with their inputs and outputs.

Step	Definition	Input	Output
Cultivation and Growth	All stages that concern agronomic practices to make olives grow and keep them healthy until harvest	Olive Trees	Olives
Harvesting	The process of gathering a ripe crop from olives fields. It can be done after natural fall, by hand, by beating the branches, with shakers, by combing (previously is commonly used to put canvases on the soil for the reception of the harvested fruits)	Olives	Olives
Postharvesting	Olives are taken from the nets on the ground and put into bins	Olives	Olives
Transport of Olives	Olives are transported to oil mill by olive grower	Olives	Olives
Olive Storage	Olives are stored in rigid and ventilated containers in a cool and dry environment	Olives	Olives
Combining Different Loads of Olives	Olives can arrive from different olive's growers and are mixed together	Olives	Olives
Cleaning	Involves defoliation and washing	Olives	Olives
Sorting	Discarding any bruised or defective fruit	Olives	Olives
Depitting	Separation of the pits from the olives	Olives	Olives

Table 2. Extract of the catalogue of parameters of interest and parameters of influence for nutritional quality, safety and authenticity/transparency concerning the virgin olive oil supply chain. The phases in italics are optional and at the choice of the food companies and "X" indicate that the process phase does not affect authenticity/traceability.

		Nutritional Quality		Safety		Authenticity/Transparency	
Phase	**Matrix**	**Parameters of Interest**	**Parameters of Influence**	**Parameters of Interest**	**Parameters of Influence**	**Parameters of Interest**	**Parameters of Influence**
Cultivation and Growth	Olives	fatty acids (FFAs, SFAs, MUFAs and PUFAs), total polyphenols, tocopherol, secoiridoids (oleuropein, hydroxytyrosol), phytosterols, pigments (carotenoids, chlorophylls), lignans, secoiridoid derivatives, 3,4-DHPEA-AC, monoglycerides and peroxides, DAGs, peroxide value, pH, total CHO, soluble solids, % in oil	climatic and pedoclimatic conditions (e.g., air composition, sun exposure, physical-chemical characteristics of soil and trees, irrigation); type and fertilisers content; pruning, pest and disease management	toxic and potentially toxic elements, Polycyclic Aromatic Hydrocarbons (PAHs), mycotoxins, radionuclides	pedoclimatic conditions (e.g., physical-chemical characteristics of soil, environmental pollution) phys-iopatological factors, biocides and plant protection products (pesticides used)	isotopic ratios, rare earth elements, micronutrients, pigments profiles, genomic profiles	cultivar, latitude, longitude, rainfall, distance from sea, sun exposition, physical-chemical characteristics of soil, fertilisers use
Harvesting			time (t), techniques applied, maturity index, detachment index	foreign matters, texture, integrity	t, harvesting system, microbiological and biological contaminants	X	X
Postharvesting			Temperature (T), t, mechanical breakages, equipment	toxic and potentially toxic elements, free acidity, peroxide, K232 value, mold, insect and microbial infection, mold, FFA, peroxide value	aeration, equipment and storage conditions (T, t), mechanical breakages, handling, foreign materials		
Transport of Olives							
Olives Storage		micronutrients content, free acidity level, peroxide, K232 value, K270 value, mold	storage conditions (T, t), processing equipment (e.g., tanks, pipes, drums, etc.)				
Arrival at the Mill		air humidity, free acidity	storage conditions (T, t)				
Combining Different Loads of Olives		micronutrients, total polyphenols, secoiridoids (oleuropein, hydroxytyrosol) phytosterols, pigments (e.g., carotenoids)	mixing ratio, content in each single load of olives	chemical residues	mixing ratio, content in each single load of olives	isotopic ratios, rare earth elements, micronutrients, pigments profiles, genomic profiles	olives loads provenance, cultivar, latitude, longitude, rainfall, distance from sea, sun exposition, physical-chemical characteristics of soil, fertilisers use
Cleaning		total polyphenols	T, t, washing water quality	toxic and potentially toxic elements, foreign matters, pesticides	T, t, washing water quality	X	X
Sorting		olive texture	X	olives texture	X		
Depitting		pits, pit dust	machines efficiency	pit dust	pit hardness		

3. Results: Traceability Search Engine Concept

The purpose of the traceability search engine concept is to support users in searching and finding available food data and information, and to provide knowledge and guidance to its users. A summary of the concept in one sentence could be the following: the traceability search engine concept uses tags having semantical meaning, groups them in dimensions, assigns each data or information resource these tags, allowing use of either a simple search or a visual space search, and offering informed guidance. In the following, some terms will be defined, the idea of dimensions are explained, and finally the visual search of data and information resources is presented.

The first term that the search engine concept uses is data or information resource which can be a dataset or a scientific publication. Such data or information resources can have any form or format, and they can be structured, semi-structured or unstructured data or information. While structured data or information such as databases or Excel files have data organised in a certain order such as rows and columns, unstructured data or information such as free text do not use an ordered structure and data and information must be extracted. Semi-structured data and information is in between structured and unstructured, and it normally combines them. An example could be a Wiki where each term is explained on a separate page and therefore using a list structure, but the content of a page is free text and therefore unstructured. This structural classification is a high-level classification, while more concrete types of data or information resources are commonly used as datasets, databases, Wikis, scientific papers, project reports or websites.

The concept does not require that data and information resources are directly included in a software that implements this concept. It is only required that a resource is described with enough information so that it can be used in the rest of the concept.

The type of a resource is a first characterisation of a data and information resource, and it is considered as a dimension in the concept. A dimension has different tags, and a tag is a term or a keyword that can be assigned to a resource, and it describes a certain aspect of the resource, see Figure 4. The dimension is therefore the group, while the tags are concrete values such as dataset, scientific publication or Wiki.

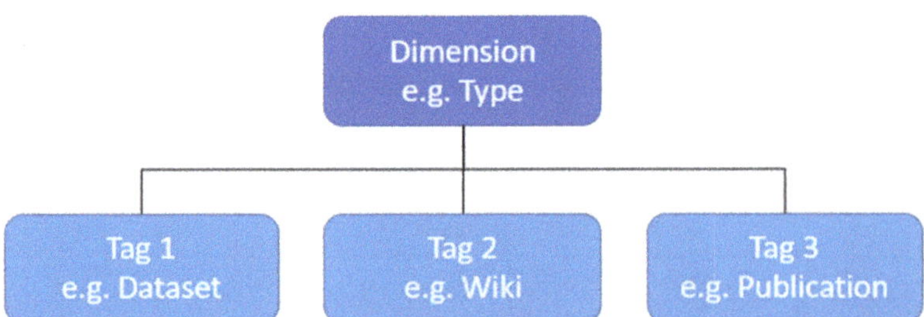

Figure 4. A dimension is a group of tags which are possible concrete terms or keywords of the dimension.

For each resource, none, one or multiple tags of a dimension can be assigned. The tags and the dimensions have a name and a description explaining their meaning and usage. Additional fields can be added such as input and output as we have seen in the former section. Tags within a dimension can also have a hierarchical structure, allowing for the use of a tree structure with multiple parents. In the food traceability search engine concept, the following dimensions are defined:

(a) Type—defines the type of the resource;
(b) Food group/matrix—describes a group of foods or a single food item;
(c) Food supply chain—describes for each food group/matrix a separate chain of phases;
(d) Country—defines the country of origin for the data or information of a resource;

(e) Main aspects of food science used in this concept—tags are nutritional quality, safety, and authenticity/traceability (based on the outcome of the former section);
(f) Research area—defines a comprehensive list of research aspects in food science which can change over time;
(g) Target audience—defines different user groups such as primary producers, processors, distributors, retailers, consumers, researches, inspection and control agencies, certification bodies, authorities and policy makers;
(h) Access mode—defines how a resource can be accessed, e.g., open access, restricted access or no access;
(i) Year—defines the year of a resource;
(j) Chemical substance—defines the list of nutrients, contaminants, and other chemical substances;
(k) Parameter of interest—describes properties of foods that are of interest when it comes to traceability in combination with nutritional quality, safety and authenticity/transparency as presented in the former section. Parameters of influence are facts that have an influence on the measurement of the parameter of interest.

The food supply chain is another example of a dimension and each step in the chain is considered a tag. Two or more dimensions can have dependencies between each other. For example, the dimension food group or matrix and the dimension food supply chain. Depending on the food group, the food supply chain can be different as presented in the last section. The concept therefore allows for the defining of dependency rules for tags and dimensions. A dependency rule for dimensions allows for defining each food group's different food supply chain. Another example is that the parameter of interest is a super dimension where all other dimensions are child dimensions; they were separated because they represent a specific aspect. A dependency rule between tags allows for defining that tags of one dimension can only occur in combination with tags of other dimensions; e.g., certain parameters of interest only occur in combination with a certain phase in the food supply chain, see Table 2.

The description of dimensions and the description of dependency rules are designed to inform users and to serve as a knowledge and documentation base in the domain of the traceability search engine. The documentation should not only allow simple description but also an advanced documentation means as shown in Figure 5. The documentation contains all the information that was collected in the former section, and it represents the current knowledge. As this is developing over time, this documentation needs to be adjusted and extended.

Possible users of such a search engine can range from laypersons to experts, and they are presented in Figure 2. Depending on the type of user, an appropriate scope of information can be made available. For more advanced users, more information can be presented while for less advanced users summaries and simplifications are enough. Therefore, the user should be able to define his/her level of expertise.

The documentation supports consumers by providing information about the production chain of a food item and allowing identification processes that can influence quality, safety, authenticity and transparency. For example, users can check the quality of milk and what affects it. Researchers on the other hand have the ability to find information about which production step influences authenticity and be able to compare their data to other datasets. Food producers also have the possibility to investigate the influences on the quality of their food production, and they can use the knowledge for improvements.

Having assigned the tags to the resources, they can be used to browse and search data and information resources. A simple search allows therefore to select one or multiple tags from one or multiple dimensions and to retrieve all resources that have these tags assigned. The result is normally presented as a list of resources. If more than one tag is used, it should be defined if the AND, OR, or both operators are used. The AND operator defines those resources must have all tags assigned while the OR operator defines that either of the tags must be assigned.

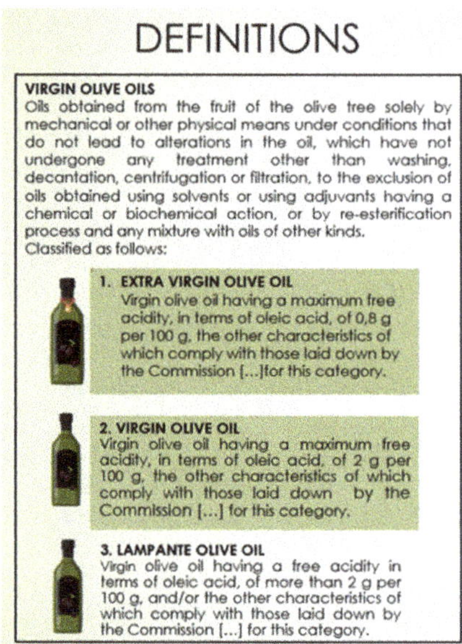

Figure 5. Example of detailed documentation of the olive oil milling process.

More interesting is the space search because the result list of resources has some limitations. The list, for instance, does not show where no resources are available and comparing tags is more or less comfortable depending on the implementation. A result list must be considered as a keyhole view where only a part is visible, while most of the room is not visible. The space search solves this issue by allowing the use of dimensions to span a result space. If two dimensions are selected, the tags of one dimension are put next to each other on one axis and the tags of the other dimension are put next to each other on the other axis. This results in a table and the data and information resource are listed in the corresponding cell. Table 3 shows a schematic example where the food supply chain was used in combination with three main aspects in food science, which are safety, quality, authenticity and transparency.

Table 3. Example of result presentation of a 2-dimensional space search.

	Primary Production	Processing	Packaging	Storage and Distribution	Retail	Final Consumption
Safety						TDS data (link) Medication concentration data (link)
Nutritional quality		Possible contamination during olive oil extraction (link)		Loss of vitamins during storage (link)	Label information (link)	Greece food composition database (link)
Authenticity/ transparency	Isotope data (link)					

The resulting table in Table 3 demonstrates how the keyhole view is removed by showing all possible combinations of two dimensions providing an overview of what data and information resources are available and where no resources are available. The example also shows that not all tags need to be mapped on an axis to reduce the number of columns and rows.

Taking into account different users and their needs, a graphic presentation of the entire supply chain is beneficial, showing its individual steps and the entire food flow process from primary production to human intake. Thanks to this, users can see the entire complexity of the process as well as obtain detailed information about the phase of the process that interests them.

The resulting cells are clickable and, when selected, another view with all resources is presented, showing more information than in the multi-dimensional result space. The idea is that the list items or the result space items represent a short summary, while more information can be found on a separate page when clicking on an item. The list or search space result presentation is called result view while the detail information page is called the detail view. How the detail view is structured and what information it contains depends on the data domain.

The space search is not limited to two dimensions, but it can combine three or more dimensions. The presentation of the result gets a challenge as multi-dimensional spaces are hard to present and maybe a decomposition in multiple 2-dimensional tables is needed.

The concept also allows for the combining of two or more dimensions on a single axis to increase the space that is spanned and to enlarge the overview of available resources. A limiting factor is the space of the computer screen, in particular if tablets and mobiles are used. In such cases, the reduction of tags mapped on axes is advisable.

Finally, the simple search and the space search can be combined. While the space search presents the results in multi-dimensional space, the dimensions that were not used to span the space can be used to further filter resources. In this way, more advanced search operations are possible and more specific results can be presented.

4. Conclusions

The traceability search engine concept and its developed model fit the purpose of collecting, organising, making available and integrating data and metadata on food quality, safety, traceability, transparency, and the authenticity of products along the food supply chain, following the FAIR approach. The developed model is helpful both to have a graphical visualisation of the entire food supply chain and to have the possibility to carry out different types of searches. Searches can be made on different dimensions alone or in combination between them (type of resource, food/matrix group, food chain, country, aspects related to food science, research area, target audience, access mode, year, chemistry, the parameter of interest) and their different tags (e.g., a step in a specific food supply chain). This model supports users in finding available food data and information, and it provides them with knowledge and guidance, according to the type of user. Indeed, depending on the expertise of the user, much more detailed information can be made available for advanced users and simplifications or summaries can be delivered for less advanced users. Sharing smart data in the network can support all actors in the food system. Thanks to the dedicated information displayed for each user category, companies, policy makers, local authorities and citizens can benefit from the model. This work integrates knowledge of food science and innovative engineering. The next step will be exploring the possibility to integrate blockchain technologies in the demonstrator to give more transparency to all users.

Author Contributions: Conceptualization, C.P., K.P. and C.Z.; Investigation, C.P., K.P., A.K., E.P. and C.Z.; Methodology, C.P., K.P., A.K., E.P. and C.Z.; Supervision, K.P. and C.Z.; Writing—original draft, C.P., K.P. and C.Z.; Writing—review & editing, C.P., K.P., A.K., E.P. and C.Z. All authors have read and agreed to the published version of the manuscript.

Funding: Food Nutrition Security Cloud (FNS-Cloud) has received funding from the European Union's Horizon 2020 Research and Innovation programme (H2020-EU.3.2.2.3.—A sustainable and competitive agrifood industry) under Grant Agreement No. 863059—www.fns-cloud.eu (accessed on 28 January 2022).

Data Availability Statement: Data is contained within the article.

Conflicts of Interest: The authors declare no conflict of interest. Premotec GmbH actively participated in the conceptualization and the writing of this paper as a partner of the FNS-Cloud project, and it has no conflict of interest.

References

1. Liu, Y.; Ma, X.; Shu, L.; Hancke, G.P.; Abu-Mahfouz, A.M. From Industry 4.0 to Agriculture 4.0: Current Status, Enabling Technologies, and Research Challenges. *IEEE Trans. Ind. Inform.* **2020**, *17*, 4322–4334. [CrossRef]
2. Cox, M.; Ellsworth, D. Application-Controlled Demand Paging for Out-of-Core Visualization. In Proceedings. Visualization'97 (Cat. No. 97CB36155). In Proceedings of the Visualization'97 (Cat. No. 97CB36155), Phoenix, AZ, USA, 24–24 October 1997; pp. 235–244.
3. Al-Mekhlal, M.; Ali Khwaja, A. A Synthesis of Big Data Definition and Characteristics. In Proceedings of the 2019 IEEE International Conference on Computational Science and Engineering (CSE) and IEEE International Conference on Embedded and Ubiquitous Computing (EUC), New York, NY, USA, 1–3 August 2009; pp. 314–322.
4. Katal, A.; Wazid, M.; Goudar, R.H. Big Data: Issues, Challenges, Tools and Good Practices. In Proceedings of the 2013 Sixth International Conference on Contemporary Computing (IC3), Noida, India, 8–10 August 2013; pp. 404–409.
5. Pollard, S.; Namazi, H.; Khaksar, R. Big Data Applications in Food Safety and Quality. In *Encyclopedia of Food Chemistry*; Elsevier: Amsterdam, The Netherlands, 2019; pp. 356–363. ISBN 978-0-12-814045-1.
6. Buckley, J. How Is Big Data Going to Change the World? Available online: https://www.weforum.org/agenda/2015/12/how-is-big-data-going-to-change-the-world/ (accessed on 16 June 2021).
7. Tao, D.; Yang, P.; Feng, H. Utilization of Text Mining as a Big Data Analysis Tool for Food Science and Nutrition. *Compr. Rev. Food Sci. Food Saf.* **2020**, *19*, 875–894. [CrossRef] [PubMed]
8. Lacam, J.-S.; Salvetat, D. Big Data and Smart Data: Two Interdependent and Synergistic Digital Policies within a Virtuous Data Exploitation Loop. *J. High Technol. Manag. Res.* **2021**, *32*, 100406. [CrossRef]
9. European Commission. Directorate General for Communications Networks, Content and Technology. In *European Open Science Cloud: A New Paradigm for Innovation and Technology*; European Commission: Luxemburg, 2019.
10. Wilkinson, M.D.; Dumontier, M.; Aalbersberg, I.J.; Appleton, G.; Axton, M.; Baak, A.; Blomberg, N.; Boiten, J.-W.; da Silva Santos, L.B.; Bourne, P.E.; et al. The FAIR Guiding Principles for Scientific Data Management and Stewardship. *Sci. Data* **2016**, *3*, 160018. [CrossRef] [PubMed]
11. European Commission Digital Economy and Society Index (DESI) 2021. Available online: https://digital-strategy.ec.europa.eu/en/library/digital-economy-and-society-index-desi-2021 (accessed on 28 January 2022).
12. Viaene, J.; Meulenberg, M.T.G. Changing Agri-Food Systems in Western Countries: A Marketing Approach. In *Innovation in Agri-Food Systems: Product Quality and Consumer Acceptance*; Wageningen Academic Publishers: Wageningen, The Netherlands, 2005; ISBN 978-90-76998-65-7.
13. Rao, S.K.; Prasad, R. Impact of 5G Technologies on Industry 4.0. *Wirel. Pers Commun.* **2018**, *100*, 145–159. [CrossRef]
14. Wolfert, S.; Ge, L.; Verdouw, C.; Bogaardt, M.-J. Big Data in Smart Farming—A Review. *Agric. Syst.* **2017**, *153*, 69–80. [CrossRef]
15. Nakamoto, S. Bitcoin P2P E-Cash Paper 2008. Available online: https://www.ussc.gov/sites/default/files/pdf/training/annual-national-training-seminar/2018/Emerging_Tech_Bitcoin_Crypto.pdf (accessed on 28 January 2022).
16. Creydt, M.; Fischer, M. Blockchain and More—Algorithm Driven Food Traceability. *Food Control* **2019**, *105*, 45–51. [CrossRef]
17. Kamilaris, A.; Fonts, A.; Prenafeta-Boldύ, F.X. The Rise of Blockchain Technology in Agriculture and Food Supply Chains. *Trends Food Sci. Technol.* **2019**, *91*, 640–652. [CrossRef]
18. Leible, S.; Schlager, S.; Schubotz, M.; Gipp, B. A Review on Blockchain Technology and Blockchain Projects Fostering Open Science. *Front. Blockchain* **2019**, *2*, 16. [CrossRef]
19. Casino, F.; Dasaklis, T.K.; Patsakis, C. A Systematic Literature Review of Blockchain-Based Applications: Current Status, Classification and Open Issues. *Telemat. Inform.* **2019**, *36*, 55–81. [CrossRef]
20. Feng, H.; Wang, X.; Duan, Y.; Zhang, J.; Zhang, X. Applying Blockchain Technology to Improve Agri-Food Traceability: A Review of Development Methods, Benefits and Challenges. *J. Clean. Prod.* **2020**, *260*, 121031. [CrossRef]
21. Serazetdinova, L.; Garratt, J.; Baylis, A.; Stergiadis, S.; Collison, M.; Davis, S. How Should We Turn Data into Decisions in AgriFood? *J. Sci. Food Agric.* **2019**, *99*, 3213–3219. [CrossRef] [PubMed]
22. FAO. WHO FAO's Strategy for a Food Chain Approach to Food Safety and Quality: A Framework Document for the Development of Future Strategic Direction. Available online: http://www.fao.org/3/y8350e/y8350e.htm (accessed on 24 June 2021).
23. Stenmarck, Å.; Jensen, C.; Quested, T.; Moates, G.; Buksti, M.; Cseh, B.; Juul, S.; Parry, A.; Politano, A.; Redlingshofer, B.; et al. *Estimates of European Food Waste Levels*; EU FUSIONS: Stockholm, Sweden, 2016; ISBN 978-91-88319-01-2.
24. World Health Organization. *WHO Estimates of the Global Burden of Foodborne Diseases*; World Health Organization, Ed.; World Health Organization: Geneva, Switzerland, 2015; ISBN 978-92-4-156516-5. Available online: https://www.eu-fusions.org/phocadownload/Publications/Estimates%20of%20European%20food%20waste%20levels.pdf (accessed on 28 January 2022).
25. Pizzuti, T.; Mirabelli, G. The Global Track&Trace System for Food: General Framework and Functioning Principles. *J. Food Eng.* **2015**, *159*, 16–35. [CrossRef]
26. Casino, F.; Kanakaris, V.; Dasaklis, T.K.; Moschuris, S.; Rachaniotis, N.P. Modeling Food Supply Chain Traceability Based on Blockchain Technology. *IFAC-PapersOnLine* **2019**, *52*, 2728–2733. [CrossRef]

27. Rychlik, M.; Zappa, G.; Añorga, L.; Belc, N.; Castanheira, I.; Donard, O.F.X.; Kouřimská, L.; Ogrinc, N.; Ocké, M.C.; Presser, K.; et al. Ensuring Food Integrity by Metrology and FAIR Data Principles. *Front. Chem.* **2018**, *6*, 49. [CrossRef] [PubMed]
28. Palocci, C.; Zoani, C. Data Sharing to Improve Food Supply Chain Management. In *The Book of Abstracts of the 5th International Conference on Metrology in Food and Nutrition—IMEKOFOODS 5*; Czech University of Life Sciences Prague: Prague, Czech Republic, 2020; pp. 118–119. ISBN 978-80-213-3036-8.
29. European Parliament; Council of the European Union Regulation (EU). No. 1308/2013 of the European Parliament and of the Council of 17 December 2013 Establishing a Common Organisation of the Markets in Agricultural Products and Repealing Council Regulations (EEC) No. 922/72, (EEC) No. 234/79, (EC) No. 1037/2001 and (EC) No. 1234/2007. (Consolidated Version 2021). Available online: https://eur-lex.europa.eu/legal-content/EN/TXT/?uri=CELEX%3A02013R1308-20211207 (accessed on 28 January 2022).
30. European Parliament; Council of the European Union Regulation (EC). No 853/2004 of the European Parliament and of the Council of 29 April 2004 Laying down Specific Hygiene Rules for Food of Animal Origin. (Consolidated Version 2021). Available online: https://eur-lex.europa.eu/legal-content/SV/TXT/?uri=CELEX:02004R0853-20211028 (accessed on 28 January 2022).

Article

A Tara Gum/Olive Mill Wastewaters Phytochemicals Conjugate as a New Ingredient for the Formulation of an Antioxidant-Enriched Pudding

Umile Gianfranco Spizzirri [1], Paolino Caputo [2], Cesare Oliviero Rossi [2], Pasquale Crupi [3], Marilena Muraglia [4], Vittoria Rago [1], Rocco Malivindi [1], Maria Lisa Clodoveo [3], Donatella Restuccia [1,*] and Francesca Aiello [1]

1 Dipartimento di Farmacia e Scienze della Salute e della Nutrizione, Dipartimento di Eccellenza 2018–2022, Università della Calabria, Ed. Polifunzionale, 87036 Rende, Italy; g.spizzirri@unical.it (U.G.S.); vittoria.rago@unical.it (V.R.); rocco.malivindi@unical.it (R.M.); francesca.aiello@unical.it (F.A.)
2 Dipartimento di Chimica e Tecnologie Chimiche, Università della Calabria & UdR INSTM della Calabria, 87036 Rende, Italy; paolino.caputo@unical.it (P.C.); cesare.oliviero@unical.it (C.O.R.)
3 Dipartimento Interdisciplinare di Medicina, Università degli Studi Aldo Moro Bari, Piazza Giulio Cesare 11, 70124 Bari, Italy; pasquale.crupi@crea.gov.it (P.C.); marialisa.clodoveo@uniba.it (M.L.C.)
4 Dipartimento di Farmacia-Scienze del Farmaco Università degli Studi di Bari, Campus Universitario E. Quagliarello Via Orabona 4, 70125 Bari, Italy; marilena.muraglia@uniba.it
* Correspondence: donatella.restuccia@unical.it; Tel.: +39-0984493298 or +39-3497839077

Citation: Spizzirri, U.G.; Caputo, P.; Oliviero Rossi, C.; Crupi, P.; Muraglia, M.; Rago, V.; Malivindi, R.; Clodoveo, M.L.; Restuccia, D.; Aiello, F. A Tara Gum/Olive Mill Wastewaters Phytochemicals Conjugate as a New Ingredient for the Formulation of an Antioxidant-Enriched Pudding. *Foods* **2022**, *11*, 158. https://doi.org/10.3390/foods11020158

Academic Editors: Mohamed Koubaa and Annalisa Tassoni

Received: 9 November 2021
Accepted: 5 January 2022
Published: 8 January 2022

Abstract: Olive mill wastewater, a high polyphenols agro-food by-product, was successfully exploited in an eco-friendly radical process to synthesize an antioxidant macromolecule, usefully engaged as a functional ingredient to prepare functional puddings. The chemical composition of lyophilized olive mill wastewaters (LOMW) was investigated by HPLC-MS/MS and ^{1}H-NMR analyses, while antioxidant profile was in vitro evaluated by colorimetric assays. Oleuropein aglycone (5.8 μg mL^{-1}) appeared as the main compound, although relevant amounts of an isomer of the 3-hydroxytyrosol glucoside (4.3 μg mL^{-1}) and quinic acid (4.1 μg mL^{-1}) were also detected. LOMW was able to greatly inhibit ABTS radical (IC$_{50}$ equal to 0.019 mg mL^{-1}), displaying, in the aqueous medium, an increase in its scavenger properties by almost one order of magnitude compared to the organic one. LOMW reactive species and tara gum chains were involved in an eco-friendly grafting reaction to synthesize a polymeric conjugate that was characterized by spectroscopic, calorimetric and toxicity studies. *In vitro* acute oral toxicity was tested against 3T3 fibroblasts and Caco-2 cells, confirming that the polymers do not have any effect on cell viability at the dietary use concentrations. Antioxidant properties of the polymeric conjugate were also evaluated, suggesting its employment as a thickening agent, in the preparation of pear puree-based pudding. High performance of consistency and relevant antioxidants features over time (28 days) were detected in the milk-based foodstuff, in comparison with its non-functional counterparts, confirming LOWM as an attractive source to achieve high performing functional foods.

Keywords: olive mill wastewater; polyphenols; antioxidant features; tara gum; pudding; rheological properties

1. Introduction

In recent years, to increase beneficial effects on the human health of many foodstuffs and beverages, some health-promoting natural substances with antioxidant activity were used as high- performing functional ingredients [1]. In particular, the exploitation of agro-industrial wastes represents a smart opportunity for sustainable growth and the scientific research so far has made considerable efforts to reuse and enhance agro-food waste as a source of bioactive compounds for large use in cosmetics, pharmaceuticals and in the agri-food field [2,3]. Our scientific experience suggests that antioxidant features of biomolecules in the olive mill wastewaters (OMW), as well as leaves, pomace and pits discharged from the EVOO production process, are partially recycled and usefully employed in the nutraceutical and pharmaceutical fields [4,5]. To this regard, literature data highlighted that the employment of macromolecular antioxidants displayed some undoubted advantages such as increased stability and slower degradation rate, compared to the compounds with low molecular weight [6]. In particular, polysaccharides are involved in the preparation of different foodstuffs specifically in a wide variety of milk-based desserts such as pudding due to their ability to influence the rheological and texture characteristics of the final product [7]. Puddings are semisolid dairy desserts that are generally milk protein-based starch pastes [8]. Nowadays, on the market are available ready-to-eat puddings or alternatively instant pudding powders able to get a gel structure by dissolving, in a relatively short period time, in cold water or milk and a subsequent cooling step [9]. The formulation of the commercial pudding powders usually consists of hydrocolloids, starch, flavors, sugars and colorants [10].

In this article, the possibility of using the polyphenolic components present in the OMW to obtain a functional polymer able to be employed in the preparation of puddings which display a high performance of consistency (mechanical properties) and relevant antioxidant features over time was investigated. For this purpose, tara gum (TG) was functionalized with reactive components of the OMW allowing the synthesis of a polymeric conjugate, representing a basic component of the pudding formulation. TG was picked up from the tara seed endosperm and based on $(1 \rightarrow 4)$-β-mannose chain with an unit of $(1 \rightarrow 6)$-α-galactose (mannose to galactose ratio equal to 3:1) [11]. In conjunction with the starch, TG allows the gelling process of the food hydrocolloid [12]. TG has been largely proposed in the food field as encapsulating material of bioactive molecules by freeze-drying and spray-drying methods [13] or in the preparation of crosslinked polymeric carriers able to deliver nutraceuticals [14]. Our challenge was to synthesize a TG-based macromolecular antioxidant by a molecular grafting reaction using a redox couple (H_2O_2/ascorbic acid) as an initiator system and involving the heteroatoms in the polymer chains of the polysaccharide and the polyphenol moieties of the active compounds in the OMW.

The polymer conjugate was used as a thickening agent in the preparation of pear puree-based pudding, which has been characterized over time by rheological tests and whose antioxidant profile has also been depicted. The employment of the antioxidant TG by increasing the total content of phenolic groups in the final product and preserving the biologically active molecules derived from fruit should guarantee a food product with a greater antioxidant capacity, practically unchanged over time. Finally, rheological analyses in asymptotic kinematics made it possible to evaluate the effects of the functionalized polysaccharide on the colloidal structure of the pudding. Specifically, the weak gel model was applied to analyze the strength of the gel and its coordination in quantitative terms.

2. Materials and Methods

2.1. Chemicals

Tara gum, pectin esterified from citrus fruits with a degree of methoxylation of 55–70%, gallic acid, (+)-hydrated catechin, L-ascorbic acid, hydrogen peroxide (H_2O_2, 30% v/v), Folin–Ciocalteu reagent, carbonate sodium (Na_2CO_3) radical 2,2′-diphenyl -1-picrylyhydrazyl (DPPH), radical 2,2′-azino-bis (3-ethylbenzothiazolin-6-sulfonic) (ABTS), potassium persulfate ($K_2S_2O_8$), ammonium molybdate tetrahydrate ($(NH_2)_2MoO_4$), sodium molybdate (Na_2MoO_4), sodium nitrite ($NaNO_2$), sodium phosphate (Na_3PO_4), aluminum chloride ($AlCl_3$), hydrochloric acid (HCl), sodium hydroxide (NaOH), absolute ethanol, methanol, Whatman No. 3 filter paper, dialysis membrane (MWCO: 12,000–14,000 Da), were purchased from Sigma Aldrich (Sigma Chemical Co., St. Louis, MO, USA). HPLC grade water, formic acid and acetonitrile were purchased from VWR (Chromasolv, VWR International Srl, Milano, Italy).

2.2. Sample Collection and Preparation

Green olives of Roggianella cultivar were harvested (October 2019) in the Northern of Calabria and processed on-site the next day (Oil mill Vinciprova srl in San Vincenzo la Costa (Cosenza, Italy)) using a semi-continuous Enorossi 150 traditional olive oil pressing system (Enoagricola Rossi, Calzolaro di Umbertide, Perugia, Italy) standardized to press a maximum of 150 kg of olives at a time. Olive mill wastewater (OMW) sample was collected and stored in 1.0 L low density polypropylene airtight containers at −18 °C until use. OMW were filtered and, subsequently, the filtrate was centrifugated (10 min at 10,000 rpm). Finally, the solution was frozen and dried by freeze-drying providing a dark colored vaporous solid (LOMW).

2.3. Chemical Characterization of Lyophilized Olive Mill Wastewaters

2.3.1. HPLC-MS Analysis of Lyophilized Olive Mill Wastewaters

HPLC 1100 system composed of a degasser, quaternary pump solvent delivery, thermostatic column compartment, auto-sampler, single wavelength UV-Vis detector and MSD triple quadrupole QQQ 6430 in a series configuration (Agilent Technologies, Palo Alto, CA, USA) was employed for the polyphenols analysis. LOMW was resuspended in 2 mL ethanol/water (1:1, *v*/*v*) to a final concentration of ~1.2 mg mL^{-1} and filtered through 0.2 µm pore size regenerated cellulose filters (VWR International Srl, Milano, Italy) and injected onto a reversed stationary phase column, Luna C_{18} (150 × 2 mm i.d., particle size 3 µm, Phenomenex, Torrance, CA, USA) protected by a C_{18} Guard Cartridge (4.0 × 2.0 mm i.d., Phenomenex). HPLC separation was carried out through a binary gradient consisting in (solvent A) H_2O/formic acid 0.1% (*v*/*v*) and (solvent B) acetonitrile: 0 min, 10% B; 1 min, 10% B; 15 min, 30% B; 22 min, 50% B; 28 min, 100% B; 34 min, 100% B; 36 min, 10% B, followed by washing and re-equilibrating the column (with ~20 column volume). The column temperature was controlled at 20 °C and the flow was maintained at 0.4 mL min^{-1}. UV-Vis detection wavelength was set at 280 nm.

Because polyphenols contain one or more hydroxyl and/or carboxylic acid groups, MS data were acquired in negative ionization mode with capillary voltage at 4000 V, using nitrogen as drying (T = 350 °C; flow rate = 9.0 L min^{-1}) and nebulizing gas (40 psi). MS and MS/MS spectra were acquired in the range between m/z 50 and 1200. All data were processed using Mass Hunter Workstation software (version B.01.04; Agilent Technologies). UV absorption, retention times (RT), elution order and mass spectra (MS and MS/MS) were compared with those from pure standards (3-hydroxytyrosol, caffeic acid and *p*-coumaric acid) and/or matched with those already reported in the literature [15–17]. Then, the main revealed compounds were quantified by multiple reaction monitoring (MRM) as 3-hydroxytyrosol (R^2 = 0.99923). Mass Hunter Optimizer software (version B.03.01; Agilent Technologies) (Table S1).

2.3.2. H-NMR Analysis of Lyophilized Olive Mill Wastewaters

The sample LOMW (16.7 mg) was dissolved in 0.6 µL of D_2O (99.9% D). ^{1}H-NMR spectrum was performed at 25 °C using a Brucker Advance 200 spectrometer of 300 MHz equipped with $^{13}C/^1H$ dual probe. The NMR experiments were recorded with a spectral width of 6983.240 Hz, an acquisition time of 10.20 s, a number of 64 scans, a relaxation time of 2 s and a pulse width of 7 s. The spectra were processed by XWIN-NMR.

2.3.3. Polyphenols Total Content

Folin–Cicocalteu method was employed to evaluate available phenolic groups (APG) [18]. Different concentrations of an aqueous solutions of LOMW (6.0 mL) were added to 1.0 mL of Folin–Ciocalteu reagent and after 3 min, 3.0 mL of Na_2CO_3 (2.0 % *w*/*v*) was also added keeping the solution under stirring in the dark (time = 2 h) and spectrophotometrically measuring at 760 nm. APG was expressed as milligrams of catechin (CT) per gram of LOMW (mg CT/g LOMW), by using an equation obtained from a calibration curve of CT at different concentrations (8.0, 16.0, 24.0, 32.0 and 40.0 µM). The method of least square was used to calculate a calibration curve. Each measure was performed in triplicate and data expressed as means (±SD). UV-Vis absorption spectra were recorded with a Jasco V-530 UV/Vis spectrometer (Jasco, Tokyo, Japan).

2.3.4. Phenolic Acid Content

Arnov test, with some modifications, was used to evaluate the phenolic acids content (PAC), expressed in milligrams of CT per gram of LOMW (mg CT/g LOMW) [19]. In a volumetric flask (10.0 mL) 1.0 mL of an aqueous solution of LOMW, 1.0 mL of HCl (0.5 mol L^{-1}), 1.0 mL of Arnov's reagent (sodium molybdate and sodium nitrite 0.1 mg mL^{-1}), 1.0 mL of NaOH (1.0 mol L^{-1}) and purified water were mixed. The absorbance of the solutions was measured by a spectrophotometer at 490 nm. Each measurement was performed in triplicate and data expressed as means (±SD).

2.3.5. Flavonoid Content

Flavonoid content (FC) in the samples was determined by a spectrophotometric method reported in literature with some modifications [20]. In a volumetric flask (5.0 mL), 0.5 mL of LOMW and 0.150 mL of $NaNO_2$ (5.0 % *w/v*) solution were mixed. After 6 min, 0.300 mL of a 6.0 % (*w/v*) $AlCl_3$ was added and after 5 min, 1.0 mL of NaOH (1.0 mol L^{-1}) was also added to the mixture by immediately measuring the absorbance (510 nm) against a control solution. FC in LOMW was expressed as milligrams of CT per gram of sample, by using a calibration curve of CT (Standard solutions concentration of CT equal to 10.0, 25.0, 50.0, 75.0, 100.0 μM). Each measurement was performed in triplicate and data expressed as means (±SD).

2.3.6. Anthocyanin Content

The anthocyanin content (AC) was determined by using the procedure reported in literature, with some modifications [21]. 1.0 g of LOMW was suspended in 10.0 mL of methanolic solution of HCl (1.0% *v/v*). The solution was incubated at 60 °C under stirring for 1 h and then filtered in a volumetric flask (10.0 mL). The absorbance was spectrophotometrically measured at 657 and 530 nm. The net absorbance was calculated according to the following equation:

$$\text{Net absorbance} = \text{Absorbance at 530 nm} - 0.25\ (\text{Absorbance at 657 nm}) \quad (1)$$

AC was calculated on the basis of cyanidin-3-glycoside (molecular weight equal to 449.1 g mol^{-1} and extinction coefficient of 29,600) [22]. AC was expressed in milligrams of anthocyanins per gram of LOMW (mg AC/g LOMW), according to the following equation:

$$\text{AC (mg/g)} = ((\text{Net absorbance})/29{,}600) \times \text{MW} \times \text{FT} \times (\text{V}/(\text{Sample weight})) \quad (2)$$

where MW is the molecular weight of cyanidin-3-glycoside, FT is the dilution factor, V is the total volume (mL). Each measurement was performed in triplicate and data expressed as means (±SD).

2.3.7. Antioxidant Properties

The antioxidant properties of LOMW were evaluated by using specific tests (inhibition capacity towards the lipophilic (DPPH) and hydrophilic (ABTS) radicals and to measure the total antioxidant activity (TCA)).

The scavenging activity of LOMW in an organic environment was evaluated in terms of decrease of the radical 2,2-diphenyl-1-picrylhydrazyl ($DPPH^{\bullet}$) concentration [23]. A total of 1.0 mL of hydro-alcoholic solutions (50:50 *v/v*) of LOMW were added to 4.0 mL of hydro-alcoholic mixture (50:50 *v/v*) and 5.0 mL of $DPPH^{\bullet}$ ethanolic solution (200 μM). The mixture was kept at 25 °C for 30 min and the residual concentration of the $DPPH^{\bullet}$

radical was spectrophotometrically evaluated at 517 nm. The percentage of inhibition of the DPPH• radical was calculated according to the following equation:

$$\text{Inhibition (\%)} = (A_0 - A_1)/A_0 \times 100 \quad (3)$$

where A_0 is the absorbance of the control solution prepared under the same conditions but without sample, while A_1 is the absorbance recorded analyzing LOMW sample. The scavenging activity of LOMW against the lipophilic radical DPPH was expressed in terms of IC_{50}. Each measure was performed in triplicate and data expressed as means (±SD).

The scavenging activity in the aqueous environment of LOMW was determined in terms of reduction of the radical 2,2′-azino-bis(3-ethylbenzothiazolin-6-sulphonic) (ABTS•) [24]. Aqueous solutions of LOMW were prepared, at different concentrations and 2.0 mL of the aqueous solution of the ABTS radical was added to 500 μL of each. The solutions were then kept in the dark for 6 min and the residual concentration of the ABTS radical was spectrophotometrically evaluated at 734 nm. The percentage of inhibition of the ABTS• was calculated according to the Equation (3), while LOMW scavenging activity was expressed in terms of IC_{50}. Each measure was performed in triplicate and data expressed as means (±SD).

Total antioxidant capacity (TAC) of LOMW was evaluated by mixing 300 μL of LOMW hydro-alcoholic solutions (50:50 v/v) with 1.2 mL of the reagent solution (28.0 mmol L^{-1} Na_3PO_4, 4.0 mmol L^{-1} $(NH_4)_2MoO_4$, 0.6 mol L^{-1} H_2SO_4) [25]. The solutions were kept at 95 °C in the dark for 150 min and then by measuring the adsorbance at 695 nm. A calibration curve was constructed by the method of least square, using 8.0, 16.0, 24.0, 32.0 and 40.0 μM hydro-alcoholic solutions (50:50 v/v) of CT. The TAC was expressed in milligrams of CT per gram of LOMW (mg CT/g LOMW). Each measure was performed in triplicate and data expressed as means (±SD).

2.4. *Synthesis of the Tara Gum Conjugate by Grafting Procedure*

The synthesis of polymeric conjugate was carried out according to the methods reported in literature, with some modifications [26]. In a 100 mL glass flask, 0.250 g of tara gum was dissolved in 37.5 mL of purified water, then 12.5 mL of H_2O_2 (120 vol) and 0.3 g of ascorbic acid were added at 25 °C. After 2 h, LOMW (equivalent to 0.035 g of CT) was dissolved into the reaction flask. After 24 h, the mixture was introduced into dialysis tubes (MWCO: 12,000–14,000 Dalton) and dipped into a glass vessel containing distilled water for 48 h. The conjugate (PLOMW) was checked to be free of unreacted antioxidant and any other compounds by LC analysis after purification step. LC analysis was performed on a Knauer (Asi Advanced Scientific Instruments, Berlin, Germany) system equipped with two pumps Smartiline Pump 1000, a Rheodyne injection valve (20 μL) and a photodiode array detector equipped with a semi-microcell. The resulting solution was frozen and dried with "freezing-drying apparatus" to provide a vaporous solid. A control polymer, blank tara gum (BTG), was also prepared under the same conditions but without LOMW.

2.5. *Characterization of Tara Gum Conjugate*

2.5.1. H-NMR Analysis of Tara Gum Conjugate

The samples of PLOMW and BTG (5.8 mg) were dissolved in 0.6 μL of D_2O (99.9% D). ^{1}H-NMR spectra were performed at 25 °C using a Brucker Advance 200 spectrometer of 300 MHz equipped with a $^{13}C/^1H$ dual probe. The NMR experiments were recorded with a spectral width of 6983.240 Hz, an acquisition time of 10.20 s, a number of 64 scans, a relaxation time of 2 s and, a pulse width of 7 s. The spectra were processed by XWIN-NMR.

2.5.2. Differential Scansion Calorimetry (DSC)

The DSC studies were performed using a SETARAM 131 instrument. The amount of each sample was around 3–10 mg. Analyses were performed from 25 to 650 °C at a temperature scan rate of 10 °C min^{-1}, under nitrogen flux.

The method used is set out in more detail below:

1. Isotherm at 25 °C for 20 min;
2. Heating from 25 to 650 °C at 10 °C min^{-1}.

2.5.3. Antioxidant Properties of the Tara Gum Conjugate

PLOMW conjugate was characterized in terms of available phenolic groups, phenolic acids and flavonoids content by the methodologies previously described, expressing the results as mg of CT per gram of PLOMW. Similarly, antioxidant performances were recorded by evaluation of TAC (expressed as mg of CT per gram of PLOMW) and scavenging activity if both aqueous (against ABTS radical) and organic (against DPPH radical) environments.

2.5.4. Toxicity of the Tara Gum Conjugate

Cell Culture

Balb/c 3T3 clone A31 and Caco-2 were purchased from ATCC, Manassas, VA, USA. Balb/c 3T3 clone A31 cells were maintained in DMEM with 10% Calf Bovine Serum (CBS; ATCC, Manassas, VA, USA) and 1% penicillin-streptomycin (10,000 unit/mL) at 37 °C in a 5% CO_2 atmosphere. Caco-2 cells were cultured in DMEM with FBS 10% (*w*/*w*), non-essential amino acids (1% *w*/*w*), L-glutamine (1% *w*/*w*) and penicillin–streptomycin (1% *w*/*w*). The cells were grown in a humidified atmosphere (5% CO_2) at 37 °C until 80% confluence and sub-cultured twice a week [27].

Neutral Red Uptake (NRU)

The *in vitro* 3T3 NRU test was performed as described by the ISO 10993-5:2009 "Biological evaluation of medical devices—Part 5: Tests for *in vitro* cytotoxicity". Briefly, 2.5×10^4 3T3 cells/well were incubated in 96-well plates and cultured for 24 h at 37 °C and 5% CO_2 in humidified hair with increases doses of BTG and PLOMW (0.39, 0.78, 1.56, 3.12, 6.25, 12.5 and 25 mg mL^{-1}) overnight in DMEM. Cell viability was measured by neutral red uptake (NRU) assay [28]. The NRU assay provided an incubation (3-h) with neutral red (50 µg/mL in DMEM) followed by an extraction with acetic acid, ethanol and water (1:50:49 *v*/*v*/*v*). The absorbance was measured at 540 nm in a microplate reader Epoch (BioTek, Winooski, VT, USA). A percentage of viability was calculated as follows:

$$\%\text{Viability} = \frac{(\text{Absorbance}_{540\text{nm}}\text{test material}) - \text{Absorbance}_{540\text{nm}}\text{blank})}{(\text{Absorbance}_{540\text{nm}}\text{control}) - \text{Absorbance}_{540\text{nm}}\text{blank})} \quad (4)$$

Cell Viability Assay

MTT staining assay was used to evaluate cell viability as described by Mosmann [29]. Briefly, 1×10^4 Caco-2 cells/well were incubated in 96-well plates and cultured for 24 h at 37 °C and 5% CO_2 in humidified hair with increases doses of BTG and PLOMW (3.12, 6.25, 12.5 and 25 mg mL^{-1}); then, fresh MTT was added to each well and after 2 h of incubation 1 mL of DMSO was used to solubilize the formazan products. The optical density (OD) was measured at 570 nm in a microplate reader Epoch (BioTek). A percentage of viability was calculated as follows:

$$\%\text{Viability} = \frac{(\text{OD}_{570\text{nm}}\text{test material}) - \text{OD}_{570\text{nm}}\text{blank})}{(\text{OD}_{570\text{nm}}\text{control}) - \text{OD}_{570\text{nm}}\text{blank})} \quad (5)$$

Con A/o-Pd Assay

Caco-2 cells were trypsinized, washed in saline solution and divided into three groups: (1) Control (2.5×10^5 cells in 5 mL of saline solution); (2) BGT (2.5×10^5 cells in 5 mL of BGT 12.5 mg mL^{-1}); (3) PLOMW (2.5×10^5 cells in 5 mL of PLMW 12.5 mg mL^{-1}). Each group was incubated at 30 °C for 15 min, under gentle shaking. Then, the cells were washed twice with 5 mL Phosphate Buffered Saline (PBS) and centrifugated (2000 rpm) for 5 min. Subsequently, the cells were transferred to a clean tube and given a final wash. After an additional centrifugation (2000 rpm) for 5 min, the supernatant was removed.

The pellet was stirred with a vortex mixer and washed twice with 12 mL of PBS and centrifugated (2000 rpm) for 5 min. Furthermore, the cells were transferred to a clean tube and washed prior to the addition of the next reagent (5 mL of PBS containing 1.0 mM calcium chloride and 10 mg L^{-1} biotinylated concanavalin A from Canavalia ensiformis (Con-A)). The mixture was incubated at 30 °C for 30 min under gentle shaking, centrifuged (2000 rpm) for 5 min. The cells were washed twice with PBS and transferred to a clean tube. 5 mL of PBS containing 5 mg/L streptavidin peroxidase was added and each tube was incubated at 30 °C for 60 min. Finally, the cells were washed and transferred to a clean tube and resuspended with 1 mL of o-phenylenediamine dihydrochloride (o-pd) solution (containing 0.4 mg o-pd and 0.4 µL 30% H_2O_2 in 1.0 mL 0.05 M citrate phosphate buffer). The oxidation of o-pd was stopped after 2 min with 1.0 mL of 1.0 M H_2SO_4 and the optical density measured at 490 nm (spectrophotometer Epoch, BioTek) [30].

2.6. Preparation of Puddings

Puddings were prepared using starch and PLOMW as gelling agent. Pears were purchased in a local supermarket and after washing, they were peeled. In order to prepare the puree, 50 mL of water was added to the fruit pulp (650 g) divided into smaller pieces. The pulp was then heated and stirred at low speed for 15 min, setting the temperature at 100 °C. Subsequently, after inactivation of the fruit enzymes, the heating was stopped and the pulp was ground for 1 min at high speed. The formulation of the pudding was determined after evaluating the consistency, according to the different quantity of gelling agent. Specifically, the pudding (PPLOMW) was prepared by mixing 100 mL of milk, 10 g of starch, 50 g of pear puree and 0.5 g of PLOMW. The starch was solubilized at room temperature in 75.0 mL of milk. The pear puree solubilized in the remaining volume of milk was then added under stirring at 80 °C for 5 min. The tare gum was gradually added without interrupting stirring and heating, to favor its hydration. The pudding was packaged in 60 g glass jars, immediately sealed with a metal lid. The pudding samples were stored at 4 °C in the refrigerator until the analysis carried out when the containers were opened (day 0), after seven, fourteen and twenty-eight days from the preparation. A control pudding (BCTG) was also prepared as control under the same conditions as described above, but using commercial tara gum (CTG) instead of PLOMW.

2.7. Characterization of Pudding

2.7.1. Rheological Characterization

The rheometric measurements of the food matrices PPLOMW and PBTG were carried out using a strain-controlled RFS III rheometer (Rheometric Scientific Inc. at Piscataway, NJ, USA), equipped with plate-plate geometry: Gap 2 mm, Φ 25 mm. The temperature was controlled by means of a Peltier system (uncertainty of 0.1 °C). The samples were preliminarily subjected to strain sweep tests at a frequency of 1.0 Hz to determine the region of linear viscoelasticity (region in which the G' and G'' vs strain modules are constant). The trend of the G' and G'' vs. frequency (frequency sweep tests) in the linear viscoelasticity region was determined in the frequency range of 0.1–16.0 Hz.

2.7.2. Antioxidant Performances of the Puddings

Pudding samples were analyzed, in terms of antioxidant properties, on the day they were prepared and after 7, 14 and 28 days, respectively [31]. Briefly, 10.0 g of pudding (PPLOMW or BCTG) were suspended in 20.0 mL of water by maceration for 24 h. The aqueous solution was then centrifuged at 8000 rpm for 10 min. Each extract was obtained by recovering the supernatant and was characterized in terms of APG, PAC and FC by the methodologies previously described, expressing the results as mg of CT per gram of pudding. Similarly, antioxidant performances were recorded by evaluation of TAC (expressed as mg of CT per gram of pudding) and scavenging activity of both aqueous (against ABTS radical) and organic (against DPPH radical) environments. Each measurement was performed in triplicate and data expressed as means (±SD).

2.8. Statistical Analysis

The inhibitory concentration 50 (IC_{50}) was calculated by non-linear regression with the use of Prism Graph- Pad Prism, version 4.0 for Windows (GraphPad Software). One-way analysis of variance test (ANOVA) followed by a multicomparison Dunnett's test were applied. All toxicity essays were performed in triplicate. Statistical analyses were made with Graph Pad. The data were evaluated by one-way analysis of variance followed by the Mann–Whitney U test. A value of $p < 0.005$ was considered significant.

3. Result and Discussion

3.1. Oil Mill Wastewater Treatment

Oil mill wastewater (OMW) displays a variable composition that is influenced by different issues including agronomic parameters, such as cultivar and maturation of the olive fruit, region of origin and climatic conditions [32]. In particular, OMW is an acidic and dark suspension mainly composed by water (83–94% *w*/*w*) and also containing inorganic (0.4–2.5% *w*/*w*) and organic substances (4–18 % *w*/*w*), including mucilage, lignin, tannins and pectins, as well as cation species such as magnesium, sodium, calcium and potassium. In addition, OMW contains phenolic compounds, the most popular high added-value ingredients, varying from 0.5 to 24 g/OMW), that represent about 98% (*w*/*w*) of the phenols typically present in olive fruit [33]. A high concentration of polyphenols generally involves a condensation step by ultrafiltration, thermal concentration or freeze-drying processes [34]. In this regard, OMWs from Roggianella *cv* were subjected to a freeze-drying process after preliminary filtration and centrifugation providing a vaporous solid (LOMW) that was deeply characterized by chromatographic and spectroscopic techniques, as well as in terms of antioxidant performance.

3.2. HPLC-MS/MS and ^{1}H-NMR Analyses of Lyophilized Oil Mill Wastewater

Separation of the main polyphenols in LOMW was carried out by HPLC-MS/MS analysis and their identification was based on mass measurements of deprotonated $[M–H]^-$ ions and MS/MS fragmentation patterns (Table 1). The compound at RT = 1.475 min (at concentration of 0.71 μg mL^{-1}) was assigned to a verbascoside residue lacking the rhamnose moiety due to its $[M–H]^-$ at m/z 477 together with the fragments at m/z 459 (water loss) and 161 (ascribable to the dehydrated ion of the caffeic acid unit), as already hypothesized in literature [31]. The compound at RT = 1.515 with a concentration of 0.19 μg mL^{-1}, exhibiting $[M–H]^-$ at m/z 169 and a fragment at m/z 151 probably produced by the loss $[M–H–H_2O]^-$, was tentatively identified as 3,4-dihydroxyphenylglycol [15].

Six phenolic acids and derivatives were also revealed, namely quinic acid (4.1 μg mL^{-1}), HyEDA (0.27 μg mL^{-1}) and decarboxymethyl-elenolic acid derivative (1.55 μg mL^{-1}), hydroxylated product of dialdhydic form of decarboxymethyl elenolic acid (3.3 μg mL^{-1}), caffeic acid (1.9 μg mL^{-1}) and *p*-coumaric acid (2.1 μg mL^{-1}) on the basis of their $[M–H]^-$ ions as well as the presence of $[M–H_2O]^-$ and $[M–CO–2H_2O]^-$ in their fragmentation patterns [15,17]. The compounds at RT = 2.245 min and 2.470 min (having concentrations of 4.3 and 2.9 μg mL^{-1}, respectively) were two 3-hydroxytyrosol glucoside isomers because they showed the same $[M–H]^-$ at m/z 315 and similar MS/MS spectra characterized by two main fragment ions at m/z 153, which were formed through the loss of a glucose moiety and at m/z 123 corresponding to the subsequent loss of the CH_2OH group. However, since different hydroxytyrosol glucoside isomers (i.e., hydroxytyrosol-1-O-glucoside, hydroxytyrosol-3′-O-glucoside and hydroxytyrosol-4′-O-glucoside) have been identified in *Olea europaea* [32], they were not furtherly distinguishable by MS analysis. Finally, oleuropein aglycone derivative (5.8 μg mL^{-1}) and 3-hydroxytyrosol (0.09 μg mL^{-1}) were recognized by matching chromatographic and MS characteristics to those of previous literature reports in the former case [15] and of a pure standard in the latter one.

As reported in Figure 1, the ^{1}H-NMR of LOMW revealed the main features typical of minor components in olive oil reported in Table 2, responsible for its interesting biological activities.

Table 1. Identified polyphenol compound in LOMW (in μg mL^{-1}). Data represent mean ± RSD (n = 3).

Compound	RT (min)	LOMW (μg mL^{-1})
Verbascoside residue	1.475	0.71 ± 0.06
3,4-dihydroxyphenylglycol	1.515	0.185 ± 0.017
Quinic acid	1.773	4.1±0.4
3-hydroxytyrosol glucoside isomers 1	2.245	4.3 ± 0.4
3-hydroxytyrosol glucoside isomers 2	2.470	2.9 ± 0.3
Dimer 407	2.610	1.56 ± 0.14
3-hydroxytyrosol	2.957	0.092 ± 0.008
Decarboxymethyl-elenolic acid derivative	3.542	1.55 ± 0.14
Hydroxylated product of dialdhydic form of decarboxymethyl elenolic acid	4.428	3.3 ± 0.3
Caffeic acid	7.996	1.91 ± 0.17
Decarboxymethyl-elenolic acid (HyEDA)	8.075	0.27 ± 0.02
Oleuropein aglycone derivative	8.124	5.8 ± 0.5
p-coumaric acid	10.798	2.13 ± 0.19

RT = retention time; LOMW = Lyophilized olive mill wastewater.

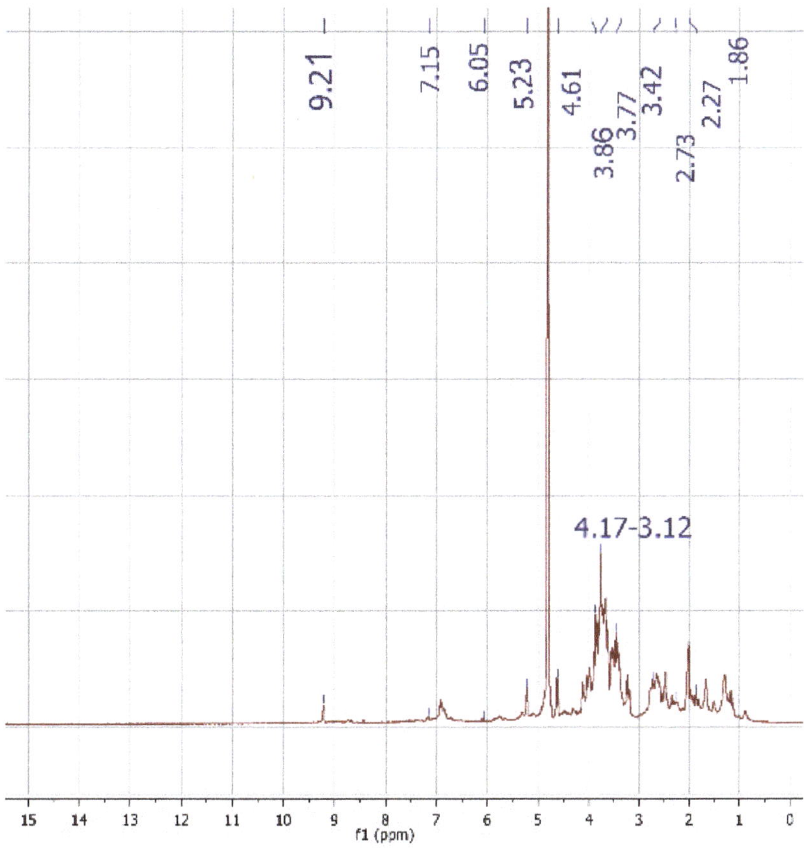

Figure 1. ^{1}H-NMR of 16.7 mg of LOMW sample in 0.6 μL of D_2O.

Table 2. Characterization of LOMW by ^{1}H-NMR spectroscopy.

Compound	Assignment	Chemical Shift (ppm)
Verbascoside	OH-18; OH-17	s 9.21; 4.61
Verbascoside	O-CH-O	d 5.23
Sugar residue of glycosides	CH-OH	m 4.17–3.12
Hydroxythyrosol	Ph-CH_2-	2.73
Hydroxythyrosol	CH_2-O	3.76
Oleuropein	CH_3	s 3.86
Oleuropein	Enantiotopic CH_2-OH	m 3.42
Caffeic acid	CH_2=CH_2	dd 6.05–7.15
Quinic acid	Enantiotopic CH_2-CH	2.27–1.86

The corrected assignment was obtained comparing the NMR spectra with those available in literature and using the Human Metabolome Database (HMDB).

Many signals belonging to sugar residues derived from glycosides and the OH group of Verbascoside, together with Ph-CH_2 and CH_2-O of Hydroxythyrosol and the CH_3 and the enantiotopic CH_2-OH of Oleuropein are detected. Furthermore, at 0.89, 1.3, 1.64, 2.02, 2.35 ppm, the signals of oleic acid are visible.

3.3. *Antioxidant Properties of Lyophilized Oil Mill Wastewater*

LOMW was characterized by evaluation of APG, FC, PAC and AC in order to provide a straight measure of the antioxidant potential of this by-product and the results are reported in Table 3.

Table 3. Antioxidant characterization of LOMW. Data represent mean ± RSD (n = 3).

Sample	APG (mg CT/g)	PAC (mg CT/g)	FC (mg CT/g)	AC (mg CT/g)	TAC (mg CT/g)	IC_{50} (mg mL^{-1})	
						DPPH Radical	ABTS Radical
LOMW	75.0 ± 0.7	50.8 ± 0.4	34.0 ± 0.4	0.15 ± 0.01	1.10 ± 0.05	0.095 ± 0.003	0.019 ± 0.001

LOMW = Lyophilized olive mill wastewater; APG = Available phenolic groups; PAC = Phenolic acids content; FC = Flavonoid content; AC = Anthocyanin content; TAC = Total antioxidant activity; DPPH = 2,2-diphenyl-1-picrylhydrazyl radical; ABTS = 2,2′-azino-bis(3-ethylbenzothiazolin-6-sulphonic radical; CT = catechin.

By-products from the olive oil extraction process are found to be particularly rich in phenolic compounds [35,36]. The available phenolic groups of LOMW was 75.0 mg CT per gram. This value appears in the same magnitude of other studies reported in literature, showing high concentration of phenolic compounds present in the OMWs [37,38]. Typically, during the extraction process, the highest (98%) amount of polyphenols in olive fruit can be found in the OMWs (0.5–24 g L^{-1} of OMW), while only 2% of them is in the oil phase [30]. A correlation of this parameter with literature data appears quite difficult because total polyphenol concentration is strictly related to many factors, such as type and region of origin, maturity of olives, method of extraction, climatic conditions and cultivation and processing techniques [39].

The analysis of FC in LOMW was carried out by using $AlCl_3$ reagent and the results were expressed as milligrams of CT per gram of sample (Table 2). In plants, the number of glycoside fractions of these compounds can vary from one to three. In particular, flavonoids are found glycosylated with carbohydrates such as glucose or rhamnose, but they can also be found linked to glucose units such as galactose, arabinose or other sugars [40]. The results showed that phenolic compounds with a flavonoid structure in LOMW are equal to 34.0 mg CT per gram, corresponding to 45.3% of the APG.

The evaluation of PAC in LOMW was carried out using the Arnov's method by expressing the results as milligrams of CT per gram of LOMW. Such compounds are hardly found in the free form, due to their ability to link quinic and tartaric acids forming esters and/or glycosylated derivatives [41]. LOMW sample provided high amounts of PAC (50.8 mg CT per gram), showing values that were equal to 67.7% of the total polyphenols.

As can be seen from the data obtained by LC-MS analysis, the phenolic acid content was mainly related to the amounts of quinic, coumaric and caffeic acids.

Finally, a colorimetric assay was employed to quantified anthocyanin compounds, a class natural pigments responsible for the coloring of most fruits and vegetables [42]. Recorded results displayed that AC in LOMW was equal to 0.15 mg CT g^{-1}, almost two orders of magnitude lower than FC and PAC.

Antioxidant properties of the food matrix were deeply investigated by specific tests, including total antioxidant capacity and scavenging activity of the LOMW against DPPH and ABTS radicals. TAC of LOMW was determined using $(NH_4)_2MoO_4$ reagent and by expressing the results as mg CT per gram of matrix. According to Prieto et al. (1999) [43], this reagent can be systematically applied for the evaluation, both in aqueous and organic environments, of the antioxidant activity of matrices with a complex composition.

The recorded results (Table 3) depicted as a high APG value was not always related to a significant antioxidant capacity according to literature data [44], highlighting as the class of phenolics deeply influenced the total antioxidant capacity of the food matrix.

In order to determine scavenger activity both in aqueous and in organic environments, LOMW was tested against DPPH and ABTS radicals. The ability of LOMW to inhibit these reactive species, was expressed in terms of IC_{50} (mg mL^{-1}), as reported in Table 3. In particular, the ability to inhibit the ABTS radical is a highly used parameter for the determination of antioxidant activity in food and biological samples [45]. Based on the inhibition kinetics of the ABTS radical, recorded LOMW IC_{50} value was 0.019 mg mL^{-1}, almost five times less compared to the activity against DPPH, suggesting that LOMW is a matrix rich in highly hydrophilic moieties.

3.4. *Synthesis of the Tara Gum Conjugate*

An innovative strategy to synthesize versatile materials with antioxidant features involved the covalent linkage of the biomolecules of LOMW on the tara gum (TG) chain. Literature data largely proposed TG as starting materials in the synthesis of innovative polymeric carries of nutraceuticals able to be employed in the food industry [13,14,46], but to the best of our knowledge covalent grafting of bioactive molecules on TG chain was not investigated. This approach allowed the preparation of a high molecular weight antioxidant compounds showing improved chemical stability, as well as lower degradation rate compared to the low molecular weight antioxidants [47]. Conjugate polymers with antioxidant features were synthesized by employing an eco-compatible, radical initiated grafting procedure. Antioxidant conjugate was synthesized by anchoring the LOMW reactive species to TG chains, using a water soluble in a radical reaction initiated by a biocompatible redox pair (H_2O_2/ascorbic acid). This redox couple showed numerous advantages, such as the opportunity of inducing polymerization reaction at room temperature drastically reducing the risks of degradation of phenolic compounds and avoiding the generation of any type of toxic products [48]. A specific LOMW/TG ratio (w/w) was used in the polymerization mixture. In particular, a quantity of LOMW equivalent to 35 mg of catechin (calculated by APG value) for each gram of commercial TG was found to be optimal. After 24 h, to remove the unreacted molecules, the conjugate was subjected to a purification process by dialysis and the resulting solution was lyophilized which led to obtaining a vaporous solid (PLOMW) whose antioxidant properties were investigated. To verify the antioxidant performance of the conjugate, a control polymer (labelled BTG), was prepared in the same conditions, but in the absence of LOMW.

3.5. *^{1}H-NMR Analysis of the Polymers*

The ^{1}H-NMR spectra of BTG (Figure 2A) and PLOMW (Figure 2B) showed the main residues of galactomannan [49], scaffold of TG and some fragments of LOMW (close to 1.12–1.40 ppm and 8.32 ppm) that confirm the conjugation.

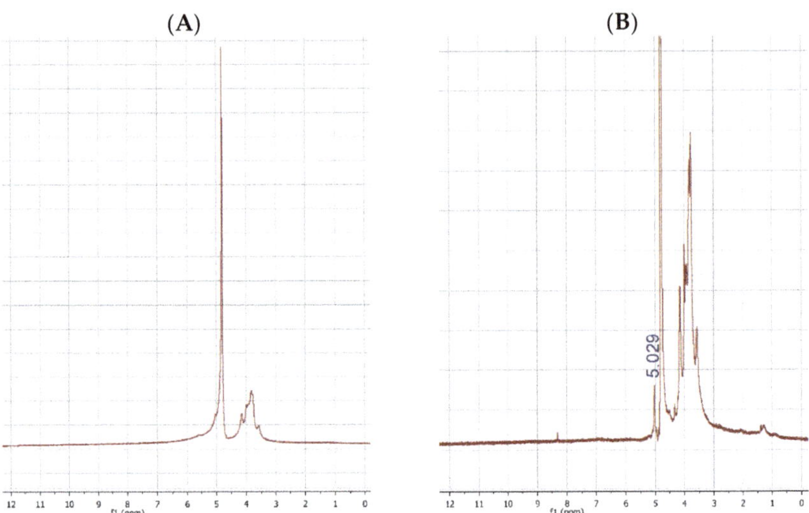

Figure 2. Panel (**A**): ^{1}H-NMR spectrum of BTG; Panel (**B**): ^{1}HNMR spectrum of PLOMW. In the panel (**B**), the signal at 5.029 ppm belongs to H_1 of α-D-galactopyranose which in the panel (**A**) collapsed into the D_2O signal.

3.6. Calorimetric Analysis of the Polymers

The Differential Scanning Calorimetry analysis of PLOMW, BTG and the commercial tara gum (CTG) was performed subjecting the samples to an isotherm at 25 °C for 20 min and heat ramp from 25 to 650 °C at a temperature scanning speed of 10 °C min^{-1} under nitrogen flow. The results obtained (Figure 3) allow to highlight the presence of two significant peaks in all the samples: an exothermic peak (100–150 °C) and an endothermic peak (around 300 °C).

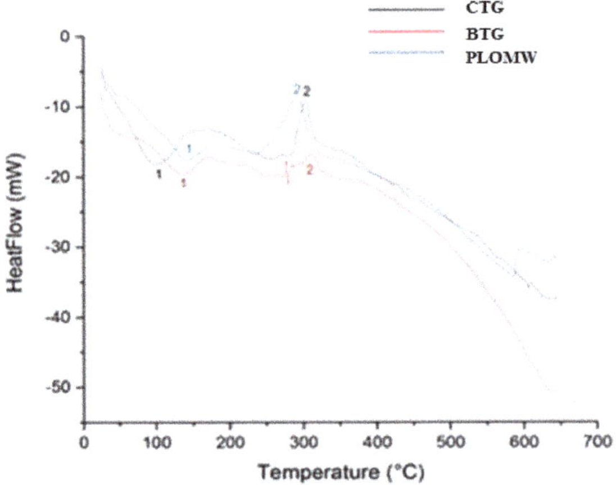

Figure 3. Differential scanning calorimetry (DSC) of PLOMW, BTG and CTG.

Specifically, the graph shows that the two polymer samples show the same peaks as TG sample, but the values are slightly shifted with respect to the reference (CTG) and this phenomenon is most likely due to the different molecular structure.

The enthalpy was then calculated for each peak and the results were reported in Table 4.

Table 4. Enthalpy and temperature values of the polymers and commercial tare rubber peaks.

Sample	T Center Peak 1 (°C)	T Center Peak 2 (°C)	Enthalpy Peak 1 (J g^{-1})	Enthalpy Peak 2 (J g^{-1})
PLOMW	141.0	311.8	125.4	−154.3
BTG	137.7	291.4	76.0	−11.6
CTG	100.4	302.6	261.4	−83.4

PLOMW = Polymer conjugate lyophilized olive mill wastewater and tara gum; BTG = Blank tara gum; CTG = commercial tara gum.

Measuring the enthalpy of fusion allows to calculate the degree of crystallinity of a substance. Therefore, a higher enthalpy corresponds to a greater interaction between the molecules. The results show that as regards the enthalpy values measured on the peak (1), these are significantly lower for the PLOMW and BTG polymers. This behavior denotes a lower level of interaction between the polymer molecules than the commercial tare rubber sample. On the other hand, the values recorded for the peak (2) show how the blank polymer has a higher enthalpy value (−11.6 J g^{-1}) and, therefore, has a greater degree of interaction between its molecules than CTG and PLOMW polymer. In particular, the latter, having the lowest enthalpy value (−154.3 J g^{-1}), is the one with the least interaction between its molecules.

3.7. Toxicity Evaluation of the Conjugate

3.7.1. NRU Assay

The effect of BGT and PLOMW on 3T3 fibroblasts was measured by *in vitro* acute toxicity assay through the neutral red uptake (NRU, Figure 4). The treatment with increased concentrations of BGT and PLOMW non-altered cell viability compared to the control, in both treatments.

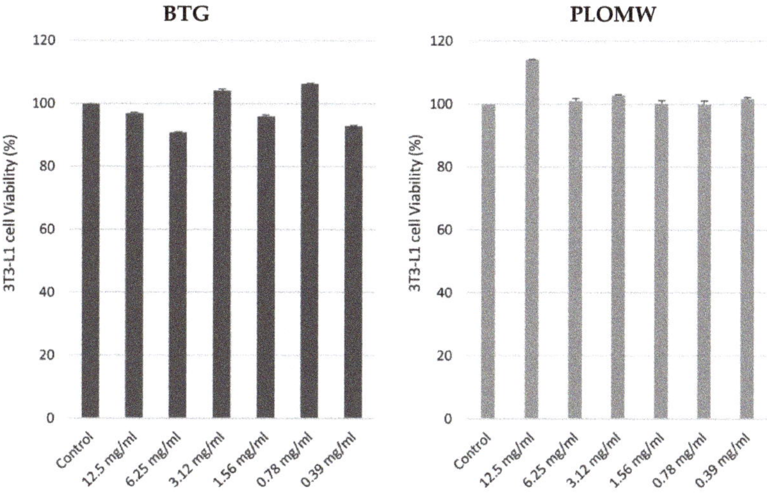

Figure 4. 3T3 cells viability (%) measured through NRU cytotoxicity assay upon treatment with increased concentrations of BGT and PLOMW (mg/mL). Each column represents the mean + SD of 3 wells/group.

3.7.2. Cell Viability by MTT Assay

To evaluate the cytotoxic and adverse cellular effects of BGT and PLOMW, was used the MTT assay. This test assesses the activity of mitochondrial enzymes in healthy cells by measuring the absorbance of purple formazan, formed through an NADP-dependent reaction catalyzed by succinate dehydrogenase in metabolically active cells [50].

As shown in Figure 5, the treatments with increased concentrations of BTG and PLOMW for 24 h, not altered Caco-2 cells viability.

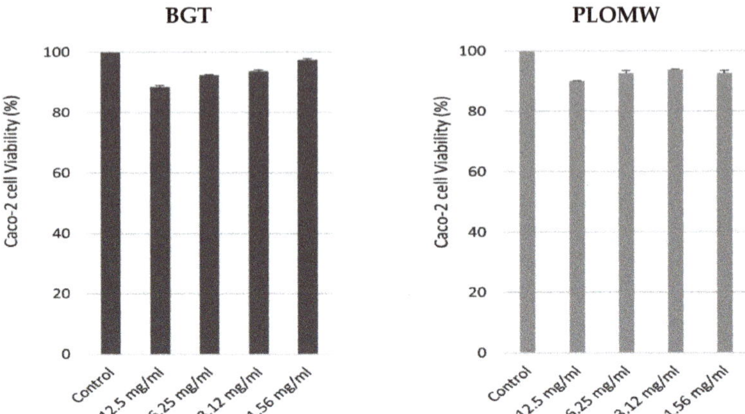

Figure 5. Effect of BTG and PLOMW on Caco-2 cells viability.

3.7.3. Bio-Adhesive Ability Assay

In this study, to evaluate the bio-adhesive ability of active substances was used a technique described by Rizza [30]. Our results showed a similar muco-adhesivity of BTG and PLOMW (Figure 6) at 12.5 mg mL^{-1} concentration. This concentration represents the final dose of dietary use.

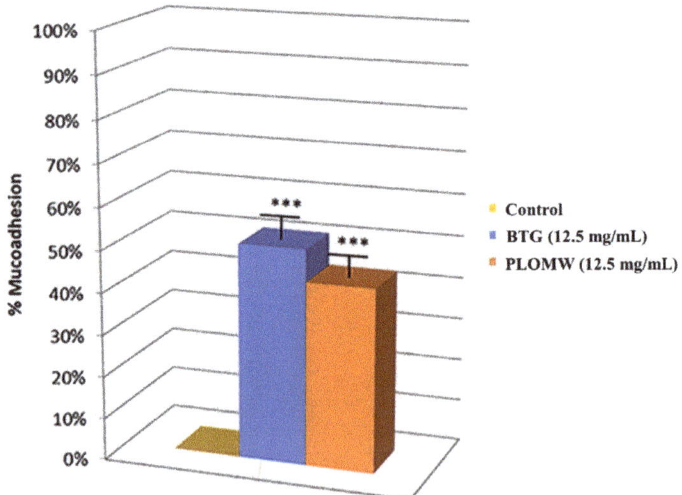

Figure 6. Muco-adhesion of BTG and PLOMW determined as reduction (%) of lectin binding on Caco-2 cells at different concentrations; *** $p < 0.001$ vs. control.

3.8. Antioxidant Performances

Antioxidant properties of PLOMW were explored by confirming that the grafting reaction has taken place and also to verify whether the reaction conditions were harmful to the active compounds in LOMW. Total phenolic groups, available phenolic acids and antioxidant activity of the conjugate have been reported in Table 5.

Table 5. Total polyphenols, phenolic acid contents and antioxidant activity of the conjugate polymer. Data represent mean ± RSD (n = 3).

Sample	APG (mg CT/g)	PAC (mg CT/g)	TAC (mg CT/g)	IC_{50} (mg mL^{-1})	
				DPPH Radical	ABTS Radical
PLOMW	16.3 ± 0.5	15.7 ± 0.4	2.26 ± 0.11	0.322 ± 0.005	0.106 ± 0.005
BTG	-	-	-	-	-

PLOMW = Tara gum grafted with lyophilized olive mill wastewater; APG = Available phenolic groups; PAC = Phenolic acids content; TAC = Total antioxidant activity; DPPH = 2,2-diphenyl-1-picrylhydrazyl radical; ABTS = 2,2′-azino-bis(3-ethylbenzothiazolin-6-sulphonic radical; CT = catechin.

Data confirmed that the grafting reaction avoids the loss of the antioxidant activity of LOMW, while BTG did not demonstrate any interference. Specifically, the analyses of the inhibition kinetics against radical species underlined the good performance of PLOMW, both in organic and aqueous environments. However, the response towards the radical ABTS appears almost three times higher, as confirmed by the IC_{50} values. The collected data clearly indicates as PLOMW is able to guarantee good results and that when introduced into a food, it can impart significant biological properties from a nutritional point of view, producing beneficial effects on the health of consumers.

3.9. *Pudding Preparation and Evaluation of the Antioxidant Properties*

Gelling properties of the conjugate suggest the employment of this macromolecular system for the preparation of suitable functional foods with high added value. In particular, the PLOMW was used for the production of a fruit-based pudding. The fruit gives numerous benefits to the human body often related to the presence of phenolic compounds. Different studies have correlated the high intake of fruit with the lower incidence of cardiovascular, neurodegenerative and chronic diseases, such as cancer and diabetes [51]. To highlight the effect and benefits derived from using PLOMW in the pudding's preparation, the pear, not particularly rich in antioxidant molecules, was chosen as raw material [52]. Although it can contain in the range 27–41 mg of phenols per 100 g of pulp, the antioxidant profile of pear remains distant from other fruits such as blackberry, raspberry, blueberry, strawberry, cherry and grape [53]. At the same time, this fruit is very digestible and tasty, especially in the preparation of fruit juices for children or hospitalized patients. In this context, puddings based on pear puree were prepared using PLOMW and CTG as gelling agent and their antioxidant capacity was investigated as a function of time. For this purpose, each pudding jar was opened at set time intervals and subjected to an extraction process according to a procedure reported in literature, with some modifications [31]. Antioxidant properties of PPLOMW were analyzed at the opening day (t = 0), after 7, 14 and 28 days. The data of total phenolic groups, phenolic acids and the scavenging properties in aqueous environment are reported in Table 6.

Available phenolic groups (t = 0 days), expressed as mg of CT per gram of pudding, confirm that the addition of the antioxidant polymer PLOMW as gelling agent, significantly increases (almost two times) APG value of PCTG. Similarly, PPOMW at time zero displayed an increased amount of available phenolic acids and more performing scavenging capacity against the hydrophilic radical ABTS. The analysis of the trend of these parameters over 28 days, in terms of APG, highlighted the slight decrease observed for PPOMW (equal to 11% after 28 days). On the contrary, APG decrease equal to 30% after 28 days was recorded with the pudding prepared using CTG. This finding was mainly interesting because it highlights that the employment of the conjugate increased phenolic groups in the functional food and maintained their concentration over time. The same trend was recorded in the content of available phenolic acids, with a decrease after 28 days of 76% for PPLOMW matrix, which increased to 82% in PCTG.

Finally, by recording the inhibition profiles towards the ABTS of the puddings over time, it was observed that the enriched pudding proved to be more effective against this radical specie, compared to the control (Figure 7). IC_{50} values recorded for the PPLOMW underwent a slight decrease over time (0.0124 mg mL^{-1}, after 28 days). On the contrary, the IC_{50} values recorded for the pudding based on CTG, underwent a greater decrease (Figure 7).

Table 6. Total polyphenols, phenolic acid contents and antioxidant activity of puddings based on PLOMW and CTG. Data represent mean ± RSD (n = 3).

Time (Days)	Pudding	APG (mg CT/g Pudding)	PAC (mg CT/g Budino)	IC_{50} (mg mL^{-1}) ABTS Radical
0	PPLOMW	0.220 ± 0.009 [a]	0.261 ± 0.011 [a]	0.0060 ± 0.0003 [a]
	PCTG	0.113 ± 0.005 [b]	0.244 ± 0.009 [b]	0.0324 ± 0.0011 [b]
7	PPLOMW	0.193 ± 0.007 [c]	0.136 ± 0.005 [c]	0.0082 ± 0.0003 [c]
	PCTG	0.101 ± 0.004 [d]	0.070 ± 0.002 [d]	0.0345 ± 0.0015 [b]
14	PPLOMW	0.194 ± 0.006 [c]	0.075 ± 0.002 [e]	0.0118 ± 0.0005 [d]
	PCTG	0.089 ± 0.003 [d]	0.056 ± 0.002 [f]	0.0397 ± 0.0018 [e]
28	PPLOMW	0.196 ± 0.006 [c]	0.062 ± 0.002 [g]	0.0124 ± 0.0006 [d]
	PCTG	0.080 ± 0.002 [e]	0.042 ± 0.001 [h]	0.0453 ± 0.0021 [f]

PPLOMW = Pudding based on tara gum grafted with lyophilized olive mill wastewater; PCTG = Pudding based on commercial tara gum; APG = Available phenolic groups; PAC = Phenolic acids content; ABTS = 2,2′-azino-bis(3-ethylbenzothiazolin-6-sulphonic radical; CT = catechin. Different letters express significant differences ($p < 0.05$).

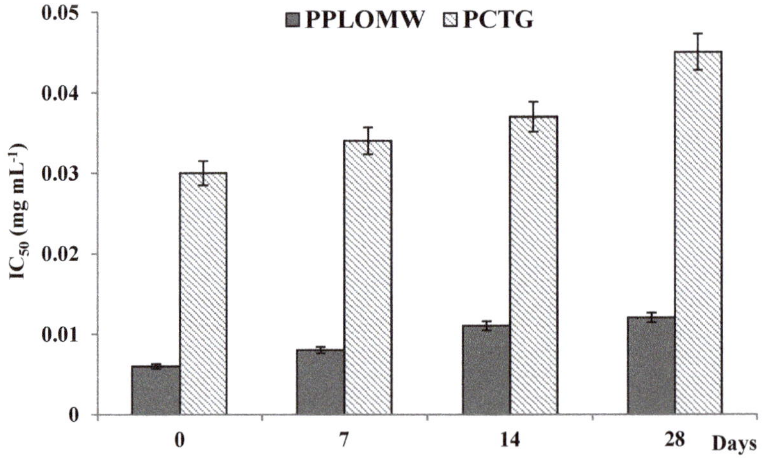

Figure 7. IC_{50} trend as function of the time of PPLOMW and PCTG.

Ultimately, the polymeric conjugate, synthesized using TG and compounds with a polyphenolic structure obtained from waste products of the oil extraction process, inserted in a food matrix such as pudding with pear, has guaranteed it a greater antioxidant capacity and the possibility of keeping it almost unchanged over time.

3.10. Rheological Analysis of Puddings

Rheology is the science that studies the deformation and flow of matter in liquid or solid form, i.e., the response of materials to mechanical stress in terms of viscosity and elasticity [54]. The rheometric measurements of the food matrices PPLOMW and PCTG were carried out by recording the frequency sweeps of the G′ and G″ modules as a function of the frequency in the linear viscoelasticity region. The trend of G′ and G″ vs. the frequency is also called the mechanical spectrum and allows a quantitative rheological characterization of the materials.

Such a linear viscoelastic behavior has been previously observed in gel systems. All mixtures show a typical gel-like response in which G′ is higher than G″. Higher values of G′ may reflect the stronger interactions existing among the domains which favor the formation of highly elastic lattices. In order to obtain quantitative information, the weak gel model was applied, which models the system as consisting of connected rheological units. This allowed to determine the values of "A" and "z", where "A" represents the interaction force between the rheological units and "z" the coordination number, that is, the number of rheological units interacting with a reference unit. The "A" and "z" values of the PPLOMW and PCGT food matrices determined by frequency sweep test are shown in Table 7.

Table 7. Frequency sweep test for PPLOMW food matrices and PCGT over time.

Sample	A ± 100	z ± 1
5 °C PCGT (t = 28 days)	8800	18
5 °C PCGT (t = 14 days)	8200	12
5 °C PCGT (t = 7 days)	5600	11
5 °C PCGT (t = 0 days)	2800	11
25 °C PCGT (t = 28 days)	7200	45
25 °C PCGT (t = 14 days)	7300	12
25 °C PCGT (t = 7 days)	4800	11
25 °C PCGT (t = 0 days)	2000	11
5 °C PPLOMW (t = 28 days)	9600	18
5 °C PPLOMW (t = 14 days)	8900	10
5 °C PPLOMW (t = 7 days)	5100	11
5 °C PPLOMW (t = 0 days)	3400	14
25 °C PPLOMW (t = 28 days)	8200	32
25 °C PPLOMW (t = 14 days)	7200	12
25 °C PPLOMW (t = 7 days)	4400	12
25 °C PPLOMW (t = 0 days)	1100	7

PPLOMW = Pudding by polymer conjugate; PCTG = pudding by commercial tara gum.

As can be seen from the data reported in Table 7, the values of "A" recorded for PCTG, both at 5 and 25 °C increase as functions of time, while the values of "z" are very similar in the first fourteen days and then significantly increased after twenty-eight days. This indicates that initially, the network formed increases the strength between the links and then expands three-dimensionally over time.

Finally, for each sample, temperature ramp tests or time cures were carried out to analyze the behavior of the system as a function of temperature and the results are shown in Figure 8A,B.

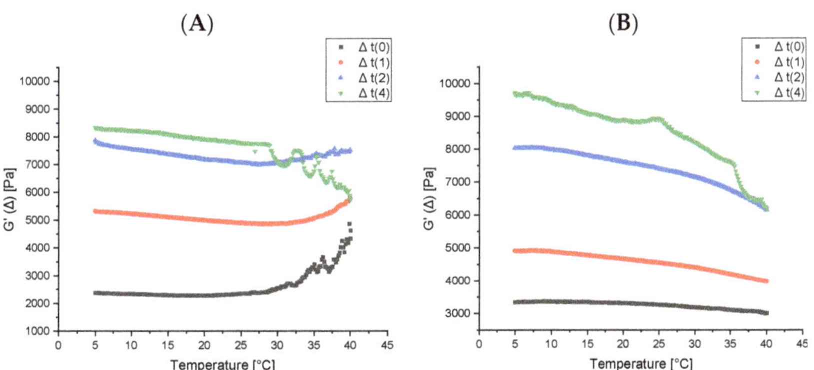

Figure 8. Time Cure test for the PCGT (**A**) and PPLOMW (**B**) food matrix over time (weeks).

In this type of test, G′ and G″ are recorded as a function of temperature at a frequency of 1 Hz within the linear viscoelastic region. The temperature range investigated was 5–40 °C, with a scanning speed of 1 °C min^{-1}. Figure 8 shows the G′ comparison (elastic modulus) of the pudding samples prepared with the two polymers investigated (PPLOMW and PCGT) over time. The recorded data indicate that at "time zero" the samples have very different values of elastic modulus (G′). In fact, the G′ values recorded for the PCGT sample are lower than those observed for PPLOMW. However, with the passage of time, the difference between the modules tends to decrease and overlap when both samples reach two weeks of maturation. It is important to note that as the maturation time increases (beyond two weeks), the elastic values of the PPLOMW sample are greater. In fact, the pudding prepared with PPLOMW at "time 4 weeks" shows the highest elastic modulus. This rheological behavior can be attributed to a greater structuring effect induced by the conjugated polymer. Preliminary data suggest that water may play a central role in the structuring through hydrogen bonds, a hypothesis that must necessarily be proved by means of spectroscopic techniques.

4. Conclusions

Olive mill wastewaters (OMW) were explored as inexpensive, precious and valuable sources of bioactive molecules to be employed in the production of antioxidant-enriched milk-based products. In this regard, lyophilized OMW (LOMW) were involved in a radical grafting reaction to synthesize a tara gum conjugate, suitable as a thickening agent in the preparation of pear puree-based pudding. Chemical composition of LOMW was evaluated by ^{1}H-NMR and HPLC-MS/MS analyses and oleuropein aglycone derivative (5.8 μg mL^{-1}) was detected as the main compound. Additionally, this olive process by-product was characterized by evaluation of the total phenolic content, flavonoids, phenolic acids and anthocyanins amount, providing a straight measure of its antioxidant features. LOMW was able to inhibit ABTS radical, displaying in the aqueous medium, scavenger properties almost one order of magnitude increased compared to the organic one. LOMW reactive species and tara gum chains were involved in an eco-friendly grafting reaction to synthesize a polymeric conjugate that was characterized by spectroscopic, calorimetric and toxicity studies. Antioxidant properties of the polymeric conjugate were also evaluated, suggesting its employing as a thickening agent in the preparation of pear puree-based pudding. Milk-based foodstuff showed high performance of consistency and relevant antioxidant features over time (28 days) in comparison with its non-functional counterparts confirming LOWM as an attractive source for the development of functional foods.

Supplementary Materials: The following supporting information can be downloaded at: https://www.mdpi.com/article/10.3390/foods11020158/s1, Table S1: Acquisition parameters for MRM HPLC-MS/MS analyses.

Author Contributions: Conceptualization: U.G.S., D.R., F.A. and M.L.C.; Data curation: U.G.S., P.C. (Pasquale Crupi), M.M., C.O.R. and V.R.; Formal analysis: U.G.S., P.C. (Paolino Caputo), M.M., P.C. (Pasquale Crupi) and R.M.; Funding acquisition: M.L.C., F.A., D.R. and C.O.R.; Investigation: U.G.S., P.C. (Paolino Caputo), P.C. (Pasquale Crupi), V.R. and F.A.; Methodology; U.G.S., P.C. (Pasquale Crupi), C.O.R., F.A. and V.R.; Supervision: M.L.C., C.O.R. and D.R.; Writing—original draft: U.G.S., P.C. (Paolino Caputo), P.C. (Pasquale Crupi), R.M. and F.A.; Writing—review & editing: D.R., C.O.R., U.G.S. and M.L.C. All authors have read and agreed to the published version of the manuscript.

Funding: This work was financially supported by AGER 2 Project, grant n. 2016-0174, AGER Foundation—Olive Tree and Oil: Competitive-Claims of olive oil to improve the market value of the product, EU project 820587—OLIVE-SOUND—Ultrasound reactor—The solution for a continuous olive oil extraction process H2020-EU.2.1.—INDUSTRIAL LEADERSHIP-EIC-FTI-2018–2020-Fast Track to Innovation (FTI)—European Union's Horizon 2020 research and innovation program under grant agreement No. 820587. University founds and POR Calabria FESR-FSE 2014/2020-Linea (B) Azione 10.5.12.

Institutional Review Board Statement: Not applicable.

Informed Consent Statement: Not applicable.

Data Availability Statement: Not applicable.

Conflicts of Interest: The authors declare no conflict of interest.

References

1. Spizzirri, U.G.; Aiello, F.; Carullo, G.; Facente, A.; Restuccia, D. Nanotechnologies: An Innovative Tool to Release Natural Extracts with Antimicrobial Properties. *Pharmaceutics* **2021**, *13*, 230. [CrossRef] [PubMed]
2. Roselli, L.; Cicia, G.; Cavallo, C.; Del Giudice, T.; Carlucci, D.; Clodoveo, M.L.; De Gennaro, B.C. Consumers' willingness to buy innovative traditional food products: The case of extra-virgin olive oil extracted by ultrasound. *Food Res. Int.* **2018**, *108*, 482–490. [CrossRef] [PubMed]
3. Mallamaci, R.; Budriesi, R.; Clodoveo, M.L.; Biotti, G.; Micucci, M.; Ragusa, A.; Curci, F.; Muraglia, M.; Corbo, F.; Franchini, C. Olive Tree in Circular Economy as a Source of Secondary Metabolites Active for Human and Animal Health Beyond Oxidative Stress and Inflammation. *Molecules* **2021**, *26*, 1072. [CrossRef] [PubMed]
4. De Luca, M.; Restuccia, D.; Clodoveo, M.L.; Puoci, F.; Ragno, G. Chemometric analysis for discrimination of extra virgin olive oils from whole and stoned olive pastes. *Food Chem.* **2016**, *202*, 432–437. [CrossRef] [PubMed]
5. Amirante, P.; Clodoveo, M.L.; Tamborrino, A.; Leone, A.; Paice, A.G. Influence of the crushing system: Phenol content in virgin olive oil produced from whole and de-stoned pastes. In *Olives and Olive Oil in Health and Disease Prevention*; Academic Press: London, UK, 2010; pp. 69–76.
6. Cirillo, G.; Curcio, M.; Spizzirri, U.G.; Vittorio, O.; Valli, E.; Farfalla, A.; Leggio, A.; Nicoletta, F.P.; Iemma, F. Chitosan-Quercetin Bioconjugate as Multi-Functional Component of Antioxidants and Dual-Responsive Hydrogel Networks. *Macromol. Mater. Eng.* **2019**, *304*, 1800728. [CrossRef]
7. Protte, K.; Weiss, J.; Hinrichs, J.; Knaapila, A. Thermally stabilised whey protein-pectin complexes modulate thethermodynamic incompatibility in hydrocolloid matrixes: A feasibility-study on sensory and rheological characteristics in dairy desserts. *LWT-Food Sci. Technol.* **2019**, *105*, 336–343. [CrossRef]
8. Lim, H.S.; Narsimhan, G. Pasting and rheological behavior of soy protein-based pudding. *LWT-Food Sci. Technol.* **2006**, *39*, 344–350. [CrossRef]
9. Alamprese, C.; Mariotti, M. Effects of different milk substitutes on pasting, rheological and textural properties of puddings. *LWT-Food Sci. Technol.* **2011**, *44*, 2019–2025. [CrossRef]
10. Ares, G.; Baixauli, R.; Sanz, T.; Varela, P. New functional fibre in milk puddings: Effect on sensory properties and consumers' acceptability. *LWT-Food Sci. Technol.* **2009**, *42*, 710–716. [CrossRef]
11. Wu, Y.; Ding, W.; He, Q. The gelation properties of tara gum blended with κ-carrageenan or xanthan. *Food Hydrocoll.* **2018**, *77*, 764–771. [CrossRef]
12. Huamaní-Meléndez, V.J.; Mauro, M.A.; Darros-Barbosa, R. Physicochemical and rheological properties of aqueous Tara gum solutions. *Food Hydrocoll.* **2021**, *111*, 106195. [CrossRef]
13. Menegon Rosário, F.; Biduski, B.; Fernando dos Santos, D.; Hadlish, E.V.; Tormen, L.; Fidelis dos Santos, G.H.; Zanella Pinto, V. Red araçá pulp microencapsulation by hydrolyzed pinhão starch, and tara and arabic gums. *J. Sci. Food Agric.* **2021**, *101*, 2052–2062. [CrossRef] [PubMed]
14. Ma, Q.; Ren, Y.; Wang, L. Investigation of antioxidant activity and release kinetics of curcumin from tara gum/polyvinyl alcohol active film. *Food Hydrocoll.* **2017**, *70*, 286–292. [CrossRef]
15. D'Antuono, I.; Kontogianni, V.G.; Kotsiou, K.; Linsalata, V.; Logrieco, A.F.; Tasioula-Margari, M.; Cardinali, A. Polyphenolic characterization of olive mill wastewaters, coming from Italian and Greek olive cultivars, after membrane technology. *Food Res. Int.* **2014**, *65*, 301–310. [CrossRef]
16. Leouifoudi, I.; Harnafi, H.; Zyad, A. Olive mill waste extracts: Polyphenols content, antioxidant, and antimicrobial activities. *Adv. Pharmacol. Sci.* **2015**, *2015*, 714138. [CrossRef]
17. De Santis, S.; Liso, M.; Verna, G.; Curci, F.; Milani, G.; Faienza, M.F.; Franchini, C.; Moschetta, A.; Chieppa, M.; Clodoveo, M.L.; et al. Extra Virgin Olive Oil Extracts Modulate the Inflammatory Ability of Murine Dendritic Cells Based on Their Polyphenols Pattern: Correlation between Chemical Composition and Biological Function. *Antioxidants* **2021**, *10*, 1016. [CrossRef] [PubMed]
18. Carullo, G.; Scarpelli, F.; Belsito, E.L.; Caputo, P.; Rossi Oliverio, C.; Mincione, A.; Leggio, A.; Crispini, A.; Restuccia, D.; Spizzirri, U.G.; et al. Formulation of New Baking (+)-Catechin Based Leavening Agents: Effects on Rheology, Sensory and Antioxidant Features during Muffin Preparation. *Foods* **2020**, *9*, 1569. [CrossRef]
19. Gawlik-Dziki, U.; Dziki, D.; Baraniak, B.; Lin, R. The Effect of Simulated Digestion In Vitro on Bioactivity of Wheat Bread with Tartary Buckwheat Flavones Addition. *LWT-Food Sci. Technol.* **2009**, *42*, 137–143. [CrossRef]
20. Ardestani, A.; Yazdanparast, R. Antioxidant and free radical scavenging potential of Achillea santolina extracts. *Food Chem.* **2007**, *104*, 21–29. [CrossRef]

21. Rababah, T.M.; Al-Omoush, M.; Brewer, S.; Alhamad, M.; Yang, W.; Alrababah, M.; Al-Ghzawi, A.A.-M.; Al-U'datt, M.; Ereifej, K.; Alsheyab, F.; et al. Total phenol, antioxidant activity, flavonoids, anthocyanins and color of honey as affected by floral origin found in the arid and semiarid mediterranean areas. *J. Food Process. Preserv.* **2014**, *38*, 1119–1128. [CrossRef]
22. Giusti, M.M.; Wrolstad, R.E. Anthocyanins: Characterization and measurement with W-visible spectroscopy. In *Current Protocols in Food Analytical Chemistry*; Wrolstad, R.E., Ed.; Wiley: New York, NY, USA, 2001.
23. Spizzirri, U.G.; Carullo, G.; Aiello, F.; Paolino, D.; Restuccia, D. Valorisation of olive oil pomace extracts for a functional pear beverage formulation. *Int. J. Food Sci. Technol.* **2021**, *56*, 5497–5505. [CrossRef]
24. Spizzirri, U.G.; Carullo, G.; De Cicco, L.; Crispini, A.; Scarpelli, F.; Restuccia, D.; Aiello, F. Synthesis and characterization of a (+)-catechin and L-(+)-ascorbic acid cocrystal as a new functional ingredient for tea drinks. *Heliyon* **2019**, *5*, e02291. [CrossRef] [PubMed]
25. Kellett, M.E.; Greenspan, P.; Pegg, R.B. Modification of the cellular antioxidant activity (CAA) assay to study phenolic antioxidants in a Caco-2 cell line. *Food Chem.* **2018**, *244*, 359–363. [CrossRef] [PubMed]
26. Restuccia, D.; Giorgi, G.; Spizzirri, U.G.; Sciubba, F.; Capuani, G.; Rago, V.; Carullo, G.; Aiello, F. Autochthonous white grape pomaces as bioactive source for functional jams. *Int. J. Food Sci. Technol.* **2019**, *54*, 1313–1320. [CrossRef]
27. Spizzirri, U.G.; Altimari, I.; Puoci, F.; Parisi, O.I.; Iemma, F.; Picci, N. Innovative antioxidant thermo-responsive hydrogels by radical grafting of catechin on inulin chain. *Carbohydr. Polym.* **2011**, *84*, 517–523. [CrossRef]
28. Stokes, W.S.; Casati, S.; Strickland, J.; Paris, M. Neutral Red Uptake Cytotoxicity Tests for Estimating Starting Doses for Acute Oral Toxicity Tests. *Curr. Protoc. Toxicol.* **2008**, *36*, 20.4.1–20.4.20. [CrossRef]
29. Mosmann, T. Rapid colorimetric assay for cellular growth and survival: Application to proliferation and cytotoxicity assays. *J. Immunol. Methods* **1983**, *65*, 55–63. [CrossRef]
30. Rizza, L.; Frasca, G.; Nicholls, M.; Puglia, C.; Cardile, V. Caco-2 cell line as a model to evaluate mucoprotective proprieties. *Int. J. Pharm.* **2012**, *422*, 318–322. [CrossRef]
31. Sun, Y.; Hayakawa, S.; Ogawa, M.; Izumori, K. Antioxidant properties of custard pudding dessert containing rare hexose, D-psicose. *Food Control* **2007**, *18*, 220–227. [CrossRef]
32. Paraskeva, P.; Diamadopoulos, E. Technologies for olive mill wastewater (OMW) treatment: A review. *J. Chem. Technol. Biotechnol.* **2006**, *81*, 1475–1485. [CrossRef]
33. He, J.; Alister-Briggs, M.; Lyster, T.D.; Jones, G.P. Stability and antioxidant potential of purified olive mill wastewater extracts. *Food Chem.* **2012**, *131*, 1312–1321. [CrossRef]
34. Rahmanian, N.; Jafari, S.M.; Galanakis, C.M. Recovery and Removal of Phenolic Compounds from Olive Mill Wastewater. *J. Am. Oil Chem. Soc.* **2013**, *91*, 1–18. [CrossRef]
35. Cardinali, A.; Pati, S.; Minervini, F.; D'Antuono, I.; Linsalata, V.; Lattanzio, V. Verbascoside, isoverbascoside, and their derivatives recovered from olive mill wastewater as possible food antioxidants. *J. Agric. Food Chem.* **2012**, *60*, 1822–1829. [CrossRef]
36. Obied, H.K.; Bedgood, D.R., Jr.; Prenzler, P.D.; Robards, K. Chemical screening of olive biophenol extracts by hyphenated liquid chromatography. *Anal. Chim. Acta* **2007**, *603*, 176–189. [CrossRef]
37. Kiritsakis, K.; Rodríguez-Pérez, C.; Gerasopoulos, D.; Segura- Carretero, A. Olive oil enrichment in phenolic compounds during malaxation in the presence of olive leaves or olive mill wastewater extracts. *Eur. J. Lip. Sci. Technol.* **2017**, *119*, 1600425. [CrossRef]
38. Mattonai, M.; Vinci, A.; Degano, I.; Ribechini, E.; Franceschi, M.; Modugno, F. Olive mill wastewaters: Quantitation of the phenolic content and profiling of elenolic acid derivatives using HPLC-DAD and HPLC/MS^2 with an embedded polar group stationary phase. *Nat. Prod. Res.* **2019**, *33*, 3171–3175. [CrossRef]
39. Ghanbari, R.; Anwar, F.; Alkharfy, K.M.; Gilani, A.H.; Saari, N. Valuable Nutrients and Functional Bioactives in Different Parts of Olive (*Olea europaea* L.). *Int. J. Mol. Sci.* **2012**, *13*, 3291–3340. [CrossRef] [PubMed]
40. Andrés-Lacueva, C.; Medina-Remon, A.; Llorach, H.; Urpi-Sarda, R.; Khan, M.; Chiva-Blanch, N.; Zamora-Ros, G.; Rotches-Ribalta, R.; Lamuela-Raventòs, R.M. Phenolic Compounds: Chemistry and Occurrence in Fruits and Vegetables. In *Fruit and Vegetable Phytochemicals: Chemistry, Nutritional Value and Stability*; De la Rosa, L., Alvarez-Parrilla, E., Gonzalez-Aguilar, G.A., Eds.; John Wiley and Sons: London, UK, 2010; pp. 53–88.
41. Manach, C.; Scalbert, A.; Morand, C.; Rémésy, C.; Jiménez, L. Polyphenols: Food sources and bioavailability. *Am. J. Clin. Nutr.* **2004**, *79*, 727–747. [CrossRef] [PubMed]
42. De Pascual-Teresa, S.; Moreno, D.A.; Garcìa-Viguera, C. Flavanols and Anthocyanins in Cardiovascular Health: A Review of Current Evidence. *Int. J. Mol. Sci.* **2010**, *11*, 1679–1703. [CrossRef] [PubMed]
43. Prieto, P.; Pineda, M.; Aguilar, M. Spectrophotometric quantitation of antioxidant capacity through the formation of a phosphomolybdenum complex: Specific application to the determination of vitamin E. *Anal. Biochem.* **1999**, *269*, 337–341. [CrossRef]
44. Giovanelli, G.; Brenna, O.V. Evolution of some phenolic components, carotenoids and chlorophylls during ripening of three Italian grape varieties. *Eur. Food Res. Technol.* **2007**, *225*, 145–150. [CrossRef]
45. Floegel, A.; Kim, D.O.; Chung, S.J.; Koo, S.I.; Chun, O.K. Comparison of ABTS/DPPH assays to measure antioxidant capacity in popular antioxidant-rich US foods. *J. Food Compost. Anal.* **2011**, *24*, 1043–1048. [CrossRef]
46. Rutz, J.K.; Zambiazi, R.C.; Borges, C.D.; Krumreich, F.D.; Da Luz, S.R.; Hartwig, N.; Da Rosa, C.G. Microencapsulation of purple Brazilian cherry juice in xanthan, tara gums and xanthan-tara hydrogel matrixes. *Carbohydr. Polym.* **2013**, *98*, 1256–1265. [CrossRef] [PubMed]

47. Kurisawa, M.; Chung, J.E.; Uyama, H.; Kobayashi, S. Enzymatic synthesis and antioxidant properties of poly(rutin). *Biomacromolecules* **2003**, *4*, 1394–1399. [CrossRef]
48. Toti, U.S.; Aminabhavi, T.M. Synthesis and characterization of polyacrylamidegrafted sodium alginate membranes for pervaporation separation of water + isopropanol mixtures. *J. Appl. Polym. Sci.* **2004**, *92*, 2030–2037. [CrossRef]
49. Chouana, T.; Pierre, G.; Vial, C.; Gardarin, C.; Wadouachi, A.; Cailleu, D.; Le Cerf, D.; Boual, Z.; Ould El Hadj, M.D.; Michaud, P.; et al. Structural characterization and rheological properties of a galactomannan from Astragalus gombo Bunge seeds harvested in Algerian Sahara. *Carbohydr. Polym.* **2017**, *175*, 387–394. [CrossRef]
50. Liu, Y.; Nair, M.G. An efficient and economical MTT assay for determining the antioxidant activity of plant natural product extracts and pure compounds. *J. Nat. Prod.* **2010**, *73*, 1193–1195. [CrossRef]
51. Salta, J.; Martins, A.; Santos, R.G.; Neng, N.R.; Nogueira, J.M.F.; Justino, J.; Rauter, A.P. Phenolic composition and antioxidant activity of Rocha pear and other pear cultivars—A comparative study. *J. Funct. Foods* **2010**, *2*, 153–157. [CrossRef]
52. Reiland, H.; Slavin, J. Systematic Review of Pears and Health. *Nutr. Today* **2015**, *50*, 301–305. [CrossRef]
53. Marinova, D.; Ribarova, F.; Atanassova, M. Total phenolics and total flavonoids in bulgarian fruits and vegetables. *J. Univ. Chem. Technol. Metall.* **2005**, *40*, 255–260.
54. Barnes, H.A.; Hutton, J.F.; Walters, K. *An Introduction to Rheology*; Elsevier Science Publishers: Amsterdam, The Netherlands, 1989; p. 199.

Article

Dairy Products with Certification Marks: The Role of Territoriality and Safety Perception on Intention to Buy

Vincenzo Russo [1,2], Margherita Zito [1,2], Marco Bilucaglia [2], Riccardo Circi [2,*], Mara Bellati [2,3], Laura Emma Milani Marin [2], Elisabetta Catania [2] and Giuseppe Licitra [4]

1 Department of Business, Law, Economics and Consumer Behaviour "Carlo A. Ricciardi", IULM University, 20143 Milan, Italy; vincenzo.russo@iulm.it (V.R.); margherita.zito@iulm.it (M.Z.)

2 Behavior and Brain Lab IULM, Center of Research on Neuromarketing, IULM University, 20143 Milan, Italy; marco.bilucaglia@iulm.it (M.B.); mara.bellati@ibba.cnr.it (M.B.); laura.emma.milani@gmail.com (L.E.M.M.); sissicatania1995@gmail.com (E.C.)

3 CNR National Research Council of Italy, 20133 Milan, Italy

4 Dipartimento Agricoltura Alimentazione e Ambiente (Di3A) Università di Catania, 95123 Catania, Italy; glicitra@unict.it

* Correspondence: riccardo.circi@studenti.iulm.it

Abstract: Over the years, the territorial origins of agri-food products have become a consolidated marketing model which stand as an alternative to mass production. References to territory, whether on packaging or in advertising, have become an increasingly popular way for marketers to differentiate products, by attributing specific characteristics to them, derived from specific cultural identities and traditions. The aim of this study is to capture the possible differences between two groups, Italian and French, in the perception and intention to buy products with certification marks. We tested a multi-group structural equations model, assessing the mediation of the Perceived Product Safety (PPS) between Packaging with reference to Territoriality (PT) and Intention to Buy (IB). Our findings show that in both groups PT has a positive association with IB and PPS and that PPS has a positive association with IB. The difference is the mediation of PPS, present only in the Italian group. This opens important considerations on the role of the perception of safety, particularly in the pandemic period, in the presentation of products, particularly in products with certification marks linked to sustainability and territoriality.

Keywords: dairy products; territoriality; safety perception; intention to buy; multi-group analyses

Citation: Russo, V.; Zito, M.; Bilucaglia, M.; Circi, R.; Bellati, M.; Marin, L.E.M.; Catania, E.; Licitra, G. Dairy Products with Certification Marks: The Role of Territoriality and Safety Perception on Intention to Buy. *Foods* **2021**, *10*, 2352. https://doi.org/10.3390/foods10102352

Academic Editor: Maria Lisa Clodoveo

Received: 24 August 2021
Accepted: 30 September 2021
Published: 2 October 2021

Publisher's Note: MDPI stays neutral with regard to jurisdictional claims in published maps and institutional affiliations.

1. Introduction

In recent years, there has been a progressive increase in consumer awareness as consumers have become more informed and more demanding with regard to the quality of agri-food products. This has led to the emergence of a growing market for products with a strong territorial identification [1]. The data of the XVIII Ismea-Qualivita Report [2] show that the demand for local or traditional foods is increasing, as they are often perceived to be of higher quality [3], more sustainable, [4] and bearers of a strong cultural identity [5].

Over the years, the emphasis on the territorial origins of agri-food products has become a consolidated marketing model, posing as an alternative to mass production [6–8]. References to territory, whether on packaging or in advertising, have become an increasingly popular way for marketers to differentiate products, by attributing specific characteristics to them, derived from specific cultural identities and traditions. Indeed, as stated by Bryła [7], "It is possible to copy all aspects of a food product, but it is impossible to change its history". Thus, the geographical origin of the product becomes an added value that enables small and medium-sized enterprises to compete with large international companies [9].

References to territoriality can be considered as a driver for the purchase of food products [10–12]. The added value of references to territorial origins has led the European

Union to adopt a package of legislation (EC Regulations 2081/92 and 2082/92) which provides protection of food names according to their origins: the Protected Designations of Origin (PDO), the Protected Geographical Indications (PGI), and the Traditional Specialities Guaranteed (TSG) [13]. On the consumers' side, these labels represent a guarantee of quality, since references to territoriality are evocative of concepts that encourage the choice of these products. First, regional products are linked to the concept of tradition, understood as the transmission of knowledge from one generation to the next [14]. Guerrero and colleagues [15] defined a traditional food product as "a product frequently consumed or associated with specific celebrations and/or seasons, normally transmitted from one generation to another, made accurately in a specific way according to the gastronomic heritage, with little or no processing/manipulation, distinguished and known because of its sensory properties, and associated with a certain local area, region, or country". The limited production area and the specificity of the territory contribute to endow the product with special characteristics in the eyes of consumers [9]. Another concept associated with local or traditional foods is that of authenticity [16,17], considered as one of the main drivers in consumers' attitudes towards brands and products [18,19]. Moreover, another key aspect determining the appeal of local products is their sustainability, that is, the use of production processes that are able to respect the environment and to provide forms of support and jobs for local communities [20,21]. Finally, another function of controlled indications of origin is to reduce the information asymmetry between producers and consumers, so that the latter can always be aware of fundamental aspects regarding the origin and the production of food [22]. For the reasons listed above, brands with certification marks are perceived by the consumer as natural, authentic, safe, and controlled. Van Dijk and colleagues [23] confirmed in their research the fundamental importance of the certification marks as a symbol of protection connected to the territory.

The role of packaging is essential to communicate the product visually and its connection with the territory. Indeed, packaging is the first visual element that puts consumers in contact with the product, a pre-requisite for information processing [23]. It is, therefore, important to investigate the importance of packaging in communicating aspects such as tradition and territoriality [10]. Indeed, packaging could play an important role in the challenge to communicate the abstract concept of "traditional" to new targets, such as young generations.

Until now, research has mostly focused on the perceptual characteristics that packaging must have to convey quality and safety. Specifically, the study by Simmonds et al. [24] suggested that transparent packaging "increased willingness to purchase, expected freshness, and expected quality, as compared to packaging that used food imagery instead. In addition, people expected the products to be tastier, to be more innovative, and were more liked overall in several of the product categories that were assessed." Chandran et al. [25] found that transparent packaging increased the product trust. However, as specified by Simmonds et al. [24], to have a positive effect on purchase intentions, the product contained in transparent packaging must be visually attractive in order to avoid the opposite effect.

Few empirical studies have instead focused on the role that references to territoriality on packaging have on the willingness to buy and on the psychological mechanisms underlying this choice [10]. Since it is recognized that consumers are willing to pay more for better quality and healthier products [26–29], in this study we investigate the role played by references to territoriality on packaging of dairy products with a certification mark by hypothesizing that:

Hypothesis 1a (H1a). Reference to territoriality on packaging has a direct and positive association with intention to buy.

Among the drivers for people to buy territorial products is the sense of safety associated with food, namely the access to healthy food with no risks to human health and no contaminants [30]. According to Espejel and colleagues [31], consumers infer from PDO labels a safety badge, due to the strict controls to which products under the protection are submitted by the regulatory councils. So, we hypothesize:

Hypothesis 1b (H1b). A positive and direct association of packaging with reference to territoriality with food safety perception.

Moreover, according to Grunert [32], the perception of safety plays a mediating role between the demand and supply of agri-food products. Therefore, according to [33] (but see also [23,31]), we hypothesize:

Hypothesis 2 (H2). A positive and direct association of food safety perception with intention to buy and that:

Hypothesis 3 (H3). Perceived product safety has a mediating role between the sense of territoriality evoked by packaging and the intention to buy.

For our research question we have chosen to focus on the specific category of products from the dairy sector. We chose to focus on dairy products for two reasons. First, dairy products are characterized by a higher contribution to climate change with respect to vegetable foodstuff production, and the relationship between perceived sustainability of dairy products and willingness to pay has already been investigated in the literature [11]. Second, in the European Union, 231 cheeses have a designation of origin [34]. In particular, it is the southern European countries (e.g., Italy, France, Spain) that have the largest number of products that are candidates for registration as PDO or PGI, and these countries are comparable to each other in their familiarity with controlled origin products [14,35]. This makes dairy products suitable for conducting a multi-group survey involving subjects from different European countries, with the aim of testing a first pilot general model or identifying any differences worth further investigation. Herein, we carried out a multi-group study, basing our survey on a sample of Italian subjects and one of French subjects.

The following, Figure 1, shows the theoretical model and the expected relationships through the hypotheses.

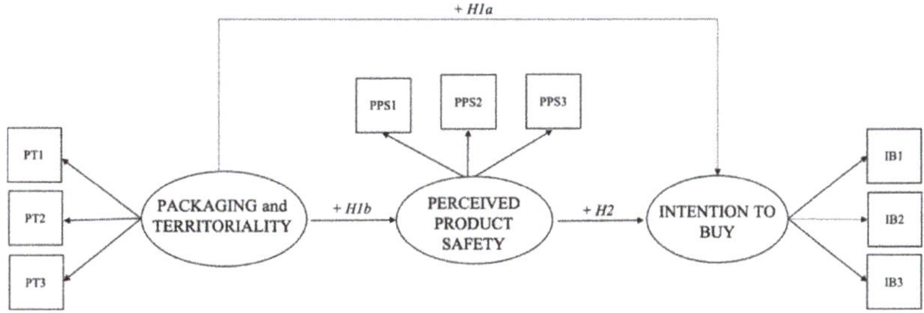

Figure 1. The hypothesized theoretical model.

2. Materials and Methods

2.1. Participants and Procedure

The study involved cheese consumers from Italy (IT; N = 400) and from France (FR; N = 200). These two countries were selected on account of the production of dairy products with certification marks. Participants completed a questionnaire placed on an online platform (Google Moduli) to which researchers added a note with instructions to fill in the questionnaire and a note to ensure anonymity. The questionnaire was administered between May and June 2020. Before the questionnaire, participants were asked to answer to some screening questions related to the consumption of cheese (that is, if they bought and ate cheese), the frequency of consumption and purchase, and whether they had food allergies, in particular, cheese allergies. Participants not satisfying the criteria of real consumption of cheese (buying, eating, and the possibility of consuming cheese without allergies) were not considered in the study. Moreover, all participants provided their informed consent in a specific box before filling in the questionnaire.

The IT sample included 60% females and 40% males, with an average age of 45 years (SD = 10.64). Among them, 11% lived alone, 72.8% were married, and 16.3% lived with their family of origin, and they had an average number of children of 0.780 (SD = 0.932). Their overall cheese consumption (coded as: 2 = rarely, 3 = sometimes, 4 = often) was 3.655 (SD = 0.563).

The FR sample included 60% females and 40% males, with an average age of 46 years (SD = 12.39). Among them, 19% lived alone, 78% were married, and 4% lived with their family of origin, and they had an average number of children of 0.915 (SD = 1.069). Their overall cheese consumption was 3.710 (SD = 0.536).

The IT and FR samples did not significantly differ in mean age, as shown by the two-sample t-test (χ^2 (598) = 0.918, p = 0.359), nor in number of children (W = 42,020.5, p = 0.274) or average cheese consumption (W = 4197.0, p = 0.211), as shown by the Mann–Whitney U tests.

As this study used convenience samples, we compared the demographic data of the two samples. According to ISTAT data, in Italy the updated distribution of females and males in the considered range of age is, respectively, 50.4% and 49.6%, whereas in France it is 51.6% and 48.4%. Even if these data are more gender distributed than the samples of the study (they are more equally distributed than the data of the study, which have a slightly higher percentage of females), they reflect a convergence between the two compared populations, making them comparable (that is, balanced distribution of female and male with slight predominance of females). This balance was also reflected in the samples of the present study, respecting the proportion of the general population and balancing, therefore, the contribution of type/gender in the study as well.

2.2. Measures

On the basis of the literature above mentioned [26–29,31,32], the study detected three main measures to assess consumers' attitudes towards the dairy products with certification marks (see the complete list of items in Table 1).

Table 1. List of measures and items.

Measure	Items
Packaging and Territoriality	1. The packaging must reflect the tradition of production 2. The packaging must recall the territory 3. The packaging must clarify the place of production
Perceived Product Safety	1. It is important that the products with the certification mark I choose are healthy 2. It is important that the products with the certification mark I choose are nutritious 3. It is important that the products with the certification mark I choose are without additives
Intention to Buy	1. Thinking about the certification mark, I am willing to pay more 2. Thinking about the certification mark, if it weren't present in one shop, I would look for it in another 3. Thinking about the certification mark, I would recommend purchasing it

The first measure is *Packaging and Territoriality* (PT). Starting from the considerations on the role of packaging and the relationship with territoriality [26–29], we formulated three items as follows: "The packaging must reflect the tradition of production"; "The packaging must recall the territory"; and "The packaging must clarify the place of production".

Participants had to indicate their agreement or disagreement with each item using a 7-point Likert scale ranging from 1 (strongly disagree) to 7 (strongly agree). This measure obtained a very satisfactory reliability, with a Cronbach's alpha coefficient (α) of 0.82 (IT = 0.86; FR = 0.80).

The second measure is *Perceived Product Safety* (PPS). This measure focuses on the perceptions of consumers of how healthy and nutritious a product is. Three items were, therefore, formulated on the basis of the literature considering the need for safe products, particularly when considering dairy products with certification marks and the considerations of aware consumers [31,32]. The items were as follows: "It is important that the products with the certification mark I choose are healthy"; "It is important that the products with the certification mark I choose are nutritious"; and "It is important that the products with the certification mark I choose are without additives". Participants had to indicate their agreement or disagreement with each item using a 7-point Likert scale ranging from 1 (strongly disagree) to 7 (strongly agree). This measure obtained a satisfactory reliability, with a Cronbach's alpha coefficient (α) of 0.72 (IT = 0.70; FR = 0.75).

The third measure is *Intention to Buy* (IB), understood as the positive attitude towards buying dairy products with certification marks. This measure is based on the consideration of the quality that a product should have and the possibility for consumers to place their trust in that product [23,31,32]. Also in this case, three items were formulated: "Thinking about the certification mark, I am willing to pay more"; "Thinking about the certification mark, if it weren't present in one shop, I would look for it in another"; and "Thinking about the certification mark, I would recommend purchasing it". Participants had to indicate their agreement or disagreement with each item using a 7-point Likert scale ranging from 1 (strongly disagree) to 7 (strongly agree). This measure obtained a satisfactory reliability, with a Cronbach's alpha coefficient (α) of 0.78 (IT = 0.80; FR = 0.72).

2.3. Data Analyses

Data analyses were performed through SPSS 27 for descriptive statistics, correlations (Pearson's r), and alpha reliabilities (α) for each scale. These analyses were performed on each group considered in the study. The multi-group structural equations model (SEM) was estimated with MPLUS 8 in order to simultaneously test in both groups, Italian and French, the relationship between the detected variables and the possible presence of a mediation by the perception of product security between the packaging with territoriality links and intentions to buy products with certification marks. It has to be specified that the hypotheses were specified a priori and a partial mediation model was performed [33]. The goodness of fit of the model was evaluated by the chi-square value (χ^2), the comparative fit index (CFI), the Tucker–Lewis index (TLI), the root mean square error of approximation (RMSEA), and the standardized root mean square residual (SRMR).

To assess possible effects of common method bias, Harman's single-factor test was performed [36] through a confirmatory factor analysis. Results obtained with MPLUS 8 showed the following fit indices: $\chi^2(28) = 709.499$, $p < 0.001$, CFI = 0.67, TLI = 0.58, RMSEA = 0.20, SRMR = 0.17. These indices show that the model could not be identified, thus indicating that one single factor did not account for the variance in the data and suggesting that common method bias was unlikely.

3. Results

From a psychometric standpoint, all the assessed variables in the study showed satisfactory Cronbach's alpha values, ranging between 0.70 and 0.90, meeting the criterion of 0.70 [37].

Correlations are shown, for each group, in Table 2 (IT) and in Table 3 (FR), with descriptive statistics of the detected measures. Both samples show high levels of PT and PPS, with means over the central point of the scale (higher levels for the IT sample), whereas the IB variable shows lower levels (with higher levels for the FR sample). As for correlations, the two samples show similar trends in correlation values, in particular for the correlation between PT and PPS ($_{r\text{IT}}$ = 0.63; $_{r\text{FR}}$ = 0.62), with higher values in the FR sample as for the correlation between PT and IB ($_{r\text{IT}}$ = 0.45; $_{r\text{FR}}$ = 0.56) and between PPS and IB ($_{r\text{IT}}$ = 0.42; $_{r\text{FR}}$ = 0.47).

Table 2. Means, Standard Deviations, and Correlations (Pearson's r) of the IT group.

	M	SD	1	2	3
1. PT	5.90	1.28	(0.86)		
2. PPS	5.63	1.08	0.63 **	(0.70)	
3. IB	3.65	0.97	0.45 **	0.42 **	(0.80)

Note: ** $p < 0.01$. Cronbach's alphas are on the diagonal (in brackets).

Table 3. Means, Standard Deviations, and Correlations (Pearson's r) of the FR group.

	M	SD	1	2	3
1. PT	5.44	1.31	(0.80)		
2. PPS	5.43	1.05	0.62 **	(0.75)	
3. IB	3.74	0.83	0.56 **	0.47 **	(0.72)

Note: ** $p < 0.01$. Cronbach's alphas are on the diagonal (in brackets).

An analysis of variance between the IT and FR samples showed two main significant differences. The IT sample perceived higher levels of PPS ($t(-2.234) = 410.453$ $p < 0.05$), whereas the FR sample showed higher levels in IB ($t(2.303) = 598$ $p < 0.05$).

The estimated multi-group SEM (Figure 2) showed satisfactory fit indices, which confirmed the goodness of the model fit: $\chi^2(52)$ = 132.991 (contribution $\chi^2{}_{IT}$ = 86.119; $\chi^2{}_{FR}$ = 46.872), $p < 0.00$, CFI = 0.96, TLI = 0.95, RMSEA = 0.07 (95% C.I.: 05; 07); SRMR = 0.05. Moreover, the multi-group SEM showed significant and good item loadings ($p < 0.001$) in each group, suggesting a good structure of the latent variables created with these groups of items. In this model, PT showed a direct, positive, and strong association with PPS ($\beta_{\text{IT}} = 0.76$; $\beta_{\text{FR}} = 0.77$) and IB ($\beta_{\text{IT}} = 0.30$; $\beta_{\text{FR}} = 0.52$), particularly in the FR group, confirming H1a and H1b. Moreover, PPS showed a positive and significant association with IB ($\beta_{\text{IT}} = 0.33$; $\beta_{\text{FR}} = 0.34$), in agreement with H2.

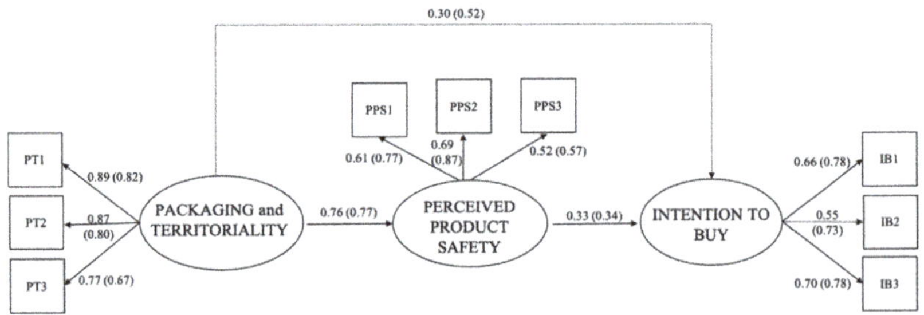

Figure 2. Results of the multi-group structural equations model. Note: Group 1 = IT—outside brackets; Group 2 = (FR)—in brackets.

Moreover, the model detected the mediating role of PPS between PT and IB. As shown in Table 4, a significant mediation was found only in the IT group, whereas in the FR group this mediation was not significant ($\beta_{\text{IT}} = 0.25$; $\beta_{\text{FR}} = n.s.$), partially confirming H3.

Table 4. Indirect effects of the estimated multi-group SEM.

Indirect Effects	Standardized Indirect Effects		
	Est.	s.e.	*p*
PT→PPS→IB	0.25 (*n.s*)	0.06 (0.14)	0.00 (0.08)

Note: Group 1 = IT—outside brackets; Group 2 = (FR)—in brackets.

4. Discussion and Conclusions

In the present study we conducted a multi-group survey with the aim of investigating the relationship between the constructs of PT, PPS, and IB. Based on previous literature, we hypothesized that the perception of rootedness in the territory, evoked by packaging, directly and positively influenced the perception of product safety and the intention to purchase dairy products (H1a and H1b). As underlined, packaging has a key role in communicating the product and in capturing the interest and the attention of consumers [10]. This is linked with the fact that the packaging is a visual stimulus, creating a first contact with the product, a pre-requisite for the processing of information [23]. As for the visual stimuli and the information processing, studies highlight that the visual element is crucial, since the majority of information processed in the brain is mainly visual [38,39]. Moreover, visual elements are considered more significant and reliable, and the use of visual stimuli can help the process of meaning building. From a practical implication standpoint, it would be useful to suggest to producers the use of immediate images and, in this case, the use of clear images and elements linked to territoriality, in order to evoke the interest and traditionality of a specific territory. As highlighted above, these elements can contribute to endow products with special characteristics in the eyes of consumers, such as authenticity [16,17] and both economic and social sustainability [20,21]. In addition, regulations on controlled designations of origin reduce the information gap between producer and consumer [22]. All these elements increase the sense of protection associated with products being rooted in a specific territory. In light of these considerations and of the obtained results, we can argue that the presence of territoriality can enhance the interest of the consumer who evaluates the product as safe, controlled, and traditional, becoming a driver for the decision to buy that product showing a reference to a specific territory [10–12]. We also hypothesized that the perception of safety directly and positively influenced purchase intention (H2). By means of a recursive SEM model, we confirmed these research hypotheses both on a sample of Italian subjects and on a sample of French subjects. This would confirm the consumers' perception of safety as a driver to buy. This is particularly important if territorial products are considered, since safe products are associated with healthy food with no risks to human health and without elements that can contaminate products [30]. In this sense, there emerges the role of communicating the element of safety also through territoriality, so strongly associated with intention to buy, both in the literature and in this study. This also opens an interesting element of deepening the role of culture, considering the mediation results found in this study.

In fact, we also hypothesized that PPS played a mediating role between PT and IB (H3). This research hypothesis was confirmed only in the Italian sample. There are several possible explanations for this result. Firstly, Italy was one of the countries hardest hit by the pandemic, which had a higher mortality rate there than the average for other countries [36]. This situation may have resulted in the Italian sample being more sensitive to the issue of perceived safety, taking into account the period in which the survey was conducted. This would be in line with the result related to the association between PT and IB. In fact, looking at this impact, in the French sample the impact of PT on IB was greater than in the Italian sample, in which the mediation of PPS was present. This would be consistent in explaining the need for safety information in the Italian sample.

Secondly, an explanation for the difference may be found in culturally different associations with the concept of "traditional food". In fact, Guerrero et al. [35] reported different associations that Italian and French consumers made after hearing the word "traditional". The French connected it with the words "tasty", "family", and "dinner", while Italians connected it with "home-made", "natural", and "old". The associations made by the French sample refer more to a concept of food quality and moments of conviviality in which food is consumed; the associations made by the Italian sample refer more to production processes and to aspects which the literature mentioned in this paper has associated with the concept of safety. Finally, it should be noted that both samples were shown packaging for Italian products. The issue of ethnocentrism [40] may have influenced the different perception of safety and the role this played in determining the propensity to buy. Future studies could investigate this further by showing both samples regional products typical of both nationalities.

Barcaccia and colleagues [41] reported how the pandemic severely affected the market for typical agri-food products in Italy, generating a surplus of supply and a drastic drop in prices. Our results suggest that a strategy to help the sector cope with the severe economic crisis could be to emphasize references to territoriality, since the perception of territoriality is directly linked to the propensity to buy. Such a strategy could create a virtuous circle with beneficial effects for local food companies and the economy as a whole. Indeed, local/regional tourism uses local food or beverages both to enhance the tourism experience and to support the tradition of local food/beverage production [42].

In conclusion, our results show that the perception of territoriality leads to a greater propensity to buy and a greater sense of perceived safety. In the Italian sample, the safety perception was shown to play a mediating role between the perception of territoriality and the propensity to buy.

A first limitation of this study was the use of only two groups. Future studies should investigate the influence of territoriality and of the perception of safety, in relation to health, on other groups as well. This would allow the determination of differences and fluctuations among different cultures and markets in order to give the right positioning to dairy products with certification marks. Moreover, another limitation of this study was related to the use of a cross-sectional design, as this did not allow us to define causal relationships between variables. Future studies would use neuromarketing techniques to capture the gap between the rational side and the experience and the emotional experience of the subject in real time by having reliable results not mediated by cognitive processes [43]. These techniques in particular would be functional in capturing the role of territoriality in conveying important and useful elements by optimizing the communication and the presentation of the products [10]. This would have a double advantage: producers would have more elements to sell their products and to tell their story and territory (with important consequences for the territory to which they belong [10,42]), and consumers would have more information, easy to find, and a better perception of health and safety. Furthermore, as stated above, other studies aiming at investigating cultural differences in the perception of territoriality on packaging should control for the effect of ethnocentrism by administering products of both nationalities to both samples.

Another limitation was the use of a convenience sampling method. This procedure allowed us to collect honest and open answers from participants [44] but not to generalize the data. Future studies should provide specific samples both from cultural and from sociodemographic standpoints. However, from a methodological point of view, the sample sizes were adequate to perform a structural equations model, according to the methodological advice suggesting a minimum sample size of 200 [45,46].

Moreover, further studies are needed to test the generalizability of our findings to other types of traditional food products. Furthermore, our results suggest the need to further investigate the role played by the perception of safety in determining the propensity to buy typical local products. Future research should also investigate the links between the concept of safety associated with food and the other drivers mentioned above (namely, heritage, authenticity, and sustainability).

Author Contributions: Conceptualization, V.R., M.Z., M.B. (Marco Bilucaglia), R.C., M.B. (Mara Bellati), L.E.M.M., E.C. and G.L.; methodology, M.Z. and M.B. (Marco Bilucaglia); formal analysis, M.Z. and M.B. (Marco Bilucaglia); investigation, V.R., R.C., M.B. (Marco Bilucaglia) and L.E.M.M.; writing—original draft preparation, V.R., M.Z, M.B. (Marco Bilucaglia), R.C., M.B. (Mara Bellati), L.E.M.M. and G.L.; writing—review and editing, V.R., M.Z., M.B. (Marco Bilucaglia), R.C., M.B. (Mara Bellati) L.E.M.M., E.C. and G.L.; visualization, V.R. and M.Z.; supervision, V.R., M.Z. and G.L.; project administration, V.R. and G.L.; funding acquisition, V.R. and G.L. All authors have read and agreed to the published version of the manuscript.

Funding: This study was developed within the project "Development of a synergy model aimed to qualify and valorize the Natural Historic Cheese of southern Italy in the Sicilian, Sardinia, Calabria, Basilicata, and Campania regions—Canestrum Casei", funded by "Progetto AGER—Agroalimentare e Ricerca 2" (RIF. 2017-1144).

Data Availability Statement: All data available upon request.

Conflicts of Interest: The authors declare no conflict of interest.

References

1. Loureiro, M.; Mccluskey, J. Assessing Consumers Response to Protected Geographical Indication Labeling. *Agribusiness* **2000**, *16*, 309–320. [CrossRef]
2. Rapporto Ismea-Qualivita on Italian PDO IGP TSG Agri-Food and Wine Production. 2020. Available online: https://www.qualivita.it/attivita/rapporto-ismea-qualivita-2020/ (accessed on 10 July 2021).
3. Wägeli, S.; Janssen, M.; Hamm, U. Organic consumers' preferences and willingness-to-pay for locally produced animal products. *Int. J. Consum. Stud.* **2016**, *40*, 357–367. [CrossRef]
4. Pilone, V.; De Lucia, C.; Del Nobile, M.A.; Contò, F. Policy developments of consumer's acceptance of traditional products innovation: The case of environmental sustainability and shelf life extension of a PGI Italian cheese. *Trends Food Sci. Technol.* **2015**, *41*, 83–94. [CrossRef]
5. Van Loo, E.J.; Grebitus, C.; Roosen, J. Explaining attention and choice for origin labeled cheese by means of consumer ethnocentrism. *Food Qual. Prefer.* **2019**, *78*, 103716. [CrossRef]
6. Domański, T.; Bryła, P. *Marketing Produktów Regionalnych na Europejskim. Rynku Zywności*; Lodz University Press: Lodz, Poland, 2013.
7. Bryła, P. The role of appeals to tradition in origin food marketing. A survey among Polish consumers. *Appetite* **2015**, *91*, 302–310. [CrossRef] [PubMed]
8. Sims, R. Food, place and authenticity: Local food and the sustainable tourism experience. *J. Sustain. Tour.* **2009**, *17*, 321–336. [CrossRef]
9. Sirieix, L.; Fort, F.; Aurier, P. Exploring terroir product meaning for the consumer. *Anthropol. Food* **2005**, *4*, 22–38. [CrossRef]
10. Russo, V.; Milani Marin, L.E.; Fici, A.; Bilucaglia, M.; Circi, R.; Rivetti, F.; Bellati, M.; Zito, M. Strategic Communication and Neuromarketing in the Fisheries Sector: Generating Ideas From the Territory. *Front. Commun.* **2021**, *6*, 659484. [CrossRef]
11. Bell, D.; Valentine, G. *Consuming Geographies: We Are Where We Ea*; Routledge: London, UK, 1997.
12. Cacciolatti, L.A.; Garcia, C.C.; Kalantzakis, M. Traditional food products: The effect of consumers' characteristics, product knowledge, and perceived value on actual purchase. *J. Int. Food Agribus. Market.* **2015**, *27*, 155–176. [CrossRef]
13. Duvaleix-Treguer, S.; Emlinger, C.; Gaigné, C.; Latouche, K. On the competitiveness effects of quality labels: Evidence from French cheese industry. In Proceedings of the 30 International Conférence of Agricultural Economists, International Association of Agricultural Economists (IAAE), Vancouver, BC, Canada, 28 July–2 August; 2018; Volume 21, p. hal-02787541.
14. Pieniak, Z.; Verbeke, W.; Vanhonacker, F.; Guerrero, L.; Hersleth, M. Association between traditional food consumption and motives for food choice in six European countries. *Appetite* **2009**, *53*, 101–108. [CrossRef]
15. Guerrero, L.; Claret, A.; Verbeke, W.; Vanhonaker, F.; Enderli, G.; Sulmont-Rosse, C.; Guerrero, L.; Guàrdia, M.D.; Xicola, J.; Verbeke, W.; et al. Consumer-driven definition of traditional food products and innovation in traditional foods. A qualitative cross- cultural study. *Appetite* **2009**, *52*, 345–354. [CrossRef]
16. Tregear, A.; Arfini, F.; Belletti, G.; Marescotti, A. The impact of territorial product qualification processes on the rural development potential of small-scale food productions. In Proceedings of the 11 World Congress of Rural Sociology, Trondheim, Norway, 25–30 July 2004.
17. Philippidis, G.; Sanjuán, A.I. Territorial food product perceptions in Greece and Spain. *J. Food Prod. Market.* **2006**, *11*, 41–62. [CrossRef]
18. Napoli, J.; Dickinson, S.; Beverland, M.; Farrelly, F. Measuring consumer-based brand authenticity. *J. Bus. Res.* **2014**, *67*, 1090–1098. [CrossRef]
19. Holt, D.B. Why do brands cause trouble? A dialectical theory of consumer culture and branding. *J. Consum. Res.* **2002**, *29*, 70–90. [CrossRef]
20. Allen, P.; Hinrichs, C. Buying into "buy local": Engagements of United States local food initiatives. *Altern. Food Geogr. Represent. Pract.* **2007**, *1*, 255–272.
21. Jia, S. Local Food Campaign in a Globalization Context: A Systematic Review. *Sustainability* **2021**, *13*, 7487. [CrossRef]
22. Bimbo, F.; Roselli, L.; Carlucci, D.; De Gennaro, B. Consumer Misuse of Country-of-Origin Label: Insights from the Italian Extra-Virgin Olive Oil Market. *Nutrients* **2020**, *12*, 2150. [CrossRef]
23. Van Dijk, H.; Houghton, J.; van Kleef, E.; van der Lans, I.; Rowe, G.; Van Ittersum, K.; Meulenberg, M.T.; Van Trijp, H.C.; Candel, M.J. Consumers' appreciation of regional certification labels: A PanEuropean study. *J. Agric. Econ.* **2007**, *58*, 1–23.
24. Simmonds, G.; Woods, A.T.; Spence, C. 'Show me the Goods': Assessing the Effectiveness of Transparent Packaging vs. Product Imagery on Product Evaluation. *Food Qual. Prefer.* **2017**, *63*, 18–27. [CrossRef]

25. Chandran, S.; Batra, R.; Lawrence, B. Is seeing believing? Consumer responses to opacity of product packaging. In *Advances in Consumer Research*; McGill, A.L., Shavitt, S., Eds.; Association for Consumer Research: Duluth, MN, USA, 2009; Volume 36, pp. 970–971.
26. Grunert, K.G. *Innovation in Agri-Food Systems: Product Quality and Consumer Acceptance*; Wageningen Academic Publishers: Wageningen, The Netherlands, 2005.
27. Garcia Martinez, M.; Poole, N. The development of private fresh produce safety standards: Implications for developing Mediterranean exporting countries. *Food Policy* **2004**, *29*, 229–255. [CrossRef]
28. Cayot, N. Sensory quality of traditional foods. *Food Chem.* **2007**, *101*, 154–162. [CrossRef]
29. Contò, F.; Antonazzo, A.P.; Conte, A.; Cafarelli, B. Consumers perception of traditional sustainable food: An exploratory study on pasta made from native ancient durum wheat varieties. *Ital. Rev. Agric. Econ.* **2016**, *71*, 325–337. [CrossRef]
30. Shamsi, K.; Compagnoni, A.; Timpanaro, G.; Cosentino, S.; Guarnaccia, P. A Sustainable Organic Production Model for "Food Sovereignty" in the United Arab Emirates and Sicily-Italy. *Sustainability* **2018**, *10*, 620. [CrossRef]
31. Espejel Blanco, J.E.; Fandos Herrera, C.; Flavián Blanco, C. Consumer satisfaction: A key factor of consumer loyalty and buying intention of a PDO food product. *Br. Food J.* **2008**, *110*, 865. [CrossRef]
32. Grunert, K.G. Food quality and safety: Consumer perception and demand. *Eur. Rev. Agric. Econ.* **2005**, *32*, 369–391. [CrossRef]
33. James, L.R.; Mulaik, S.A.; Brett, J.M. A tale of two methods. *Organ. Res. Methods* **2006**, *9*, 233–244. [CrossRef]
34. Anedda, R.; Melis, R.; Curti, E. Quality Control in Fiore Sardo PDO Cheese: Detection of Heat Treatment Application and Production Chain by MRI Relaxometry and Image Analysis. *Dairy* **2021**, *2*, 270–287. [CrossRef]
35. Guerrero, L.; Claret, A.; Verbeke, W.; Enderl, G.; Zakowska-Biemans, S.; Vanhonacker, F.; Issanchou, S.; Sajdakowska, M.; Signe Granli, B.; Scalvedi, L.; et al. Perception of traditional food products in six European regions using free word association. *Food Qual. Prefer.* **2009**, *21*, 225–233. [CrossRef]
36. Posdakoff, P.M.; MacKenzie, S.B.; Lee, J.Y.; Podsakoff, N.P. Common method biases in behavioural research: A critical review of the literature recommended remedies. *J. Appl. Psychol.* **2003**, *88*, 879–903.
37. Nunnally, J.C.; Bernstein, I.H. *Psychometric Theory*; McGraw-Hill: New York, NY, USA, 1994.
38. Sharma, A.; Bhosle, A.; Chaudhary, B. Consumer perception and attitude towards visual elements in social campaign advertisement. *J. Bus Manag.* **2012**, *3*, 6–17. [CrossRef]
39. Zito, M.; Fici, A.; Bilucaglia, M.; Ambrogetti, F.S.; Russo, V. Assessing the Emotional Response in Social Communication: The Role of Neuromarketing. *Front. Psychol.* **2021**, *12*, 625570. [CrossRef] [PubMed]
40. Bryła, P. Regional ethnocentrism on the food market as a pattern of sustainable consumption. *Sustainability* **2019**, *11*, 6408. [CrossRef]
41. Barcaccia, G.; D'Agostino, V.; Zotti, A.; Cozzi, B. Impact of the SARS-CoV-2 on the Italian Agri-Food Sector: An Analysis of the Quarter of Pandemic Lockdown and Clues for a Socio-Economic and Territorial Restart. *Sustainability* **2020**, *12*, 5651. [CrossRef]
42. Boyne, S.; Hall, D.; Williams, F. Policy, support and promotion for food-related tourism initiatives. *J. Travel Tour. Market.* **2003**, *14*, 131–154. [CrossRef]
43. Poels, K.; DeWitte, S. How to capture the heart? Reviewing 20 years of emotion measurement in advertising. *J. Advert. Res.* **2006**, *46*, 18–37. [CrossRef]
44. Molino, M.; Ingusci, E.; Signore, F.; Manuti, A.; Giancaspro, M.L.; Russo, V.; Zito, M.; Cortese, C.G. Wellbeing costs of technology use during Covid-19 remote working: An investigation using Italian translation of the technostress creators scale. *Sustainability* **2020**, *12*, 5911. [CrossRef]
45. Boomsma, A. Nonconvergence, improper solutions, and starting values in LISREL maximum likelihood estimation. *Psychometrika* **1985**, *50*, 229–242. [CrossRef]
46. Boomsma, A. The robustness of maximum likelihood estimation in structural equation models. In *Structural Modeling by Example: Applications in Educational, Sociological, and Behavioral Research*; Cuttance, P., Ecob, R., Eds.; Cambridge University Press: New York, NY, USA, 1987; pp. 160–188.

MDPI

Article

Application Research: Big Data in Food Industry

Qi Tao [1], Hongwei Ding [1], Huixia Wang [2] and Xiaohui Cui [1,*]

[1] Key Laboratory of Aerospace Information Security and Trusted Computing, Ministry of Education, School of Cyber Science and Engineering, Wuhan University, Wuhan 430072, China; qtao17@whu.edu.cn (Q.T.); hwding@whu.edu.cn (H.D.)
[2] Hubei Provincial Institute for Food Supervision and Test, Wuhan 430223, China; whx097@126.com
* Correspondence: xcui@whu.edu.cn

Abstract: A huge amount of data is being produced in the food industry, but the application of big data—regulatory, food enterprise, and food-related media data—is still in its infancy. Each data source has the potential to develop the food industry, and big data has broad application prospects in areas like social co-governance, exploit of consumption markets, quantitative production, new dishes, take-out services, precise nutrition and health management. However, there are urgent problems in technology, health and sustainable development that need to be solved to enable the application of big data to the food industry.

Keywords: food industry; big data; application prospect; sustainable development

Citation: Tao, Q.; Ding, H.; Wang, H.; Cui, X. Application Research: Big Data in Food Industry. *Foods* **2021**, *10*, 2203. https://doi.org/10.3390/foods10092203

Academic Editor: Maria Lisa Clodoveo

Received: 19 August 2021
Accepted: 11 September 2021
Published: 17 September 2021

Publisher's Note: MDPI stays neutral with regard to jurisdictional claims in published maps and institutional affiliations.

1. Introduction

Consumers are no longer satisfied with having enough to eat; food quality has become a key factor in determining consumer choice [1], and their demands and preferences change with the season, time, weather, mood, and other factors [2]. However, food choice is a luxury not every person enjoys. The Food and Agriculture Organization (FAO) of the United Nations reported that 88% of countries face a serious malnutrition burden and so has issued healthy dietary guidelines that cover a wide range of food and nutrition (http://www.fao.org/nutrition/education/food-dietary-guidelines/regions/countries/united-states-of-america/en/ (accessed on 2 September 2021)). Traditional food science has been unable to satisfy increasing demand for food in a world where "healthy nutrition" has overtaken "well-fed" as the predominant paradigm of consumption (https://baijiahao.baidu.com/s?id=1681293145359468742&wfr=spider&for=pc (accessed on 2 September 2021)) [3]. However, big data offers food science a new means of scientific analysis [4].

The food supply chain is composed of economic stakeholders from primary producers to consumers. It has the characteristics of large volume, many links, wide distribution, diverse types, and scattered data, and it is becoming more complex. Millions of tons of food move around the world every year, so no enterprise can promise that every risk node on the production line is absolutely safe. Any flaw in the supply chain could bring a disaster and huge regulatory difficulties for government departments. However, big data provides a solution to regulatory difficulties [5] by helping enterprises understand consumer demand better and uncover food industry trends through big data analysis. The food industry collects large datasets through real-time monitoring and can improve food safety if analyzed in conjunction with sample data [6]. When industry data is combined with data on consumer dietary behavior, food enterprises can optimize their investment and adjust the direction of research and development in a timely manner. [7].

This paper uses bibliometrics to analyze the research progress of big data in the food field. According to Bradford's Law, a small number of core journals collect enough information to reflect the latest and most important advances in science and technology. The database of Web of Science Core Collection contains more than 12,000 core journals from more than 250 subject areas. It defined the search topic "Food & Big Data" and selected 1672 papers from its database. The research progress of big data on food is shown in Figure 1 It has increased significantly since 2014 because USD 35.8 billion was invested in global agrifood from 2010 to 2019, and after 2014 the scale of financing grown rapidly (https://agfunder.com/research/agfunder-agrifood-tech-investing-report-2019/ (accessed on 2 September 2021)). This increased capital investment promoted research into big food data, and the rapidly rising trend is from 2010 to 2021 is shown in Figure 2. China's food industry has attracted global attention since 2012 because the country's new government leaders stressed that they would pay more attention to food safety (http://www.xinhuanet.com/politics/2017-01/03/c_1120239001.htm (accessed on 2 September 2021)). As the second largest economy and the world's largest trading country, China has great international influence. (https://www.brookings.edu/research/chinas-influence-on-the-global-middle-class/ (accessed on 2 September 2021)). Big data has been one of the focuses of research since 2013, mainly in food safety, food security and agriculture. Its application to food safety may still be in its infancy, but it is affecting the entire supply chain. The literature contains analyses on the feasibility and need for big data in the food industry [4,6], but there are no in-depth analyses. Therefore, this paper will mainly discuss the following three aspects in depth.

- The major sources of big data in food industry and its challenges.
- The market application trends of big data in the food industry.
- The main challenges to applying big data.

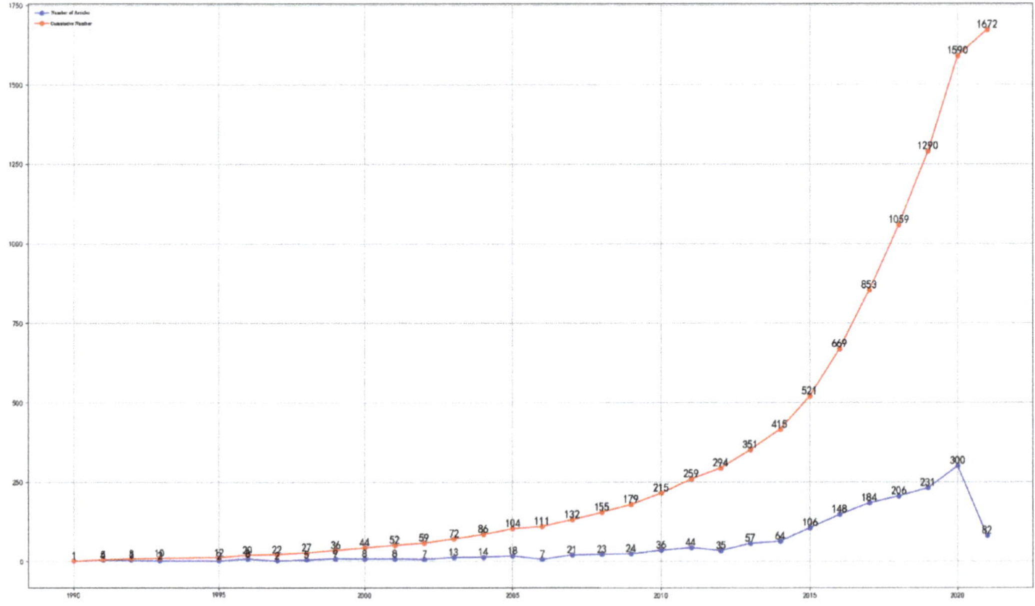

Figure 1. Research progress of data in the food field. From 1990 to 2010, the research papers of big data on food grew at a rate of 100% every five years, and since 2010 it has grown by nearly 300% every five years.

Figure 2. Analysis of the research direction of the data from 2010 to 2021 in the food field. Since 2013, food security and big data have become a focus of researchers interested in the potential value of big data on food. The research focuses on IoT-based data collection and its application to smart farming, supply chain management, food nutrition, and sustainable development.

The authors of this paper hope to help researchers develop a deeper understanding of the research progress of big data in the food field and to provide guidance for further research.

2. Big Data in Food Industry

Big data sources of food mainly include regulatory, food enterprise (including data generated at every link of the industrial chain from planting to restaurants) , and media data (including food-related news, video, pictures and audio). High-quality big data analysis can help develop the food industry, whereas analyses from low-quality data can adversely affect managers' prediction of market demand [8], and social stability [9].

2.1. Food Regulatory Data

Food regulatory data usually includes department regulatory and product sampling data. Marvin [4] has detailed public information about food safety supervision and sampling inspection in various countries. This information includes, reports on animal and plant disease monitoring, hazards, food-borne diseases, which provided support for researchers of deep-risk information. Rapid Alert System of Food and Feed (RASFF) is a commonly used online food safety database for industry and scientific research in the European Union (EU). Food safety databases in other countries include the Import Rejection Report (IRR) and the Inspection Classification Database (ICD) in the U.S. and the State Administration for Market Regulation (SAMR) alerts in China [5]. With the increasingly close connection between countries, the trend of "table globalization" has become increasingly prominent. In 2017, the amount of food China imported from Australia, the United States, Japan, Germany, Southeast Asia, and other countries exceeded RMB 1.5 trillion (https://www.askci.com/news/chanye/20171212/084457113784.shtml (accessed on 2 May 2021)). As shown in Table 1, the SAMR usually shares its sampling inspection results of imported and exported food on government websites, which allows consumers to know the quality of food on the market. The U.S. government shares food sample analysis reports through the FSIS system. The EFSA database contains data on food consumption habits and patterns across the European Union. Such statistical data allows users to quickly screen long-term and acute exposures to potentially hazardous substances in the food chain. In addition, the World Health Organization (WHO) established the Global Environmental Monitoring System (GEMS/Food) in 1976, in which participating institutions submit data on food pollutant concentrations and set up data centers to help governments, the Codex Alimentarius Commission(CAC) and other institutions to assess trends in food contaminants [10]. In 2015, the WHO integrated data from the fields of agriculture, food, public health and economics to build a big data services platform for food safety (https://www.who.int/foodsafety/foscollab/en/ (accessed on 2 May 2021)) to improve risk monitoring.

Table 1. The public regulatory database.

Database	Database Type	Data Description	Country	Organization	Link/Source
Import and export food sampling	Alerts/notifications	Results of food sampling	China	NMPA	https://www.cfsn.cn/
Food sampling report	Alerts/notifications	Food sampling report	USA	FSIS	https://www.fsis.usda.gov/science-data/sampling-program
European Food Consumption Database	Alerts/notifications	European food consumption habits	European	EFSA	https://www.efsa.europa.eu/en/data-report/efsa-food-composition-db
Import food sampling	Alerts/notifications	Results of food sampling	Japan	MHLW	https://www.mhlw.go.jp/content/11135200/000663987.pdf

Challenges. The sharing and circulation of data among food regulatory departments is conducive to the construction of intelligent supervision of the food supply chain [11]. However, there are several challenges.

- **Limited shared data.** Government departments have not been able to fully disclose detailed monitoring and sampling data, leading to the repeated testing of product quality indicators by enterprises, which has resulted in serious waste of social resources and increased operating costs. How to encourage various departments to share data is an important research direction.
- **Lack of system standards.** Due to a lack of system standards, the independence of departments leads to the relative independence of food supervision system. In addition, the limitation of responsibilities further intensifies the independence of departments. So, it is urgent that a new mode of interdepartmental data sharing be explored.
- Due to inconsist food standards, **there are differences in the names and categories of the same food**, which is an obstacle to data sharing. The Estonian government has proposed X-Road architecture of data sharing among the basic sectors [12], and a few European governments will also be involved in international data sharing [13].

2.2. Food Enterprise Data

The food industry chain is composed of enterprises from agriculture, fishing, processing and restaurant and is characterized by many links and wide distribution. At present, all agricultural machinery is electronically controlled to improve operational performance [14,15]. Cloud computing, the Internet of Things, big data and blockchain can integrate isolated production lines in the food supply chain into data-driven interconnected intelligent systems. Through semantic active technology, each operation is automatically integrated, improving the efficiency of precision agriculture and enterprise management [16]. Using sensors and drones to collect data on weather, geography, and animal and crop behavior can help farmers optimize crop planting and animal growth cycles. Intelligent devices capture actionable data and make decisions that reduce equipment downtime [17].

In recent years, research on the IoT in the food industry has promoted the diversification of the IoT platform to address market needs, [18,19] different monitoring models [20], and unbalanced energy consumption [21]. IoT-integrated applications will help food companies create new data sources. Industry 4.0 not only promotes the rapid development of Agriculture 4.0, but also enables enterprises to transmit real-time information to identify and meet the changing demands of stakeholders [15]. According to the Eurostat report (https://www.brookings.edu/research/chinas-influence-on-the-global-middle-class/ (accessed on 2 September 2021)), the application of smart agriculture will save 4–6% of agricultural costs and increase market value by 3% by 2026. The application of big data can not only enable businesses to deal with challenges in food production, but also to obtain more affordable raw materials to reduce production costs [14]. It also promotes the development of smart agriculture, which helps save water [22], preserve soil, limit carbon emissions [23] and improveproductivity [24]. Smart agriculture provides an opportunity for farmers, service providers, government and other stakeholders (such as financial institutions, investors, traders) to share their experiences in optimizing the agricultural supply chain with the production sustainability [25].

Challenges. The food industry can benefit from big data services, but there are challenges that need to be addressed, including data fairness such as the searchability, accessibility, interoperability, and reusability of shared data, and a lack of information standards and data processing technology.

- **Lack of Information standards.** (a) The lack of standardized protocols has created incompatibilities among information management systems [26]; (b) The development of the IoT is still in its initial stages. As IoT manufacturers develop independently, the data generated may be difficult to interpret and share. In addition, there are other problems such as IoT security in food safety. Any insecure IoT node in the food supply chain can be a weak link in the entire system.
- **Immature processing of food big data.** Although cloud computing has been used by many organizations, its appplication to big data regarding food safety is still in its infancy. There as also problems such as system scalability, data fairness, data security, and legal issues, which have not been adequately addressed. Blockchain technology is expected to bring a safer and more transparent food supply chain, but it is still immature and difficult to apply. Currently, the blockchain applicationa to food safety is limited to traceability, and issues such as data integrity and data governance still need research.
- **Improved supply-chain decision-making.** There is still a need to help farmers make effective decisions in Agriculture 4.0, maintain effective connections among different complex networks, and identifythe dynamic needs of stakeholders.

2.3. Media Data

Social media has become the main way for users to obtain and share food information [27,28]. According to a statistical report on China's Internet Network Information Center, the number of Internet users in China has reached 854 million, and 88.8% [29]. Social networks have gradually become the mainstream platform for disseminating information, a constant strem of videos, news, and other types of data [30]. Purcell et al. [31] found that two-thirds of Internet users get their news from Facebook and share news through social media. Through research into the generation, and promotion of social events on the Internet, the mode and characteristics of information transmission can be discovered, which provides support for practical application scenarios. In 2009, Google successfully predicted the spread of the H1N1 virus based on query data in its search engine and brought the public valuable time to prevent an outbreak [32]. Combining the real-time advantage of big data with the conventional and available advantages of traditional data will enable effective response to the transmission of public health events such as COVID-19 [33]. Singh et al. discovered supply chain management problems by using Twitter data to improve supply chain management in food industry [34].

The public participates in the discussion of events by different media, and it expresses clear opinions and attitudes in the form of public opinion [35]. The report (https://baijiahao.baidu.com/s?id=1617643364060321280&wfr=spider&for=pc (accessed on 2 September 2021)) shows that food safety and food rumors were first among hot food events in 2018, and the topic has become one of the prime targerts for media rumors, and social media's intensification of rumors can create a widespread crisis [35]. By analyzing and understanding the trend of food incidents based on social media data, regulators can formulate timely countermeasures like enhancing public awareness through science education and shaping public opinion [36–38]. However, the field of media data has its challenges that need to be overcome.

- **Multi-source heterogeneous data fusion.** Media data on Twitter, Facebook, YouTube and various information portals have complex formats and multiple sources, and there is a lack of technology to identify relevant data in one sources and link it to others. In the absence of a "fusion technology" for multi-source heterogeneous data further study is urgently needed to address social media rumors and their negative impact on public security.

- **Rumor detection.** Network rumors seriously impair the public's ability to recognize authentic of network information. The generation, influence, and propagation mechanism of network rumors have been studied [35–38], but there is still no answer to the problem of improving the public's abiity to evaluate network information. The ability to perceive a risk has a positive impact on users' attitudes, but social media undermines it. Authoritative information on refuting rumors, such as government notices and mainstream media news, have a significant effect on reducing the public's acceptance of network rumors and improving the willingness to identify network rumors.
- **Rumor control.** Many studies based on communication science and psychology analyze food rumors by the way they are transmitted, but there is a lack of relevant research on government management. Existing food rumors detection technology is mostly simulation, which rarely considers the responsibilities of government institutions. It will be necessary to consider the role of the government, social media, and human networks to establish an efficient, and authoritative information platform to dispel rumors in time. This will improve public risk and prevention awareness, strengthen and punishment mechanisms for rumormongers, and strengthen legal education and behavioral guidance for the public [35]. The study of social network interaction patterns will be used to explore how to promote and change the public's perception, attitude, and behavior on rumors concerning food, health or other fields.

3. Application of Big Data in Food Industry

The food supply chain is from farm to table, where the main links are planting and breeding, storage, processing, circulation (transportation) and consumption [39]. The discovery of value information on original data needs to go through a continuous cycle: of "discrete data—integrated data—knowledge understanding—mechanism extraction—application effect analysis", from which the potential value of data sets can be mined. The processing system of big data application in food industry is shown in Figure 3. It is composed of five modules: big data collection of food industry, big data processing and fusion, big data mining and analysis, big data view and big data security. Each module is closely connected, and its functions are briefly described as follows.

- **Big data collection of the food industry.** Based on sensors, web crawlers, near-infrared detection instruments, food-related data is collected from different sources.
- **Multi-source data processing and fusion.** The collected data contaiins a considerable anount of redundant and dirty data [40], but through cleaning and conversion they can be removd and the data can be put into a standardized format. Then data features are extracted, and fusion is performed using probability statistics [41], logical reasoning and machine learning [42].
- **Big data mining and analysis.** This is a discovery mode of mining valuable knowledge from massive data that can accuratetly predict activities through scaled data [43]. Some big data technology, including SVM(Support Vector Machine) [44], Random Forest [45] and Naive Bayes [46], are used to analyze fused dataset to discover potential patterns to create social value [47,48].
- **Big data view.** Due to the characteristics of the complexity and multidimensional of data, it is necessary to generate a data view that can be easily expressed and understood by users. These are usually parallel coordinate, scatter graph, and scatter graph matrix methods [49,50].
- **Big data security.** In the lifecycle of big data on food, there are security risks in each processing stage [51], so research into big data security technology has become an

important research topic. This module provides security technical support for all big data processing to ensure the safety, reliability, and controllability of data.

3.1. Social Co-Governance in the Food Industry

Social co-governance in the food industry provides a feasible solution to the issues of food security and food quality by using public wisdom [52]. Social co-governance is usually based on crowdsourcing to cooperate with consumers or experts to create value [53]. The failure rate of new food product development exceeds 40%, and the failure of new products usually affects the continued operation of small and medium-sized enterprises [3]. Large food companies have tried to collect consumer preference data through crowdsourcing, and have decided the direction of product development based on an analysis of big data. The Danone company encouraged consumers to vote for a creamy dessert flavor, and the 400,000 participants in 2006 more than doubled to 900,000 in 2011. Lay's used the wisdom of crowds to develop more than 245,825 flavors of potato chip [54]. Procter & Gamble, Starbucks and Unilever sought better product design based on collective intelligence [55]. Employees often have a wealth of heterogeneous expertise, and companies can gain insight from their workforce to help improve economic performance. In addition, crowdfunding is another form of social co-governance, sharing business risks and alleviating capital pressure through mutual assistance [56]. Social co-governance has great potential for food security. Combining the mobile data of consumer groups with food shelf life, the intelligent control of food inventory can be realized to prevent food spoilage and waste [57]. Social co-governance can also be applied to monitoring foodborne diseases [32], identfying contaminated products, reducing the risk rate of food rumors and enhancing food safety [58].

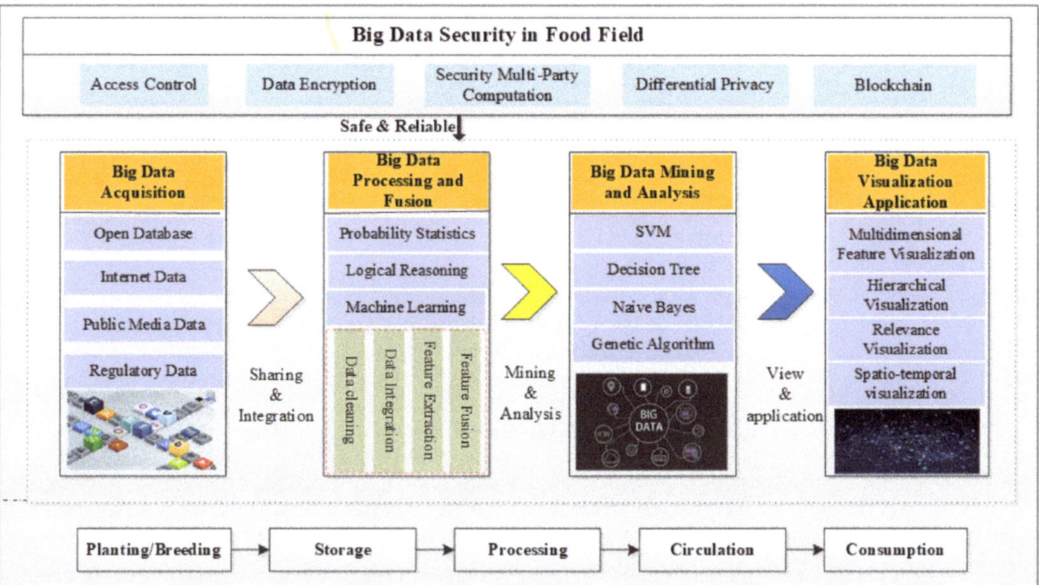

Figure 3. The processing model of big data in food.

Although social co-governance can enable food enterprises to obtain consumer demand information through diversified channels, it is still difficult to obtain effective information in time due to the limitation of enterprise resources [53]. In addition, there is a lack of an incentive or fair evaluation method in the food industry to convince consumers to participate.

3.2. Exploit Consumption Markets

There is a huge amount of food-related data both inside and outside the food supply chain, and the collection and analysis can promote enterprises to expand their markets [59]: (1) By collecting commodity and retail information for analysis, they can appraise the market situation, grasp the business dynamics of their competitors, and define the market positioning of products, thereby grasping market opportunity. (2) Collecting consumer information (purchase lists and channels, commodity preferences, usage cycle, family information, working condition, values) will establish a customer database that can give enterprises portraits of their customers that reveal their preferences, consumption tendency, value orientation and commodity reputation. With this information, enterprises can develop efficient marketing strategies and develop trust, so they can continue to compete effectively. (3) Data clustering analysis of consumers' food evaluations (advantages and disadvantages, quality, nutritional value) from social platforms such as Facebook, Twitter and Sina Weibo, allows enterprises to anticipate potential problems and optimize the quality of goods and services.

3.3. Quantitative Production

By predicting future commercial demand based on historical sales' data, agricultural and livestock products can be planned to reduce the probability of "cheap vegetables hurting farmers". In addition, big data analysis can help predict weather more accurately, helping farmers and herdsmen to prepare for natural disasters. Analyses based on consumption and crop growth data help farmers decide which crops varieties to increase and which to reduce, improve crop yield, facilitate rapid sales,and achieve a return on capital. Big data can also optimize grazing area for local herdsmen and improve the usage rate of pasture. Fishermen can scientifically arrange fishing moratoria and locate fishing areas based on the results of big data analysis.

Consumption trends and habits provided by big data, enable governments to provide accurate guidance for agricultural and animal husbandry production, suggest production levels according to demand, and avoid unnecessary waste of resources and social wealth caused by excess capacity. When combined with drones, big data can promote the development of precision agriculture by allowing farmers to collect information on the growth of crops, diseases, and pests, at a much lower cost and with much higher accuracy than by hired aircraft [60].

3.4. New Dishes New Experience

In 1992, chefs Heston Blumenthal and Francois Benzi were deciding which ingredients with similar flavors would work well together, when someone created a combination of white chocolate and caviar. Due to the chemical differences, it tasted terrible. Today, because of food sciecne, there is a large amount of information about food chemicals [61] and how they make food taste. Consequently, Ahn et al. [62,63] developed a flavor network of ingredients connected by shared flavor compounds, in which flavors were limited by the type of raw materials. Garg et al. developed a flavor database with richer food materials [64]. Simas et al. promoted a flavor network and constructed the food-bridging network [65]. However, some well-known food combinations such as red wine and beef, do not share chemical compositions or flavor compounds, yet they are still very popular. Therefore, food pairings need to be seen in a broader spctrum, not just based on flavor compounds or chemical composition.

The future of food flavor design could be traced to 2019 when the company McCormick partnered with IBM to use Artificial Intelligence and big data to generate new flavor combinations by analyzing data from millions of datasets to meet the changing consumer demand.

3.5. Take-Out Service

In China, the number of online takeout users accounts for more than 44% of sales, and the scale has exceeded 398 million people (https://www.qianzhan.com/analyst/detail/220/200512-65621d53.html (accessed on 2 September 2021)). The take-out markets with its large number of users and rapid growth has generated a huge amount of takeout data. The takeout big data service platform not only helps the government supervise the industry, but also creates huge economic and social value. First, it predicts and informs customers of the delivery time, thereby avoiding disrupting consumers' daily plans and helping restaurants establish a good reputation. Second, it helps the take-out enterprise understand consumer demand. Third, the take-out big data platform promotes the transparency of the supply chain, which is conducive to establishing and improving customers' trust. Fourth, the overall running of the city can be clearly understood by analyzing the take-out dataset [66].

Since take-out data involves sensitive private information (the customer's location, preference, bank, identity, and communication), ensuring data security in the take-oout big data platform is a serious challenge.

3.6. Precise Nutrition and Health Management

The development of big data provides technical support for the processing of massive data, and scientific guidance for human nutrition and health management. In the past, people usually learned nutrition information from experts, books, and the Internet. However, there was a lack of accurate nutrition and health management for individuals because of the difference in individual health conditions [67]. In an example of applying big data to the people's daily diet Teng et al. proposed to use a recipe recommendation algorithm to determine which food ingredients were necessary [68]. Grace et al. combined case-based reasoning and a deep learning algorithm to generate new recipes [69,70]. However, it may also generate "dark cuisine" due to the the uncertain factor of deep learning. Some scholars, like Freyne et al., focused on a diet therapy. They developed a personalized recipe recommendation system for obese people based on the suggestions from medical professionals and research on obese people [71]. In anoher instance, Yoshida et al. proposed a personalized recipe recommendation based on users' food preferences [72]. Zeevi et al. broke with traditional experience-based nutritional recommendations by using machine-learning algorithms to combine data (e.g., blood parameters, dietary habits, and gut microbiota) to formulate personalized diets that optimize postprandial glucose levels and metabolites [73]. The combination of big data with Artificial Intelligence will provide a new approach for the research of precision nutrition.

4. Challenges

While the food supply chain can benefit from big data, the following challenges need to be addressed.

- **Low data collection efficiency and poor data quality.** Due to the lack of effective data collection technology, there are problems with applying big data to food: missing or insufficient data, and difficulties with data forensics. Therefore, it is of great significance to study the collection and verification methods of multisourced big data on food. In the future, edge computing technology will be used to link the food supply chain thereby superseding traditional collection, which is hard to apply to a complex collection environment and collection requirements. Intelligent web crawler technology can be used to collect public opinion data from the Internet, and associate it with data from the physical world to form a complete big data view of food safety. It will solve the dilemma of the separation between physical world and public opinion data found in traditional methods.
- **Data islands.** There is one are in the food industry where data is not shared: government. Since regulatory data usually involve state secrets, and enterprise data may involve trade secrets, they reluctance to share these data seriously hinders the application and promotion of big data. Moreover, the independent management systems of

enterprises and government creates poor system compatibility, which inhibits data circulation. Therefore,a new business model is urgently needed, for example: the establishment of an incentive mechanism to explore new methods of data fusion to establish an effective privacy protection method and encourage data owners to share data.

- **Data security.** Private information, such as consumer activity and social relationships, is hidden in food data. For example, based on the user's location, there are connections to tracking data (such as order and logistical data and a location picture), so consumers can be tracked through association analysis, which may represent a risk to the their security [74]. In the life cycle of big data, there are security threats and privacy leakage risks at each processing stage [51,75,76]. The security protection analysis of big data for food is shown in Table 2. Data encryption is one of the effective methods for protecting data security. Rivest [77] proposed a privacy homomorphic encryption that would delegate part of a complex operations to a third party, directly operating on the ciphertext, who would return the results without disclosing of information. Yao et al. [78] proposed a secure multiparty computing scheme (SMC), in which all participants realize collaborative computing, and the privacy of all parties is protected, which eliminates the need for trusted third parties. However, ciphertexts have a serious impact on data readability. The strongly robust blockchain has anti-network attack, anti-eavesdropping, anti-tampering and anonymity features, that can protect the security and reliability of data [79]. Combining blockchain with an access control algorithm, Uchibeke et al. [80] proposed that data owners independently manage their data and prevent data leakage. However, blockchain technology still needs to be studied for security and work efficiency [81]; thus, research into big data security and privacy protection technology has become an important research area in the food field.

Table 2. Analysis of data security protection technology in the whole life cycle of food data.

Life Cycle	Challenges	Protection Technology
Data collection	Data corruption, data loss, data leakage and data forgery	Data encryption [82]
Data storage	Illegal intrusion and data disclosure	Storage encryption [83], blockchain [79]
Data transmission	Data leakage and data corruption	Data encryption [84], privacy protection [85], blockchain [81]
Data usage	Information leakage and data abuse	Access control [86], SMC [78], data encryption [77], differential privacy protection [87]
Data Destruction	Privacy disclosure, destruct the data media	Data trusted deletion [88]

- **Security risks of crowdsourcing services.** On one hand, food enterprises have a large amount of enterprise data, but it is difficult to discover the potential value of the information. On the other hand, in crowdsourcing, consumers are not enthusiastic enough to participate, and there is a risk that the crowd will get out of control. In the take-out market, unlicensed crowdsourcing deliverers have become the biggest uncontrollable factor in takeout distribution, and it is a serious risk for take-out food safety. (1) Since the equipment of deliverers is self-regulated, it is difficult to control whether delivery safety standards can be achieved. (2) It is difficult to guarantee the hygiene and health of the people involved in the delivery. If a deliverer carries an infectious disease or the food is contaminated on the way, food safety cannot be guaranteed at all. (3) Take-out deliverers increase the management difficulty of the platform. Recently, the platform can only discipline deliverers through a user account ban. But the deliverer only needs to register with a new mobile phone number and borrow a person's ID card for real name authentication, and then he can continue to deliver. Therefore, the establishment of a fair, incentive, and perfect crowdsourcing service mechanism will be one of the important research topics.
- **Healthy and sustainable development of the food industry.** By 2050, the world's population will exceed 9 billion, and only through a healthy and sustainable food in-

dustry can global food demand be met. [89]. However, resource waste and foodborne diseases are key factors restricting this sustainable development: excessive chemical residue in crops from the of chemical fertilizers and pesticides [90]; perishable food loss in developing countries and enormous food waste in developed countries [91]; high energy consumption and pollution from food processing and transportation; and the discarding of potentially contaminated food because of the iinability to quickly and efficiently trace the source of a contamination. Therefore, the economic and environmental sustainability of stakeholders are the core factors for promoting the sustainable development of the food industry chain.

5. Conclusions

Because big data can provide a large amount of effective business information, the development of data-driven industries has attracted attention from all countries [92]. In 2012, the United States promoted big data as a national strategy to promote the formation of new economic growth and enhance national competitiveness [93]. Subsequently, EU member states formulated big data development strategies to transform traditional national governance models [94]. Big data to promote the development of the food industry has become a main research topic. This paper introduced the application of big data in food industry and showed that the main data sources are regulatory, enterprise, and media data. The results showed the great potential for big data for the food industry. Big data has particularly broad application prospects in social co-governance of the food industry, quantitative production, exploitation of consumption markets, new dishes, take-out services, and precise nutrition and health management. But, to exploit this full potential of big data, technical, social, and health and sustainable development issues require further research.

Author Contributions: Q.T.: Conceptualization, Methodology, Software, Validation and Writing—Original Draft & Review & Editing. H.D.: Investigation. H.W.: Validation and Writing—Review. X.C.: Supervision, and Funding acquisition. All authors have read and agreed to the published version of the manuscript.

Funding: The authors would like to acknowledge the support provided by the National Key R&D Program of China (No. 2018YFC1604000).

Institutional Review Board Statement: Not applicable.

Informed Consent Statement: Not applicable.

Data Availability Statement: Data sharing is not applicable for this article.

Conflicts of Interest: The authors have no competing interest to declare.

References

1. Angowski, M.; Jarosz-Angowska, A. Importance of Regional and Traditional EU Quality Schemes in Young Consumer Food Purchasing Decisions. *Eur. Res. Stud. J.* **2020**, *23*, 916–927. [CrossRef]
2. Kearney, J. Food consumption trends and drivers. *Food Consum. Trends Drivers* **2010**, *365*, 2793–2807. [CrossRef]
3. Palacios, M.; Martinez-Corral, A.; Nisar, A.; Grijalvo, M. Crowdsourcing and organizational forms: Emerging trends and research implications. *J. Bus. Res.* **2016**, *69*, 1834–1839. [CrossRef]
4. Marvin, H.J.; Janssen, E.M.; Bouzembrak, Y.; Hendriksen, P.J.; Staats, M. Big data in food safety: An overview. *Crit. Rev. Food Sci. Nutr.* **2017**, *57*, 2286–2295. [CrossRef]
5. Jin, C.; Bouzembrak, Y.; Zhou, J.; Liang, Q.; van den Bulk, L.M.; Gavai, A.; Liu, N.; van den Heuvel, L.J.; Hoenderdaal, W.; Marvin, H.J. Big Data in food safety-A review. *Curr. Opin. Food Sci.* **2020**, *2020*, 24–32. [CrossRef]
6. Wilesmith, J.W.; Ryan, J.; Atkinson, M. Bovine spongiform encephalopathy: Epidemiological studies on the origin. *Vet. Rec.* **1991**, *128*, 199–203. [CrossRef]
7. Srinivasan, K.; Nirmala, R. A Study on Consumer Behavior towards Instant Food Products (With Special References to Kanchipurm Town). *IOSR J. Bus. Manag.* **2014**, *16*, 17–21. [CrossRef]
8. Garcia-Bernardo, J.; Takes, F.W. The effects of data quality on the analysis of corporate board interlock networks. *Inf. Syst.* **2018**, *78*, 164–172. [CrossRef]

9. Hu, Y.; Pan, Q.; Hou, W.; He, M. Rumor spreading model with the different attitudes towards rumors. *Phys. A Stat. Mech. Its Appl.* **2018**, *502*, 331–344. [CrossRef]
10. Moy, G.G.; Vannoort, R.W. Total Diet Studies. 2013. Available online: https://link.springer.com/content/pdf/bfm%3A978-1-4419-7689-5%2F1.pdf (accessed on 2 September 2021).
11. Tao, Q.; Cui, X.; Huang, X.; Leigh, A.M.; Gu, H. Food safety supervision system based on hierarchical multi-domain blockchain network. *IEEE Access* **2019**, *7*, 51817–51826. [CrossRef]
12. Willemson, J. Pseudonymization service for X-road eGovernment data exchange layer. In Proceedings of the International Conference on Electronic Government and the Information Systems Perspective, Toulouse, France, 29 August–2 September 2011; Springer: Berlin/Heidelberg, Germany, 2011; pp. 135–145. [CrossRef]
13. Freudenthal, M.; Willemson, J. Challenges of Federating National Data Access Infrastructures. In Proceedings of the International Conference for Information Technology and Communications, Bucharest, Romania, 8–9 June 2017; Springer: Cham, Switzerland, 2017; pp. 104–114. [CrossRef]
14. Lezoche, M.; Hernandez, J.E.; Díaz, M.d.M.E.A.; Panetto, H.; Kacprzyk, J. Agri-food 4.0: A survey of the supply chains and technologies for the future agriculture. *Comput. Ind.* **2020**, *117*, 1–10. [CrossRef]
15. Lasi, H.; Fettke, P.; Kemper, H.G.; Feld, T.; Hoffmann, M. Industry 4.0. *Bus. Inf. Syst. Eng.* **2014**, *6*, 239–242. [CrossRef]
16. Mondino, P.; González-Andújar, J.L. Evaluation of a decision support system for crop protection in apple orchards. *Comput. Ind.* **2019**, *107*, 99–103. [CrossRef]
17. Spiegel, R. Plant Data Everywhere and Anywhere. *Des. News* **2015**, *70*, 30–33.
18. Lee, J.; Park, G.I.; Shin, J.H.; Lee, J.H.; Sreenan, C.J.; Yoo, S.E. SoEasy: A software framework for easy hardware control programming for diverse IoT platforms. *Sensors* **2018**, *18*, 2162. [CrossRef]
19. Choi, J.; In, Y.; Park, C.; Seok, S.; Seo, H.; Kim, H. Secure IoT framework and 2D architecture for End-To-End security. *J. Supercomput.* **2018**, *74*, 3521–3535. [CrossRef]
20. Prathibha, S.; Hongal, A.; Jyothi, M. IoT based monitoring system in smart agriculture. In Proceedings of the 2017 IEEE International Conference on Recent Advances in Electronics and Communication Technology (ICRAECT), Bangalore, India, 16–17 March 2017; pp. 81–84. [CrossRef]
21. Ilie, M. Internet of things in agriculture. In *International Conference on Competitiveness of Agro-Food and Environmental Economy Proceedings*; The Bucharest University of Economic Studies: Bucharest, Romania, 2018.
22. O'Connor, N.; Mehta, K. Modes of greenhouse water savings. *Procedia Eng.* **2016**, *159*, 259–266. [CrossRef]
23. Ochoa, K.; Carrillo, S.; Gutierrez, L. Energy efficiency procedures for agricultural machinery used in onion cultivation (*Allium fistulosum*) as an alternative to reduce carbon emissions under the clean development mechanism at Aquitania (Colombia). *IOP Conf. Ser. Mater. Sci. Eng.* **2014**, *59*, 012008. [CrossRef]
24. Mayer, J.; Gunst, L.; Mäder, P.; Samson, M.F.; Carcea, M.; Narducci, V.; Thomsen, I.K.; Dubois, D. Productivity, quality and sustainability of winter wheat under long-term conventional and organic management in Switzerland. *Eur. J. Agron.* **2015**, *65*, 27–39. [CrossRef]
25. Annosi, M.C.; Brunetta, F.; Monti, A.; Nati, F. Is the trend your friend? An analysis of technology 4.0 investment decisions in agricultural SMEs. *Comput. Ind.* **2019**, *109*, 59–71. [CrossRef]
26. Bouzembrak, Y.; Klüche, M.; Gavai, A.; Marvin, H.J. Internet of Things in food safety: Literature review and a bibliometric analysis. *Trends Food Sci. Technol.* **2019**, *94*, 54–64. [CrossRef]
27. Aertsen, T.; Gelders, D. Differences between the public and private communication of rumors. A pilot survey in Belgium. *Public Relat. Rev.* **2011**, *37*, 281–291. [CrossRef]
28. Wang, H.; Xu, Z.; Fujita, H.; Liu, S. Towards felicitous decision making: An overview on challenges and trends of Big Data. *Inf. Sci.* **2016**, *367*, 747–765. [CrossRef]
29. CNNIC. The 44th China Statistical Report on Internet Development. 2019. Available online: http://www.cac.gov.cn/pdf/20190829/44.pdf (accessed on 12 May 2021).
30. Soon, J.M. Consumers' Awareness and Trust Toward Food Safety News on Social Media in Malaysia. *J. Food Prot.* **2020**, *83*, 452–459. [CrossRef]
31. Chua, A.Y.; Banerjee, S. To share or not to share: The role of epistemic belief in online health rumors. *Int. J. Med. Inform.* **2017**, *108*, 36–41. [CrossRef]
32. Ginsberg, J.; Mohebbi, M.H.; Patel, R.S.; Brammer, L.; Smolinski, M.S.; Brilliant, L. Detecting influenza epidemics using search engine query data. *Nature* **2009**, *457*, 1012–1014. [CrossRef]
33. Liu, T.; Jin, Y. Human Mobility and Spatio-temporal Dynamics of COVID-19 in China: Comparing Survey Data and Big Data. *Popul. Res.* **2020**, *44*, 44–59.
34. Singh, A.; Shukla, N.; Mishra, N. Social media data analytics to improve supply chain management in food industries. *Transp. Res. Part E Logist. Transp.* **2018**, *114*, 398–415. [CrossRef]
35. Li, D.; Ma, J. How the government's punishment and individual's sensitivity affect the rumor spreading in online social networks. *Phys. A Stat. Mech. Its Appl.* **2017**, *469*, 284–292. [CrossRef]
36. Yuan, J.; Lu, Y.; Cao, X.; Cui, H. Regulating wildlife conservation and food safety to prevent human exposure to novel virus. *Ecosyst. Health Sustain.* **2020**, *6*, 1–12. [CrossRef]

37. Mou, Y.; Lin, C.A. Communicating food safety via the social media: The role of knowledge and emotions on risk perception and prevention. *Sci. Commun.* **2014**, *36*, 593–616. [CrossRef]
38. Rutsaert, P.; Pieniak, Z.; Regan, Á.; McConnon, Á.; Kuttschreuter, M.; Lores, M.; Lozano, N.; Guzzon, A.; Santare, D.; Verbeke, W. Social media as a useful tool in food risk and benefit communication? A strategic orientation approach. *Food Policy* **2014**, *46*, 84–93. [CrossRef]
39. Tao, Q.; Cui, X.; Zhao, S.; Yang, W.; Yu, R. The Food Quality Safety Management System Based on Block Chain Technology and Application in Rice Traceability. *J. Chin. Cereals Oils Assoc.* **2018**, *33*, 102–110.
40. Liu, C. Maximum likelihood estimation from incomplete data via EM-type Algorithms. *Adv. Med. Stat.* **2003**, 1051–1071._0028. [CrossRef]
41. Ling, W.; Dong-Mei, F. Estimation of missing values using a weighted k-nearest neighbors algorithm. In Proceedings of the 2009 IEEE International Conference on Environmental Science and Information Application Technology, Wuhan, China, 4–5 July 2009; Volume 3, pp. 660–663.
42. Li, J.; An, Y.; Fei, R.; Wang, H.; Yan, Q. Activity recognition method based on weighted LDA data fusion. *Intell. Autom. Soft Comput.* **2017**, *23*, 509–517. [CrossRef]
43. Hill, M.; Connolly, P.; Reutemann, P.; Fletcher, D. The use of data mining to assist crop protection decisions on kiwifruit in New Zealand. *Comput. Electron. Agric.* **2014**, *108*, 250–257. [CrossRef]
44. Saunders, C.; Stitson, M.; Weston, J.; Holloway, R.; Bottou, L.; Scholkopf, B.; Smola, A. Support vector machine. *Comput. Sci.* **2002**, *1*, 1–28.
45. Arumugam, A. A predictive modeling approach for improving paddy crop productivity using data mining techniques. *Turk. J. Electr. Eng. Comput. Sci.* **2017**, *25*, 4777–4787. [CrossRef]
46. Tan, K.; Ma, W.; Wu, F.; Du, Q. Random forest–based estimation of heavy metal concentration in agricultural soils with hyperspectral sensor data. *Environ. Monit. Assess.* **2019**, *191*, 1–14. [CrossRef]
47. Li, D.; Wang, X. Dynamic supply chain decisions based on networked sensor data: An application in the chilled food retail chain. *Int. J. Prod. Res.* **2017**, *55*, 5127–5141. [CrossRef]
48. Zhang, Q.; Huang, T.; Zhu, Y.; Qiu, M. A Case Study of Sensor Data Collection and Analysis in Smart City: Provenance in Smart Food Supply Chain. *Int. J. Distrib. Sens. Netw.* **2013**, *9*, 382132. [CrossRef]
49. Inselberg, A. The plane with parallel coordinates. *Vis. Comput.* **1985**, *1*, 69–91. [CrossRef]
50. Viau, C.; McGuffin, M.J.; Chiricota, Y.; Jurisica, I. The FlowVizMenu and parallel scatterplot matrix: Hybrid multidimensional visualizations for network exploration. *IEEE Trans. Vis. Comput. Graph.* **2010**, *16*, 1100–1108. [CrossRef] [PubMed]
51. Chen, H.; Yan, Z. Security and privacy in Big data lifetime: A review. In Proceedings of the International Conference on Security, Privacy and Anonymity in Computation, Communication and Storage, Zhangjiajie, China, 16–18 November 2016; Springer: Cham, Switzerland, 2016; pp. 3–15. [CrossRef]
52. Soon, J.M.; Saguy, I.S. Crowdsourcing: A new conceptual view for food safety and quality. *Trends Food Sci. Technol.* **2017**, *66*, 63–72. [CrossRef]
53. Simula, H.; Ahola, T. A network perspective on idea and innovation crowdsourcing in industrial firms. *Ind. Mark. Manag.* **2014**, *43*, 400–408. [CrossRef]
54. Djelassi, S.; Decoopman, I. Customers' participation in product development through crowdsourcing: Issues and implications. *Ind. Mark. Manag.* **2013**, *42*, 683–692. [CrossRef]
55. Lutz, R.J. Marketing Scholarship 2.0. *J. Mark.* **2011**, *75*, 225–234. [CrossRef]
56. Castellion, G.; Markham, S.K. Perspective: New Product Failure Rates: Influence of A rgumentum ad P opulum and Self-Interest. *J. Prod. Innov. Manag.* **2013**, *30*, 976–979. [CrossRef]
57. Pollard, S.; Namazi, H.; Khaksar, R. Big data applications in food safety and quality. *Encycl. Food Chem.* **2019**, 356–363. [CrossRef]
58. Goel, P.; Datta, A.; Mannan, M.S. Application of big data analytics in process safety and risk management. In Proceedings of the 2017 IEEE International Conference on Big Data (Big Data), Boston, MA, USA, 11–14 December 2017; pp. 1143–1152. [CrossRef]
59. Lister, C. Easily accessible food composition data: Accessing strategic food information for product development and marketing. *Food N. Z.* **2018**, *18*, 1–17.
60. Murugan, D.; Garg, A.; Singh, D. Development of an Adaptive Approach for Precision Agriculture Monitoring with Drone and Satellite Data. *IEEE J. Sel. Top. Appl. Earth Obs. Remote Sens.* **2017**, *10*, 5322–5328. [CrossRef]
61. Zhang, W.; Ji, X.; Yang, Y.; Chen, J.; Gao, Z.; Qiu, X. Data fusion method based on improved DS evidence theory. In Proceedings of the 2018 IEEE International Conference on Big Data and Smart Computing (BigComp), Shanghai, China, 15–17 January 2018; pp. 760–766. [CrossRef]
62. Ahn, Y.Y.; Ahnert, S.E.; Bagrow, J.P.; Barabási, A.L. Flavor network and the principles of food pairing. *Sci. Rep.* **2011**, *1*, 1–7. [CrossRef]
63. Osowski, S.; Linh, T.H.; Brudzewski, K. Neuro-fuzzy network for flavor recognition and classification. *IEEE Trans. Instrum. Meas.* **2004**, *53*, 638–644. [CrossRef]
64. Garg, N.; Sethupathy, A.; Tuwani, R.; Nk, R.; Dokania, S.; Iyer, A.; Gupta, A.; Agrawal, S.; Singh, N.; Shukla, S.; et al. FlavorDB: A database of flavor molecules. *Nucleic Acids Res.* **2018**, *46*, 1210–1216. [CrossRef]
65. Simas, T.; Ficek, M.; Diaz-Guilera, A.; Obrador, P.; Rodriguez, P.R. Food-bridging: A new network construction to unveil the principles of cooking. *Front. ICT* **2017**, *4*, 14. [CrossRef]

66. Hao, J.; Liu, J.; Wu, W.; Tang, F.; Xian, M. Research on the Influence of O2O Take-out Chinese User Satisfaction on Continuous Intention: The Moderating Effect of User Habit. *J. China Stud.* **2018**, *21*, 101–120. [CrossRef]
67. Zeevi, D.; Korem, T.; Zmora, N.; Israeli, D.; Rothschild, D.; Weinberger, A.; Ben-Yacov, O.; Lador, D.; Avnit-Sagi, T.; Lotan-Pompan, M.; et al. Personalized Nutrition by Prediction of Glycemic Responses. *Cell* **2015**, *163*, 1079–1094. [CrossRef] [PubMed]
68. Teng, C.Y.; Lin, Y.R.; Adamic, L.A. Recipe recommendation using ingredient networks. In Proceedings of the 4th Annual ACM Web Science Conference, Evanston, IL, USA, 22–24 June 2012; pp. 298–307.
69. Grace, K.; Maher, M.L.; Wilson, D.C.; Najjar, N.A. Combining CBR and deep learning to generate surprising recipe designs. In Proceedings of the International Conference on Case-Based Reasoning, Atlanta, GA, USA, 31 October–2 November 2016; Springer: Cham, Switzerland, 2016; pp. 154–169. [CrossRef]
70. Li, H.C.; Ko, W.M.; Chung, S.C. Personalized Meal Planning Mehtod and System Thereof. U.S. Patent Application No 12/009,695, 11 June 2009.
71. Freyne, J.; Berkovsky, S. Recommending food: Reasoning on recipes and ingredients. In Proceedings of the International Conference on User Modeling, Adaptation, and Personalization, Big Island, HI, USA, 20–24 June 2010; Springer: Berlin/Heidelberg, Germany, 2010; pp. 381–386.
72. Kawahara, T.; Yoshida, M.; Tanaka, M.; Ido, S.; Nakano, H.; Adachi, N.; Abe, Y.; Nagatomo, W. Automated CD-SEM recipe creation technology for mass production using CAD data. In Proceedings of the Metrology, Inspection, and Process Control for Microlithography XXV, San Jose, CA, USA, 27 February–3 March 2011; Volume 7971, pp. 170–177.
73. Wolever, T.M.S. Personalized nutrition by prediction of glycaemic responses: Fact or fantasy? *Eur. J. Clin. Nutr.* **2016**, *70*, 411–413. [CrossRef] [PubMed]
74. Scott, T.; Jonathan, S.; Carl, L.; Agarwal, H.S.; Kohno, T. Devices That Tell On You: Privacy Trends in Consumer Ubiquitous Computing. In Proceedings of the 2007 USENIX Security Symposium, Boston, MA, USA, 6–10 August 2007; pp. 55–70.
75. Abouelmehdi, K.; Beni-Hessane, A.; Khaloufi, H. Big healthcare data: Preserving security and privacy. *J. Big Data* **2018**, *5*, 1–18. [CrossRef]
76. Mehmood, A.; Natgunanathan, I.; Xiang, Y.; Hua, G.; Guo, S. Protection of big data privacy. *IEEE Access* **2016**, *4*, 1821–1834. [CrossRef]
77. Rivest, R.L.; Adleman, L.; Dertouzos, M.L. On data banks and privacy homomorphisms. *Found. Secur. Comput.* **1978**, *4*, 169–180.
78. Yao, A.C. Protocols for secure computations. In Proceedings of the IEEE 23rd Annual Symposium on Foundations of Computer Science (sfcs 1982), Chicago, IL, USA, 3–5 November 1982; pp. 160–164.
79. Li, H.; Li, Z.; Tian, N. Resource Bottleneck Analysis of the Blockchain Based on Tron's TPS. In Proceedings of the International Conference on Natural Computation, Fuzzy Systems and Knowledge Discovery, Kunming, China, 20–22 July 2019; Springer: Cham, Switzerland, 2019; pp. 944–950. [CrossRef]
80. Uchibeke, U.U.; Schneider, K.A.; Kassani, S.H.; Deters, R. Blockchain access control Ecosystem for Big Data security. In Proceedings of the 2018 IEEE International Conference on Internet of Things (iThings) and IEEE Green Computing and Communications (GreenCom) and IEEE Cyber, Physical and Social Computing (CPSCom) and IEEE Smart Data (SmartData), Halifax, NS, Canada, 30 July–3 August 2018; pp. 1373–1378.
81. Tao, Q.; Chen, Q.; Ding, H.; Adnan, I.; Huang, X.; Cui, X. Cross-Department Secures Data Sharing in Food Industry via Blockchain-Cloud Fusion Scheme. *Secur. Commun. Netw.* **2021**, *2021*, 6668339. [CrossRef]
82. Blakley, G.R. Safeguarding cryptographic keys. *In Proceedings of the International Workshop on Managing Requirements Knowledge*; IEEE Computer Society: New York, NY, USA, 1979; p. 313.
83. Huang, X.; Tao, Q.; Qin, B.; Liu, Z. Multi-authority attribute based encryption scheme with revocation. In Proceedings of the 2015 IEEE 24th International Conference on Computer Communication and Networks (ICCCN), Las Vegas, NV, USA, 3–6 August 2015; pp. 1–5. [CrossRef]
84. Hsu, C.; Zeng, B.; Zhang, M. A novel group key transfer for big data security. *Appl. Math. Comput.* **2014**, *249*, 436–443. [CrossRef]
85. Dwork, C. Differential privacy: A survey of results. In Proceedings of the International Conference on Theory and Applications of Models of Computation, Xi'an, China, 25–29 April 2008; Springer: Berlin/Heidelberg, Germany, 2008; pp. 1–19.
86. Hu, V.C.; Grance, T.; Ferraiolo, D.F.; Kuhn, D.R. An access control scheme for big data processing. In Proceedings of the 10th IEEE International Conference on Collaborative Computing: Networking, Applications and Worksharing, Miami, FL, USA, 22–25 October 2014; pp. 1–7. [CrossRef]
87. Lv, D.; Zhu, S. Correlated Differential Privacy Protection for Big Data. In Proceedings of the 2018 IEEE 32nd International Conference on Advanced Information Networking and Applications (AINA), Krakow, Poland, 16–18 May 2018; pp. 1011–1018. [CrossRef]
88. Hao, J.; Liu, J.; Wu, W.; Tang, F.; Xian, M. Secure and fine-grained self-controlled outsourced data deletion in cloud-based IoT. *IEEE Internet Things J.* **2019**, *7*, 1140–1153. [CrossRef]
89. Godfray, H.C.J.; Beddington, J.R.; Crute, I.R.; Haddad, L.; Lawrence, D.; Muir, J.F.; Pretty, J.; Robinson, S.; Thomas, S.M.; Toulmin, C. Food security: The challenge of feeding 9 billion people. *Science* **2010**, *327*, 812–818. [CrossRef] [PubMed]
90. Walter, A.; Finger, R.; Huber, R.; Buchmann, N. Opinion: Smart farming is key to developing sustainable agriculture. *Proc. Natl. Acad. Sci. USA* **2017**, *114*, 6148–6150. [CrossRef] [PubMed]
91. Parfitt, J.; Barthel, M.; Macnaughton, S. Food waste within food supply chains: Quantification and potential for change to 2050. *Philos. Trans. R. Soc. B Biol. Sci.* **2010**, *365*, 3065–3081. [CrossRef] [PubMed]

92. Gobble, M.M. Big data: The next big thing in innovation. *Res.-Technol. Manag.* **2013**, *56*, 64–67. [CrossRef]
93. Tabesh, P.; Mousavidin, E.; Hasani, S. Implementing big data strategies: A managerial perspective. *Bus. Horizons* **2019**, *62*, 347–358. [CrossRef]
94. Salas-Vega, S.; Haimann, A.; Mossialos, E. Big data and health care: Challenges and opportunities for coordinated policy development in the EU. *Health Syst. Reform* **2015**, *1*, 285–300. [CrossRef] [PubMed]

MDPI

Review

Exploring the Future of Edible Insects in Europe

Simone Mancini [1,2], Giovanni Sogari [3,*], Salomon Espinosa Diaz [1], Davide Menozzi [3], Gisella Paci [1,2] and Roberta Moruzzo [1]

1 Department of Veterinary Sciences, University of Pisa, Viale delle Piagge 2, 56124 Pisa, Italy; simone.mancini@unipi.it (S.M.); salomon.espinosadiaz@phd.unipi.it (S.E.D.); gisella.paci@unipi.it (G.P.); roberta.moruzzo@unipi.it (R.M.)
2 Interdepartmental Research Center "Nutraceuticals and Food for Health", University of Pisa, Via del Borghetto 80, 56124 Pisa, Italy
3 Department of Food and Drug, University of Parma, Parco Area delle Scienze 27/A, 43124 Parma, Italy; davide.menozzi@unipr.it
* Correspondence: giovanni.sogari@unipr.it; Tel.: +39-0521-906-545

Abstract: The effects of population increase and food production on the environment have prompted various international organizations to focus on the future potential for more environmentally friendly and alternative protein products. One of those alternatives might be edible insects. Entomophagy, the practice of eating insects by humans, is common in some places but has traditionally been shunned in others, such as European countries. The last decade has seen a growing interest from the public and private sectors to the research in the sphere of edible insects, as well as significant steps forward from the legislative perspective. In the EU, edible insects are considered novel foods, therefore a specific request and procedure must be followed to place them in the market; in fact, until now, four requests regarding insects as a novel food have been approved. Insects could also be used as feed for livestock, helping to increase food production without burdening the environment (indirect entomophagy). Market perspectives for the middle of this decade indicate that most of the demand will be from the feed sector (as pet food or livestock feed production). Undoubtedly, this sector is gaining momentum and its potential relies not only in food, but also in feed in the context of a circular economy.

Keywords: entomophagy; novel food; mealworm; grasshopper; cricket; locust; acceptance; alternative protein; sustainability; neophobia

Citation: Mancini, S.; Sogari, G.; Espinosa Diaz, S.; Menozzi, D.; Paci, G.; Moruzzo, R. Exploring the Future of Edible Insects in Europe. *Foods* **2022**, *11*, 455. https://doi.org/10.3390/foods11030455

Academic Editor: Andrea M Liceaga

Received: 15 December 2021
Accepted: 24 January 2022
Published: 3 February 2022

1. European Market of Insects

In the last decade, the insect sector has increased worldwide, also reaching popularity via mass media communications, mostly when new regulations or outcomes are released in relation to its potential in all areas of food chain production. Indeed, insects, when compared to conventional livestock animals, require low amounts of water, land, and feed to produce the same quantity of nutritional molecules [1]. Insects, as poikilotherm animals, can convert feed (substrates) with high efficiency in their body, decreasing the feed conversion ratio. Several international agencies and organizations have identified insects as one of the players in the commercial food chain in the future and as an active actor in reducing the environmental impact while increasing animal production [1,2]. Several international organizations are supporting the sector worldwide, such as IPIFF (International Platform of Insects as Food and Feed, Brussels, Belgium), AFFIA (Asian Food and Feed Insect Association, Bangkok, Thailand), NACIA (North America Coalition for Insect Agriculture, Chicago, IL, USA), and IPAA (Insect Protein Association of Australia, Canberra, ACT). Edible insects are becoming a part of the human diet, directly or indirectly, in many regions of the world [3,4], even if they currently represent a niche market in Western societies [5]. According to the Global Market Insights report, in 2019, the size of the edible insects market exceeded 112 million USD [6]. Though the global market data for edible insects are inconsistent, also given their short market history and consumer availability, different reports point to an overall growth of the market for edible insects [6–8]. The Global Market Insights report estimated that the edible insects market will grow by 47% between 2019 and 2026 [6]. It is expected that the largest increase might happen in North America and Europe [9]. The global edible insects market is segmented not only with regard to geography, but also by insect species, product type (e.g., whole insects, insect meal, and insect powder), and application (e.g., food and feed) [10]. As regards geography, the Asia–Pacific region led by Thailand, China, and Vietnam dominated the market in 2019 with around 41% of the market share in terms of revenue, followed, respectively, by Europe (22%) (led by the UK, the Netherlands, and France), Latin America (21%), North America (13%), and the Middle East and Africa (3%) [7]. Based on the insect type, beetles dominated the global edible insects market in 2019 with around 30% of the market share in terms of revenue, followed by caterpillars and Hymenoptera, respectively [7]. Insects and insect-based products are consumed in different forms: it is possible to find whole insects, or insects processed into food ingredients (e.g., flour or powder), which are then incorporated into final products (e.g., energy bars, burgers, or compound feed) [11,12]. Insects can be eaten roasted, fried, or boiled as a whole (which is the traditional way of preparation in most tropical countries), or they can be dried and ground and can then be added to foods otherwise [13].

In the past decade, in the edible insects market, numerous entrepreneurs and companies have become active in the production of insects [5]. Several initiatives have gradually transitioned from ambitious startups to well-established operators that are about to rapidly grow their production capacity. In Europe, the insect production sector was initially based mainly on small- to medium-sized startups, which undertook insect breeding in zoological gardens for biocontrol purposes or the production of animal feed [14]. Nowadays, the situation is different, and it is possible to find a certain number of insect feed business operators (iFeedBOs) (some of them are also active in food production activities) and some insect food business operators (iFoodBOs).

In the market factsheets developed each year by the IPIFF, it is possible to find specific information about these two kinds of operators. In Europe, from the 500 tons of insect-based products (whole insects, insect ingredients, and products with added edible insects) in 2019, the market will expand to 260,000 tons by 2030 [15], with powder/insect ingredients accounting for over 75% of the total. Insect FoodBOs comprise micro-companies (81%) with fewer than 10 employees, followed by small (16%) and medium-sized (3%) companies, with the number of employees between 10 and 50 or 50 and 250, respectively. The level of investment varies a lot among the operators. For most of them (63%), the investment is under 500,000 euros, and only 3% receive investments up to 25 million [16]. With the growth of the insect sector and growing investment, more jobs will be created (not only direct jobs, but also indirect ones, such as in specialized retail, logistics, administration, or research). Insect FoodBOs are involved in the different stages of insect production: primary production, processing into insect ingredients, and preparation of insect-based food. In all, 28% of the iFoodBOs are involved in all of the stages, 19% are involved in production and processing, 12%—in processing and preparation, 36%—in preparation, 3%—in production, and almost 3%—in the selling of different edible insect products to end consumers through their respective channels (e.g., online platforms). The majority of iFoodBOs are concentrated in northern European countries, led by the United Kingdom, Germany, and Belgium, and followed by the Netherlands, France, Finland, and Denmark [12]. As a matter of fact, according to some authors [1,17], the Netherlands has to be considered as the European "hub" of early research associated with the human consumption of insects in the first decade of the 21st century. The insect FoodBOs segment their final market, choosing between the national, European, and extra-European markets. In 2020, the majority (more than 60%) of the European iFoodBOs primarily focused the sale of products in the country of production. However, by 2025, iFoodBOs are expected to concentrate most of their activity on the EU market, even if they decide to maintain their activity at both the national and international level. As regards iFeedBOs, according to the IPIFF, most of them are SMEs (i.e., small enterprises with 10–50 employees or medium-sized enterprises with 50–250 employees). Micro-enterprises (1–10 employees) also represent more than 40% of the companies active in feed production. By 2030, it is expected that almost one out of two FBOs will be a large enterprise—with more than 250 employees. Insect FeedBOs have managed to raise over one billion euros in investments—figures that may reach three billion euros by 2025 [18].

By the middle of the decade, most of the demand for insect meal will likely lie in the pet food sector (almost 40–50% of the insect meal produced) and in aquaculture feed production (reaching 25–35% in terms of the share) [19]. Indeed, until 2021, in the European Union, insects as feed were only authorized for aquaculture (Commission Regulation 2017/893) and pet nutrition. Recently, the EU member states voted positively on a regulation aimed at enabling the use of insect-processed animal proteins (PAPs) in poultry and pig nutrition (Commission Regulation 2021/1372 of 17 August 2021). Thus, according to the IPIFF forecasts [18], the next relevant market for insects as iFeedBOs in terms of the quantity of insect meal sold will be the poultry (20–30%) and pig markets (5–15%), which will see a rapid increase following entry into force of the approval of insect PAPs.

The potential use of edible insects in European countries is currently debated among many stakeholders of the food supply chain. This paper aims to provide a concise but comprehensive state of the art on how insects could be used as food in light of the new legislative framework and recent EU novel food authorizations. In addition, we underline the crucial role of consumer acceptance of edible insects and reflect on how consumers will play a role in this sector. Our discussion takes into consideration the current situation at the European level in the framework of research policy and business perspective.

2. Regulation of Insects as Food in the EU

Before insect-based food products started to attract a significant level of attention in the European food market, small-scale production and trade of edible insects had not been considered sufficiently important to be subject to legislative matters or safety standards supervision [20]. In fact, in most European countries, the consumption of edible insects is still very low and often seen as socially improper [21]. This is also because in Europe, edible insects and insect-based products are classified as novel foods (NFs) [11], that is, food products that do not have a history of human consumption within the region or country in question or, more specifically, any food that was not consumed "significantly" prior to May 1997, according to the EU regulations. However, interest in insect-based products has been increasing in the last few years, triggered especially by the potential environmental, economic, and food security benefits that insects could offer [22,23]. The novelty of using insects for human consumption, as well as the high interest from the public and the media in such products [5], poses important questions and concerns regarding the risks derived from insect production, processing, and consumption [24]. Moreover, the need in reliable guidelines has concerned producers and slowed down the growth of the sector, as also highlighted jointly in 2018 by NACIA, IPIFF, AFFIA, and IPAA [5]. Answering these questions, and therefore ensuring the safety of consumers, requires valid scientific assessments that adequately inform policymakers with the responsibility to authorize the incorporation of such products into the market.

Consequently, in 2015, the European Food Safety Authority (EFSA) published its first scientific opinion on the risks associated with the production and consumption of farmed insects as food and feed [25]. It covered considerations about potential allergenic and environmental risks, as well as chemical and biological hazards linked to external factors such as the methods used for their production, the substrates they are fed on, and the lifecycle stage at which they are harvested, among others. The EFSA concluded that as long as insects are fed with currently permitted feed materials, the potential occurrence of such hazards is expected to be similar to that of other non-processed protein sources. This means that insects can only be safely reared on substrates of vegetable origin or specific allowed animal-origin materials, preventing the possibility of using substrates containing manure and other waste materials. This opinion laid the groundwork for a new European regulation on insects as NF, i.e., Regulation (EU) 2015/2283. According to this new regulation, insect food products may be commercialized only if authorized by the European Commission (EC). The process of obtaining such authorization is very straightforward: anyone who intends to place an NF on the EU market must submit an application to the EC. After the application is verified and validated, it becomes available to all the member states, and the EFSA is requested to provide a scientific safety assessment within nine months from a valid application. Based on this opinion, authorization is granted or not. This regulation also indicates the specific insects that can be utilized as NFs, labeling recommendations for insect-based products, as well as general conditions for the inclusion of NFs in the union list, which is updated by the EC and serves as a reference for economic operators who wish to place an authorized NF on the market (Figure 1).

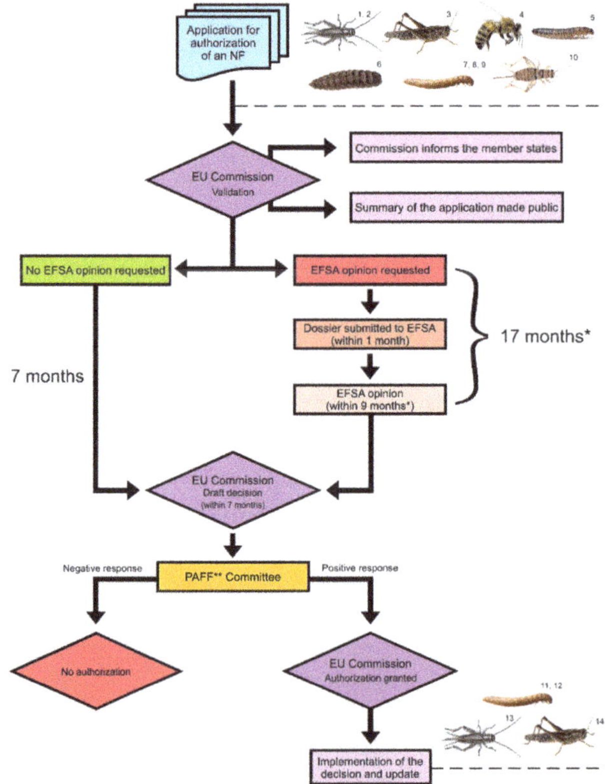

Figure 1. Novel food (NF) authorization process. * This amount of time can be longer if the EFSA requests additional information from the applicant. ** The Standing Committee on Plants, Animals, Food and Feed. [1,2] *Acheta domesticus* (Belgian Insect Industry Federation—Belgium; Cricket One—Vietnam). [3] *Locusta migratoria* (Belgian Insect Industry Federation—Belgium). [4] *Apis mellifera* (Finnish Beekeepers' Association—Finland). [5] *Alphitobius diaperinus* (Proti-Farm Holding NV—the Netherlands). [6] *Hermetia illucens* (Enorm Biofactory A/S—Denmark). [7,8,9] *Tenebrio molitor* (Nutri'Earth—France; Belgian Insect Industry Federation—Belgium; Ynsect—France). [10] *Gryllodes sigillatus* (SAS EAP Group—Saint-Orens-de-Gameville, France). [11,12] *Tenebrio molitor* (SAS EAP Group—France; Protix Company—the Netherlands). [13] *Acheta domesticus* (Protix Company, Fair Insects BV—Dongen, the Netherlands). [14] *Locusta migratoria* (Protix Company, Fair Insects BV—Dongen, the Netherlands).

More recently, Regulations (EU) 2017/2469 and (EU) 2017/893 specify some rules regarding the administrative and scientific requirements to an application to request authorization of commercialization of an NF in the EU market, as well as some amendments of the general criteria for insect and insect protein production, with particular attention to the substrate options that can be used to rear them, which are limited to those permitted for other livestock species.

Although, on the one hand, the highly restrictive regulations for the production and commercialization of insects and insect-based products in Europe allow governments to ensure the safety of consumers, on the other hand, they also slow down the development of an industry capable of offering potential environmental and economic benefits [26]. At the same time, in some countries, the lack of formal standards for production have led to uncertainty and concern among many insect farmers, providing an unstable basis for the industry (though also opening the possibility for the industry itself to innovate in the development of standards) [27]. Therefore, new scientific evidence enabling an optimized and improved assessment and the identification of specific critical control points in the entire production and processing chain of insects is essential.

3. European Authorizations

Based on the regulation of NFs, as of December 2021, four authorizations deal with insects. First, the EC authorized the placing on the EU market of dried *Tenebrio molitor* larvae (yellow mealworm) as an NF. The application requested for dried *Tenebrio molitor* larvae to be used as whole, dried insects in the form of snacks and as a food ingredient in several food products, the target population being the general population. The applicant (SAS EAP Group, France) made a request to the Commission for the protection of the proprietary data submitted in the application.

Based on the applicant's request, whole, dried larvae or larval powder were proposed to be used as ingredients in food products, defined, using the FoodEx2 hierarchy, as snacks other than chips and similar (maximum use level, ML = 100 g NF/100 g product), protein and protein components for sports people (ML = 10 g NF/100 g), biscuits (ML = 10 g NF/100 g), legumes-based dishes (ML = 10 g NF/100 g), and uncooked pasta-based dishes (ML = 10 g NF/100 g) (see Table 1 for a complete list of products) [28].

Table 1. Food products and the maximum use levels reported in the four EFSA opinions on insects as novel foods (NFs).

Food Category	**Maximum Use Level (g NF/100 g)**								
	T. molitor [1]	***L. migratoria*** [2]			***T. molitor*** [2]			***A. domesticus*** [2]	
	W/D/P	**F**	**D**	**P**	**D**	**P**	**F**	**D/P**	**F**
Beans and vegetable meal								5	15
Beer and beer-like beverages		2	2	2	1	1	1	1	1
Biscuits	10								
Biscuits, sweet, plain					8	8	30	8	30
Bread and rolls with special ingredients added								10	30
Caesar salad		15	5	5				5	15
Canned or jarred legumes		20	15	15					
Canned/jarred vegetables		20	15	15					
Cereal bars					15	15	30	15	30
Chickpeas (without pods)		40	20	20	30	30	40	25	40
Chips/crisps					20	20	40		
Chocolate and similar		30	10	10	10	10	30	10	30

Table 1. *Cont.*

Food Category	**Maximum Use Level (g NF/100 g)**								
	T. molitor [1]	*L. migratoria* [2]			*T. molitor* [2]			*A. domesticus* [2]	
	W/D/P	**F**	**D**	**P**	**D**	**P**	**F**	**D/P**	**F**
Crackers and breadsticks					10	10	30	10	30
Dried pasta					5	5	15	1	3
Dried stuffed pasta					15	15	30	15	30
Frozen yoghurt		15	5	5	5	5	15	5	15
Hummus								5	15
Legume (beans) soup		15	5	5				5	15
Legume-based dishes	10	15	5	5	5	5	15		
Meatballs					16	16	40	16	40
Meat burgers (no sandwiches)					16	16	40	16	40
Meat imitates		80	50	50	50	50	80	50	80
Mixed alcoholic drinks		2	2	2	1	1	1	1	1
Mixed vegetable soup		15	5	5					
Mixed vegetable soup (dry)		15	5	5				5	15
Multigrain bread and rolls					10	10	30		
Mushroom soup		15	5	5				5	15
Mushroom soup (dry)		15	5	5				5	15
Oilseeds		40	20	20	30	30	40	25	40
Onion soup		15	5	5				5	15
Pasta-based dishes, cooked					5	5	15		
Pasta, filled, cooked								5	15
Pasta-based dishes, uncooked	10								
Pizza and pizza-like dishes		15	5	5	5	5	15		
Pizza and similar with cheese and vegetables								5	15
Pizza and similar with cheese, vegetables, and fruits								5	15
Potato soup		15	5	5				5	15
Potato-based dishes		15	5	5	5	5	15		
Potatoes and vegetable meal								5	15
Pre-mixes (dry) for baked products					15	15	30	15	30
Prepared pasta salad		15	5	5				5	15
Primary derivatives from nuts and similar seeds		40	20	20	30	30	40	25	40
Protein and protein components for sports people	10								
Sausages		30	10	10					
Snacks other than chips and similar	100	100	100	100	100	100	100	100	100
Soups and salads					5	5	20		
Tartar sauce					10	10	30	15	30
Tomato soup		15	5	5				5	15
Tomato soup (dry)		20	5	5				5	20
Tortilla chips								20	40
Tree nuts		40	20	20	30	30	40	25	40
Unsweetened spirits and liqueurs		2	2	2	1	1	1	1	1
Whey powder					20	20	40	20	40

W, whole; F, frozen; D, dried; P, powder. Applicant: [1] SAS EAP Group [28]; [2] Fair Insects BV (Protix Company) [29–31].

The Commission also authorized the placing on the market of a second insect, *Locusta migratoria* (migratory locust), as an NF on 12 November 2021. The request dealt with frozen and dried forms of the insect to be used in different food categories (Table 1). The applicant was Fair Insects BV (Protix Company, the Netherlands). The NF consists of frozen, dried, and powder forms of the migratory locust, intended to be marketed as a snack or as a food ingredient in several food products. Moreover, Protix Company requested data protection [29]. On 8 December 2021, two requests by Fair Insects BV (Protix Company, the Netherlands) were authorized as NFs. The authorized requests dealt with frozen and dried formulations of whole *Tenebrio molitor* and frozen and dried formulations of whole *Acheta domesticus* (house cricket) (Table 1). The applicant requested protection of the data for both NFs [30,31].

The other 10 applications dealt with *Tenebrio molitor* (Nutri'Earth—France; Belgian Insect Industry Federation—Belgium; Ynsect—France), *Acheta domesticus* (Belgian Insect Industry Federation—Belgium; Cricket One—Vietnam), *Gryllodes sigillatus* (SAS EAP Group—France), *Locusta migratoria* (Belgian Insect Industry Federation—Belgium), *Alphitobius diaperinus* (Proti-Farm Holding NV—the Netherlands), *Hermetia illucens* (Enorm Biofactory A/S—Denmark), and *Apis mellifera* (Finnish Beekeepers' Association—Finland).

Based on the applicants' proposed uses, all of the NFs were listed in the "Snacks other than chips and similar" products as whole products. Thus, all of the enterprises will be able to place on the market products containing only insects that have been dried, frozen, or ground into powder (in relation to the request). These include bakery products, such as bread, biscuits, and crackers containing mealworms and crickets. Six different types of products referred to pasta-like products, all listed in the two requests made by Fair Insects BV (Protix Company) for mealworms and crickets. Moreover, three different types of pizza were listed for these requests.

Edible insects are commonly presented and positioned to consumers as meat alternatives [17,32]. This argument is not supported by research or entrepreneurial ideas, but is wrongly based on the idea that insects could be a substitute for meat products, providing significant environmental benefits. Edible insects are more than an alternative to meats; especially in relation to the nutritional–economic status of the consumers/country, insects could be positioned as a major protein or energy food, decreasing the gaps between the rich and poor nutritional diets across the developing and the developed countries (without negatively affecting the environment). Furthermore, in the developed countries, consumers are unwilling to accept the direct substitution of a "nice" slice of meat with a "strange" dish of insects (food neophobia) [33–36]. In this contest, meat-like products are only partially reported in the lists. Indeed, sausages, meatballs, and meat burgers were listed with an inclusion of approximately 30–40% (w/w) of frozen locusts, mealworms, and crickets, or approximately 10–16% (w/w) of dried/powdered insects. Contrarily, meat imitates will be intended as more insect-based, with a maximum of 50% of dried/powdered insects or 80% of frozen insects.

Interestingly, a high number of vegetable-based foods was reported in the requests. Vegetable-based dishes, meals, soups, salads, and canned/jarred items were requested by all the applicants. Moreover, oilseed and primary derivatives from nuts and similar seeds were listed in the applications with quite high maximum contents of insects (respectively, 20–30 dried or powder NF g/100 g product and 40 frozen NF g/100 g products).

It seems like producers are going to use edible insects in beverages more as a curiosity, and likewise the caterpillar of *Comadia redtenbacheri* (gusanos rojos in Spanish) in a mezcal traditional recipe [37]. The percentages in beer and beer-like, mixed alcohol drinks, as well as unsweetened spirits and liqueurs, range between 1% and 2% w/w.

Foods intended for sports people are also listed as "protein and protein components for sports people" and whey powder. In these products, the insect contents are 10%, 20%, or 40%.

4. Consumer Acceptance of Insects as Food

In the past few years, a considerable number of studies has been published on consumers' acceptance of alternative proteins, including edible insects as food [32]. Several recent up-to-date reviews on eating insects and insect-based foods provide a critical overview within the EU [4], as well as globally [38]. In addition, Sogari et al. [39] provided a comprehensive perspective on the overall state of research activity on consumer attitude and behavior toward entomophagy without date restrictions. Their results showed that the number of publications has increased substantially since 2015, after the publication of the FAO report "Edible Insects: Future Prospects for Food and Feed Security" [1,5], and the trend has maintained since then [38]. Most of these studies focused on European consumers, with only a few of them including a cross-country comparison (e.g., Italian and Dutch samples in [40]). As a result of the majority of these studies being conducted in Western countries, researchers have investigated consumers' reactions to processed insect-based foods such as snacks (e.g., biscuits, chips, and bars) rather than whole and traditional dishes with insects [4,41].

This rising publication trend calls out for a need to provide a thematic synthesis. First, compared to even few years ago, as Dagevos [38] suggested, we have gained a considerable understanding of consumers' acceptance of insect-eating. It is now clear that a high level of food neophobia (i.e., fear of new foods) implies a low inclination to consider insects as a food to try and purchase, regardless of the respondents' origin [42,43]. Another motivation of Western consumers' aversion to eating insects is disgust, a primary emotion and a major aversive reaction toward insect consumption since research on entomophagy started [5,41].

The aversion to insects as a food is also strictly linked to the food culture [24]. Several studies have shown that the stronger the gastronomic culture within the society, the greater the rejection, and vice versa [4,40]. This is also linked to other people's opinions (social norms), which could strongly affect the acceptance of consuming this novel food. On the contrary, individuals who have already had a positive experience of eating edible insects show a higher willingness to eat insects in the future [4]. Moreover, previous experience of eating insects plays a significant role in the willingness to repeat consumption due to a positive perception of tastiness (e.g., expected liking) and reduced food neophobia [38,41,44].

Several studies have analyzed the effects of information about the benefits of entomophagy on the attitudes toward insect food products, indicating that this strategy increased insects' acceptance [32,41]. For example, Mancini et al. [4] showed how an education lecture about the ecological, safety, nutritional, and taste-related aspects of insect-eating increased the willingness to try insects and insect-containing foods among a group of university students. However, consumers' response to these information treatments is strongly correlated with subjective interests in the nutritional or environmental benefits of the eaten food [38]. Thus, it is likely that young people, such as students, who are sensitive to the current challenges in food sustainability [45] will more likely be responsive to information about the health or sustainability aspects of insects as food than older consumers. In fact, early adopters of insect food products are often identified in young adults [4], especially males [32,46]. In particular, it has been noted by Jones [47] that in her study, many young people of school age welcomed the opportunity to have their preconceptions surrounding what is acceptable challenged in light of a more sustainable food system. Besides gender and age, the dietary regimen also plays a role with respect to consumers' attitudes toward entomophagy. For example, Elorinne et al. [46] showed that non-vegan vegetarians' attitudes toward eating insects were the most positive, whereas vegans held the most negative attitude, considering eating insects as immoral and irresponsible. In fact, in the Netherlands, supermarkets display insect-based convenience foods such as burgers and nuggets in the same section as vegetarian products [48].

Familiar food preparation is another driver for insect consumption [4,44]. Thus, the current market strategy is to develop highly processed insect-based foods that are familiar in Western diets such as burgers, bread, biscuits, crackers, crisps, candy bars, shakes, soups, sauces, and pasta [12,38]. This rationality of "hiding" the appearance of insects and including them in crushed, bruised, or powdered forms, rather than visible and whole, is supported by a large amount of empirical evidence [4,41]. It seems obvious that to reduce disgust-based aversion and consequently raise the acceptance toward entomophagy, insects need to be processed in a form that resembles known products (i.e., familiar-looking).

Another recent stream of consumer acceptance studies focuses on using insects as a feed source, which is sometimes defined as "indirect entomophagy" [49]. From the literature, it seems that consumers will have higher acceptance of using insects as feed for farmed animals than as food for human consumption [32]. In fact, in a recent study, Menozzi et al. [50] found that there is a potential consumer interest in considering insects as a protein substitute in the poultry sector.

5. Conclusions and Future Prospects

World population growth and the increase in food demands push scholars to research alternative protein sources for both human and animal consumption. As presented in the previous sections, the last decade has seen a growing interest from the public and private sectors in research in the sphere of edible insects, as well as significant steps forward from the legislative perspective in the EU. Moving from entomophobic countries such as most of those in Europe to more open entomophilic societies will be favored by common and shared stakeholder efforts, including national governments, the research community, and the private sector [24]. This latter will include actors across the production, supply, and consumption of food, which will be crucial to industry success [48]. Undoubtedly, this sector is gaining momentum, and its potential relies not only on food and feed, but also on the context of a circular economy. In particular, insects have the potential to convert a wide range of organic byproducts into food and feedstuffs, which then go back into the production cycle [51]. However, due to the current lack of evidence related to the safety of the final products, EU regulations still prohibit the use of waste products as a substrate for growing edible insects. Our review showed how the most edible insect species, i.e., crickets, mealworms, grasshoppers, and locusts, investigated in consumer studies [41] are those already authorized by the EFSA (i.e., *Tenebrio molitor* and *Locusta migratoria*), or currently subject to a safety evaluation by the authority. As suggested by House [17], these species are the "industry standard" food insects in Europe as the positive result of a number of technical, practical, and legislative factors.

Although some adventurous consumers and sensation-seekers could try a visible and whole insect [38], it seems clear that the marketing strategy adopted by most insect companies is to develop processed products (e.g., bakery, meat, pasta, and pizza products) in which insects are "hidden" in the form of a powder or similar. In addition, it has to be highlighted how even if the current legislation and industry are contributing to the idea that insects are a legitimate and edible (i.e., safe) food, this does not mean that they will instantly be accepted and consumed [48]. Acceptance will especially depend upon education, especially in the context of schools and young individuals, with the aim of changing mis- and preconceptions about edible insects [47].

The prospect for future research should then concentrate on different aspects: (i) The economic convenience of introducing insect meals into animals' diets, both in the developing and the developed countries. Future studies on insects as feed and food should better investigate the structure cost and profitability of insect farms. This type of analysis could facilitate the development of other farms to undertake the production of insects and also support the sustainable development of this sector from the circular economy perspective. (ii) Consumers' willingness to pay for eating insects, insect-based foods, and animal feed with insects as meal. Thus far, a considerable number of consumer-oriented studies have focused on consumers' attitudes and behavior toward entomophagy. However, studies investigating consumers' willingness to pay for such products in real settings are lacking. Future research on this aspect is also important to push farms to embrace this new sector of activity. (iii) The role of governments in supporting the insect farming sector. Governments play a decisive role in facilitating a shift toward new and sustainable food solutions. The sustainability of food systems is a global issue, and food systems will have to adapt to face diverse challenges. As mentioned in the "Farm to Fork strategy," all actors of the food chain must play their part in achieving the sustainability of the food chain. Farmers and producers need to transform their production methods more quickly and make the best use of nature-based, technological, digital, and space-based solutions to deliver better climate and environmental results, increase climate resilience, and reduce and optimize the use of inputs. However, these solutions require not only human and financial investments, but also a collective approach involving public authorities at all levels of governance that could increase participation and provide a voice for farms and producers in global processes related to food. (iv) Promoting the use of insects from the packaging and labeling standpoint. Attractive naming and descriptions, communication of health benefits, branding (logo), and product image are important to consumers' perceptions of insect foods. Future entomophagy research should therefore focus on improving marketing strategies to ensure that insects and insect-based products become more appealing. A correct communication strategy could ameliorate access to insects and reduce future generations' unfamiliarity, inappropriateness, and disgust that are the major causes of the non-acceptance of insects today.

Author Contributions: Conceptualization, S.M., G.S. and R.M.; writing—original draft preparation, S.M., G.S., R.M. and S.E.D.; writing—review and editing, S.M., G.S., R.M., S.E.D., D.M. and G.P.; funding acquisition, S.M. All authors have read and agreed to the published version of the manuscript.

Funding: The research was funded by the University of Pisa, PRA (Progetti di Ricerca di Ateneo) grant No. PRA_2020_12 (Produzione di Insetti come Feed e Food (PIFF)).

Acknowledgments: The authors kindly acknowledge the FOODS journal Award Committee who conferred to article https://www.mdpi.com/2304-8158/8/7/270 (accessed on 23 January 2022), Foods Best Paper Award 2019, granting this publication free of charge.

Conflicts of Interest: The authors declare no conflict of interest.

References

1. van Huis, A.; van Itterbeeck, J.; Klunder, H.; Mertens, E.; Halloran, A.; Muir, G.; Vantomme, P. *Edible Insects. Future Prospects for Food and Feed Security*; Food and Agriculture Organization of the United Nations: Rome, Italy, 2013; Volume 171, ISBN 978-92-5-107595-1.
2. Moruzzo, R.; Mancini, S.; Guidi, A. Edible Insects and Sustainable Development Goals. *Insects* **2021**, *12*, 557. [CrossRef]
3. Macombe, C.; le Feon, S.; Aubin, J.; Maillard, F. Marketing and Social Effects of Industrial Scale Insect Value Chains in Europe: Case of Mealworm for Feed in France. *J. Insects Food Feed.* **2019**, *5*, 215–224. [CrossRef]
4. Mancini, S.; Moruzzo, R.; Riccioli, F.; Paci, G. European Consumers' Readiness to Adopt Insects as Food. A Review. *Food Res. Int.* **2019**, *122*, 661–678. [CrossRef]
5. Payne, C.; Megido, R.C.; Dobermann, D.; Frédéric, F.; Shockley, M.; Sogari, G. Insects as Food in the Global North—The Evolution of the Entomophagy Movement. In *Edible Insects in the Food Section*; Springer: Cham, Switzerland, 2019; pp. 11–26. [CrossRef]
6. Edible Insects Market Analysis 2020–2026 | Industry Forecasts. Available online: https://www.gminsights.com/industry-analysis/edible-insects-market (accessed on 16 November 2021).
7. Edible Insects Market to Reach US$850 Million Globally by. Available online: https://www.globenewswire.com/news-release/2020/05/22/2037713/0/en/Edible-Insects-Market-to-reach-US-850-Million-globally-by-end-of-2027-Coherent-Market-Insights.html (accessed on 16 November 2021).
8. Edible Insects—Statistics & Facts | Statista. Available online: https://www.statista.com/topics/4806/edible-insects/#dossierKeyfigures (accessed on 16 November 2021).
9. Guiné, R.P.F.; Correia, P.; Coelho, C.; Costa, C.A. The Role of Edible Insects to Mitigate Challenges for Sustainability. *Open Agric.* **2021**, *6*, 24–36. [CrossRef]
10. Ebenebe, C.I.; Ibitoye, O.S.; Amobi, I.M.; Okpoko, V.O. African Edible Insect Consumption Market. In *African Edible Insects as Alternative Source of Food, Oil, Protein and Bioactive Components*; Springer: Cham, Switzerland, 2020; pp. 19–51. [CrossRef]
11. Derrien, C.; Boccuni, A. Current Status of the Insect Producing Industry in Europe. In *Edible Insects in Sustainable Food Systems*; Springer International Publishing: Cham, Switzerland, 2018; pp. 471–479.
12. Pippinato, L.; Gasco, L.; di Vita, G.; Mancuso, T. Current Scenario in the European Edible-Insect Industry: A Preliminary Study. *J. Insects Food Feed.* **2020**, *6*, 371–381. [CrossRef]
13. Daub, C.-H.; Gerhard, C. Essento Insect Food AG: How Edible Insects Evolved from an Infringement into a Sustainable Business Model. *Int. J. Entrep. Innov.* **2021**, 146575032110308. [CrossRef]
14. Halloran, A.; Flore, R.; Vantomme, P.; Roos, N. *Edible Insects in Sustainable Food Systems*; Halloran, A., Flore, R., Vantomme, P., Roos, N., Eds.; Springer International Publishing: Cham, Switzerland, 2018; ISBN 978-3-319-74010-2.

15. Skotnicka, M.; Karwowska, K.; Kłobukowski, F.; Borkowska, A.; Pieszko, M. Possibilities of the Development of Edible Insect-Based Foods in Europe. *Foods* **2021**, *10*, 766. [CrossRef]
16. IPIFF Edible Insects on the European Market. Available online: https://ipiff.org/wp-content/uploads/2020/06/10-06-2020-IPIFF-edible-insects-market-factsheet.pdf (accessed on 16 November 2021).
17. House, J. Insects Are Not 'the New Sushi': Theories of Practice and the Acceptance of Novel Foods. *Soc. Cult. Geogr.* **2019**, *20*, 1285–1306. [CrossRef]
18. IPIFF The European Insect Sector Today: Challenges, Opportunities and Regulatory Landscape. Available online: https://ipiff.org/wp-content/uploads/2019/12/2019IPIFF_VisionPaper_updated.pdf (accessed on 16 November 2021).
19. An Overview of the European Market of Insects as Feed. Available online: https://ipiff.org/wp-content/uploads/2021/04/Apr-27-2021-IPIFF_The-European-market-of-insects-as-feed.pdf (accessed on 22 January 2022).
20. Mariod, A.A. The Legislative Status of Edible Insects in the World. In *African Edible Insects as Alternative Source of Food, Oil, Protein and Bioactive Components*; Springer: Cham, Switzerland, 2020; pp. 141–148. [CrossRef]
21. Koko, M.Y.F.; Mariod, A.A. Sensory Quality of Edible Insects. In *African Edible Insects as Alternative Source of Food, Oil, Protein and Bioactive Components*; Springer: Cham, Switzerland, 2020; pp. 115–122. [CrossRef]
22. van Huis, A. Edible Insects Are the Future? *Proc. Nutr. Soc.* **2016**, *75*, 294–305. [CrossRef]
23. Mancini, S. Insects as Sustainable Feed and Food. *Acta Fytotech. Et Zootech.* **2020**, *23*, 214–216. [CrossRef]
24. Svanberg, I.; Berggren, Å. Insects as Past and Future Food in Entomophobic Europe. *Food Cult. Soc.* **2021**, *24*, 624–638. [CrossRef]
25. European Food Safety Authority. Risk Profile Related to Production and Consumption of Insects as Food and Feed. *EFSA J.* **2015**, *13*, 4257. [CrossRef]
26. Lähteenmäki-Uutela, A.; Marimuthu, S.B.; Meijer, N. Regulations on Insects as Food and Feed: A Global Comparison. *J. Insects Food Feed.* **2021**, *7*, 849–856. [CrossRef]
27. Bear, C. Making Insects Tick: Responsibility, Attentiveness and Care in Edible Insect Farming. *Environ. Plan. E Nat. Space* **2021**, *4*, 1010–1030. [CrossRef]
28. Turck, D.; Castenmiller, J.; de Henauw, S.; Hirsch-Ernst, K.I.; Kearney, J.; Maciuk, A.; Mangelsdorf, I.; McArdle, H.J.; Naska, A.; Pelaez, C.; et al. Safety of Dried Yellow Mealworm (*Tenebrio molitor* Larva) as a Novel Food Pursuant to Regulation (EU) 2015/2283. *EFSA J.* **2021**, *19*, e06343. [CrossRef]
29. Turck, D.; Castenmiller, J.; de Henauw, S.; Hirsch-Ernst, K.I.; Kearney, J.; Maciuk, A.; Mangelsdorf, I.; McArdle, H.J.; Naska, A.; Pelaez, C.; et al. Safety of Frozen and Dried Formulations from Migratory Locust (*Locusta migratoria*) as a Novel Food Pursuant to Regulation (EU) 2015/2283. *EFSA J.* **2021**, *19*, e06667. [CrossRef]
30. Turck, D.; Bohn, T.; Castenmiller, J.; de Henauw, S.; Hirsch-Ernst, K.I.; Maciuk, A.; Mangelsdorf, I.; McArdle, H.J.; Naska, A.; Pelaez, C.; et al. Safety of Frozen and Dried Formulations from Whole Yellow Mealworm (*Tenebrio molitor* Larva) as a Novel Food Pursuant to Regulation (EU) 2015/2283. *EFSA J.* **2021**, *19*, e06778. [CrossRef]
31. Turck, D.; Bohn, T.; Castenmiller, J.; de Henauw, S.; Hirsch-Ernst, K.I.; Maciuk, A.; Mangelsdorf, I.; McArdle, H.J.; Naska, A.; Pelaez, C.; et al. Safety of Frozen and Dried Formulations from Whole House Crickets (*Acheta domesticus*) as a Novel Food Pursuant to Regulation (EU) 2015/2283. *EFSA J.* **2021**, *19*, e06779. [CrossRef]
32. Onwezen, M.C.; Bouwman, E.P.; Reinders, M.J.; Dagevos, H. A Systematic Review on Consumer Acceptance of Alternative Proteins: Pulses, Algae, Insects, Plant-Based Meat Alternatives, and Cultured Meat. *Appetite* **2021**, *159*, 105058. [CrossRef]
33. de Boer, J.; Schösler, H.; Boersema, J.J. Motivational Differences in Food Orientation and the Choice of Snacks Made from Lentils, Locusts, Seaweed or "Hybrid" Meat. *Food Qual. Prefer.* **2013**, *28*, 32–35. [CrossRef]
34. Caparros Megido, R.; Gierts, C.; Blecker, C.; Brostaux, Y.; Haubruge, É.; Alabi, T.; Francis, F. Consumer Acceptance of Insect-Based Alternative Meat Products in Western Countries. *Food Qual. Prefer.* **2016**, *52*, 237–243. [CrossRef]
35. Tan, H.S.G.; Fischer, A.R.H.; van Trijp, H.C.M.; Stieger, M. Tasty but Nasty? Exploring the Role of Sensory-Liking and Food Appropriateness in the Willingness to Eat Unusual Novel Foods like Insects. *Food Qual. Prefer.* **2016**, *48*, 293–302. [CrossRef]
36. Alexander, P.; Brown, C.; Arneth, A.; Dias, C.; Finnigan, J.; Moran, D.; Rounsevell, M.D.A. Could Consumption of Insects, Cultured Meat or Imitation Meat Reduce Global Agricultural Land Use? *Glob. Food Secur.* **2017**, *15*, 22–32. [CrossRef]
37. Sogari, G.; Vantomme, P. *A Tavola Con Gli Insetti*; Mattioli 1885: Fidenza, Italy, 2014.
38. Dagevos, H. A Literature Review of Consumer Research on Edible Insects: Recent Evidence and New Vistas from 2019 Studies. *J. Insects Food Feed* **2021**, *7*, 249–259. [CrossRef]
39. Sogari, G.; Liu, A.; Li, J. Understanding Edible Insects as Food in Western and Eastern Societies. *Environ. Health Bus. Oppor. New Meat Altern. Mark.* **2018**, 166–181. [CrossRef]
40. Menozzi, D.; Sogari, G.; Veneziani, M.; Simoni, E.; Mora, C. Explaining the Intention to Consume an Insect-Based Product: A Cross-Cultural Comparison. In *Theory of Planned Behaviour: New Research*; Seyal, A.H., Al, E., Eds.; Nova Publisher: New York, NY, USA, 2017; pp. 201–215. ISBN 9781536110968.
41. Sogari, G.; Menozzi, D.; Hartmann, C.; Mora, C. How to Measure Consumers Acceptance towards Edible Insects?—A Scoping Review About Methodological Approaches. In *Edible Insects in the Food Sector*; Springer International Publishing: Cham, Switzerland, 2019; pp. 27–44. [CrossRef]
42. Liu, A.J.; Li, J.; Gómez, M.I. Factors Influencing Consumption of Edible Insects for Chinese Consumers. *Insects* **2019**, *11*, 10. [CrossRef]

43. Mancini, S.; Sogari, G.; Menozzi, D.; Nuvoloni, R.; Torracca, B.; Moruzzo, R.; Paci, G. Factors Predicting the Intention of Eating an Insect-Based Product. *Foods* **2019**, *8*, 270. [CrossRef]
44. Pambo, K.O.; Okello, J.J.; Mbeche, R.M.; Kinyuru, J.N.; Alemu, M.H. The Role of Product Information on Consumer Sensory Evaluation, Expectations, Experiences and Emotions of Cricket-Flour-Containing Buns. *Food Res. Int.* **2018**, *106*, 532–541. [CrossRef]
45. Arnaudova, M.; Brunner, T.A.; Götze, F. Examination of Students' Willingness to Change Behaviour Regarding Meat Consumption. *Meat Sci.* **2022**, *184*, 108695. [CrossRef]
46. Elorinne, A.-L.; Niva, M.; Vartiainen, O.; Väisänen, P. Insect Consumption Attitudes among Vegans, Non-Vegan Vegetarians, and Omnivores. *Nutrients* **2019**, *11*, 292. [CrossRef]
47. Jones, V. 'Just Don't Tell Them What's in It': Ethics, Edible Insects and Sustainable Food Choice in Schools. *Br. Educ. Res. J.* **2020**, *46*, 894–908. [CrossRef]
48. House, J. Insects as Food in the Netherlands: Production Networks and the Geographies of Edibility. *Geoforum* **2018**, *94*, 82–93. [CrossRef]
49. la Barbera, F.; Verneau, F.; Videbæk, P.N.; Amato, M.; Grunert, K.G. A Self-Report Measure of Attitudes toward the Eating of Insects: Construction and Validation of the Entomophagy Attitude Questionnaire. *Food Qual. Prefer.* **2020**, *79*, 103757. [CrossRef]
50. Menozzi, D.; Sogari, G.; Mora, C.; Gariglio, M.; Gasco, L.; Schiavone, A. Insects as Feed for Farmed Poultry: Are Italian Consumers Ready to Embrace This Innovation? *Insects* **2021**, *12*, 435. [CrossRef]
51. Moruzzo, R.; Riccioli, F.; Espinosa Diaz, S.; Secci, C.; Poli, G.; Mancini, S. Mealworm (*Tenebrio molitor*): Potential and Challenges to Promote Circular Economy. *Animals* **2021**, *11*, 2568. [CrossRef] [PubMed]

MDPI

Review

Overview on Innovative Packaging Methods Aimed to Increase the Shelf-Life of Cook-Chill Foods

Maria Lisa Clodoveo [1], Marilena Muraglia [2,*], Vincenzo Fino [2], Francesca Curci [2], Giuseppe Fracchiolla [2] and Filomena Faustina Rina Corbo [2]

1 Interdisciplinary Department of Medicine, University Aldo Moro Bari, 70125 Bari, Italy; marialisa.clodoveo@uniba.it

2 Department of Pharmacy-Drug Sciences, University Aldo Moro Bari, 70125 Bari, Italy; vincenzo.fino@gmail.com (V.F.); francesca.curci@uniba.it (F.C.); giuseppe.fracchiolla@uniba.it (G.F.); filomena.corbo@uniba.it (F.F.R.C.)

* Correspondence: marilena.muraglia@uniba.it

Abstract: The consumption of meals prepared, packaged, and consumed inside and outside the home is increasing globally. This is a result of rapid changes in lifestyles as well as innovations in advanced food technologies that have enabled the food industry to produce more sustainable and healthy fresh packaged convenience foods. This paper presents an overview of the technologies and compatible packaging systems that are designed to increase the shelf-life of foods prepared by cook–chill technologies. The concept of shelf-life is discussed and techniques to increase the shelf life of products are presented including active packaging strategies.

Keywords: cook–chill technology; foods; packaging; shelf-life; sustainability

Citation: Clodoveo, M.L.; Muraglia, M.; Fino, V.; Curci, F.; Fracchiolla, G.; Corbo, F.F.R. Overview on Innovative Packaging Methods Aimed to Increase the Shelf-Life of Cook-Chill Foods. *Foods* **2021**, *10*, 2086. https://doi.org/10.3390/foods10092086

Academic Editors: Elena Canellas and Marlene Cran

Received: 22 June 2021
Accepted: 30 August 2021
Published: 3 September 2021

1. Introduction

1.1. Food Away from Home (FAFH) vs. Food Packaging Technology

The consumption of food away from home (FAFH) is constantly rising in the whole industrialized society because of the increasing need to carry out activities for both children at school and adults at work [1,2]. A rational approach should take in account innovative methods for cooking food stuff and new packaging. Foods are expected to be tasty and healthy even if not cooked at the moment of consumption. A rational design for cooking and store foods could reduce waste and containing food expenditure. Food packaging technology research should help food manufacturers to reach this goal meeting the needs of consumers, the society and the manufacturers, often very different from each other.

While the consumers are oriented to products that match their own lifestyle, the society needs safer and environmentally sustainable products that means major attention also on packaging material, not only on food stuff. On the other hand, the manufacturers are interested in the use of better and more cost-effective packaging materials and technologies to meet the market demand and make profits [3,4].

1.2. Food, Energy, Environment Trilemma

In the next decades, the global community will attend to an increasing of food demand, due to the third world demographic explosion, and the displacement of food crops, due to a change of the land use that will be directed to the biofuel and bioenergy production [5]. A trend is visible in the scientific literature showing the conflict between food production and bioenergy/biofuel production is the so-called *"food, energy, environment trilemma"* [6,7]. One of the possible strategies that should be used to solve this problem is the enhancement of the food transformation efficiency intended as a reduction of waste and, therefore, a general reduction of losses in edible material during food manipulation and consumption.

Now larger and medium sized catering and hospitality companies are forced to use new technologies for cooking and storage due to business needs. Maybe in the near future, small companies, restaurants, and small communities including families will embrace this new "food preparation philosophy" in order to reduce food costs.

The cook–chill processing procedure has been widely used in catering since the early Nineties [8] coming from the practice of serving unused food the day after preparation to avoid waste. This procedure consists in rapid cooling of cooked food to 2–4 °C before or after packaging that will be stored at the same temperature until it will be served both hot (by warming at 70 °C) or cold.

The current paper presents an overview of the advances in the cook–chill technology developed to increase the shelf-life of foods without loss of their nutritional and organoleptic value.

2. The Cook–Chill Technology

The habit of consuming food is a common practice for households who have access to refrigeration. At the same time, lack of adequate preparation and storage can result in incidents of food poisoning [9]. This can also result in considerable food spoilage which contributes to increasing food wastage.

Companies involved in serving meals have implemented specific technologies of the last two or three decades able to prepare and package meals designed for reheating without loss of taste, flavor and texture. To obtain this result it is strictly necessary to follow a simple protocol that is schematically shown in Figure 1.

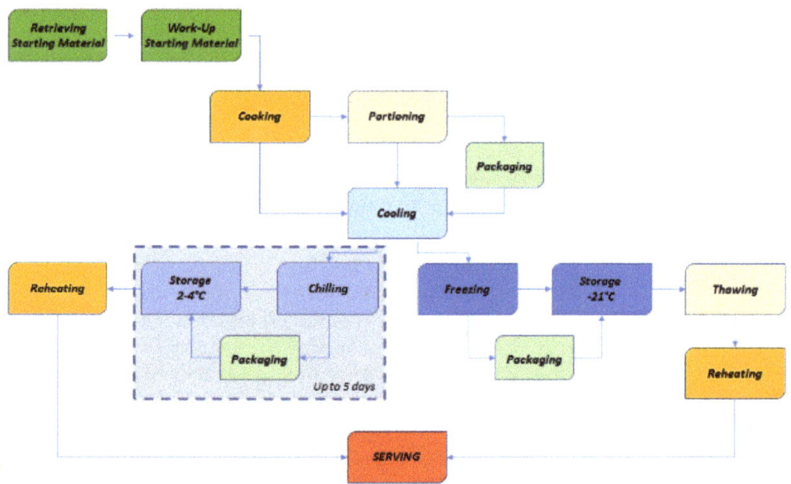

Figure 1. A block scheme of the principal steps in cook–chill procedure.

2.1. Retrieving and Work-Up of Raw Materials

Choice, storage, handling, and preparation operations of raw materials follow simple hygienic rules principally derived from good sense and coded by administration such as U.S. Food and Drug Administration (FDA) or European Food Safety Authority (EFSA) (see legislation paragraph below). The standard quality of starting materials must be as high as possible because it is not going to improve with cooking, so the systematic control on the supplier chain guarantees to keep the quality level.

After selection of raw materials, it is necessary to follow the basic food safety principles, to ensure that the proper temperature and humidity of the starting materials are kept. In addition, in preparing procedures, basic food safety principles have to be applied: separate work surfaces and dedicated utensils should be used for different types of food (fish, meat, poultry and vegetables) to prevent cross-contamination. At the best conditions, food preparation should take place in a separate area from cooking and post-cooking.

Frozen raw materials should be completely thawed out before use, rapid high temperature thawing should be avoided because it can allow the growth of pathogens [10–12] and may leave cold spots at the core of the food. Thawing with microwave ovens is generally not recommended for the same reasons [7,13,14].

2.2. Cooking

Almost all kinds of cooking procedures are compatible with the cook–chill procedure, this fact makes it a very flexible tool for food manufacturers. The main and unique problem that it must be faced is the temperature control at the core of the food to ensure the proper destruction of pathogens that may be present inside it [15].

The control of temperature depends on the type of food, its thickness, the type of pathogens that eventually present, its charge, and cooking methods [16–23].

2.3. Cooling

The cooling step of the procedure is the most important of the entire process and it can be carried out both before or after portioning and/or packaging of the cooked food.

The chilling apparatus that must be used in a cook–chill procedure should be capable of chilling the food to around 3 °C in 2–4 h in dependence of the just cooked food temperature. The rapid chill of the food avoids bacterial growth and preserves the appearance, texture, flavor and nutritional value of the food [24].

Food probes are frequently used to check the process, because there are plenty of variables that should influence the chill rate. If we only consider the type of food and the cooking procedure followed in different recipes, we should have a too large volume of information that we have to manage. Furthermore, each kind of food presents different chill problems related principally on the nutrient content, on the water amount and its quantity. Liquid preparation can be efficiently cooled by adding microbiologically pure slush ice to concentrated soup dishes, while solids foods such as meat, fish or vegetables are conveniently air-chilled [18].

Heat conduction by air circulation depends on surface area for heat exchange, gradient temperature between hot surfaces and air flow, surface heat transfer coefficient.

Each product-refrigerating system couple can be characterized by an own surface heat transfer coefficient that depends on the velocity impressed to the air flow and on its specific heat. This constant value varies in a range from 5 to 500 $W \cdot m^{-2} \cdot K^{-1}$ [25].

With the aim to improve the performance of the chilling technology a lot of computational studies were performed on fluid dynamics [26–30] that presumably will influence the next generation chilling technology.

While the improvement of the shelf-life of food using a cold chain after the cooking process is a simple idea, the industrial or semi-industrial scale-up process is not so easy. The rapid and homogeneous reduction of food temperature appear to be the main problem of the entire process involving a lot of factors ascribed to the intrinsic nature of the foodstuff and its variability, so the equipment should be able to be used to apply the procedure to the highest possible number of recipes to make the process economically favorable for foodservice industries.

3. Patent in Cook–Chill Technology

The following sections present a brief overview of the recent patent activity in food service and catering. Patents related to both equipment designed for the cook–chill process and other types of facilities useful for heat removal, cooking, sterilization or storage of low temperatures that could be used in the same process are discussed.

3.1. Cook–Chill Apparatus

This group of inventions includes both cooking apparatus for low scale food production than very large devices useful for industrial purposes. An example of a low scale apparatus is a device consisting in a plate with a lower multizonal heater that can cooks two different recipes at the same time divided in two different vessels. It has also an upper heater for grilling [31–33].

Several inventions are based on the same principle, this kind of apparatus consists in a water bath (typically a tank) in which packaged food can be cooked and rapidly chilled changing the temperature of the bath. Inventions differ between them in the warming system [34], water filling and replacement devices, and temperature control devices [35–38]. Some similar apparatus was designed also for unpackaged food [39].

3.2. Cooker/Rethermalizer Systems

Cooker or re-thermalizer devices were designed in a similar way in respect to the cooker–chiller devices for packaged food. Food fills racks in heating liquids filled tanks. The bath temperature is controlled by temperature sensor devices [40–45].

3.3. Storage

Cook–chill technologies rely on the containment of foods and would not be possible without materials that can withstand cooking and storage temperatures.

These items can be tubular bags with one end open and the other sealed with a heat seal. The bags consist of a multilayered co-extruded plastic film made of various inert polymers such as nylon, polyethylene, polyesters used alone or in various combinations between them [46–49].

3.4. Procedures

This section presents some examples of complementary inventions (devices and protocols) in work-up procedures of cook–chill processes.

One of these is a device for distributing cooled food, for large kitchens, comprising a belt conveyor, several mobile feed distributor devices and a refrigerator circuit closed with a refrigeration system. The cooling capacity of the system can be divided into several cooling points which are arranged in the longitudinal direction of the belt conveyor and have means for connecting a feed distributor apparatus [50]. A computer program for management of out-of-hospital cooking service that can on-line receive meal orders from a plurality hospital is described in a patent work [51]. This system can efficiently send information to large scale kitchens that cook and deliver daily meal to each patient that is registered on a master file containing nutritional and eat data according to medical prescriptions.

In addition, small companies such as restaurants can follow industrial protocol for preparing pre-cooked food [52] dividing the process in sequential steps: food cooking, portioning and packaging, removing oxygen by vacuum pump, pasteurization or sterilization and blast-chilling of the packages.

4. The Shelf-Life of Foods

4.1. Shelf-Life Definition

A unique definition of shelf-life does not exist. Generally, it could be defined as the amount of time in which a useable wholesome state under the expected conditions of storage using agreed upon methods and acceptance criteria [53,54].

In the last 25 years, experts and institutions have provided several definitions speaking in term of acceptable sensory and nutritional properties of foods, or general eligibility under defined environmental conditions.

Consumer preferences and lifestyles have impacted food product formulation, preparation, and consumption habits, and this too provides the impetus for renewed focus on shelf-life determination. Upon commercialization, shelf-life tests must be performed upon the first several lots of production in order to verify previously determined outcomes with prototypical samples. In some instances, challenge studies must be performed to validate the ongoing safety throughout shelf-life. Then for ongoing food production, routine shelf-life testing is an essential quality metric, albeit a lagging indicator of quality. Attributes measured include microbiological counts, chemical degradation, physical deterioration, and sensory properties [55].

In 1993 IFST (Institute of Food Science and Technology) for the first time identified the key factors that must be considered in the shelf-life assessing process: (a) safety; (b) sensory, chemical, physical and microbiological characteristics; and (c) nutritional label declaration.

In 2005, it was defined the sensory shelf-life [56] as the time during which the product keeps its sensory characteristics and performance declared by the manufacturer providing to the end users its benefits.

It is clear for all definitions that safety of the foods is out of discussion during this time, in other words, the food is always considered safe for a longer period than the shelf-life.

4.2. Law Regulation on Shelf-Life

In this paragraph, we consider the law regulation of FDA and EFSA as supranational organizations belonging respectively to the United States of America and the European Union that represent a part of the world population that may influence the food trade, analyzing contact points and main differences between them.

Law regulation on labelling of information related to the shelf-life of a food product may vary quite a lot from country to country.

USDA (United States Department of Agriculture) in the United States [57] reports that product dating on food stuff is not required by federal regulation except for infant foods, so date applying on labels is a voluntary activity done by manufacturers that follow FSIS (Food Safety and Inspection Service) regulation [58].

Close to the calendar date, eventually present on the package, there must be a phrase that explains the meaning of the date shown such as, sell by, use before, or use by.

"Use by date" is an expression used mainly for infant foods that means "the last date recommended for the use of the product while at peak quality". This is information that refers directly to the shelf-life of the product since it is not a safety date.

By the way, US-FDA (Food and Drug Administration) permits the use of this phrase on infant foods under its close control meaning that the nutritional content of the food is a safety statement for this kind of consumers and ensuring them that the formulas contain not less than the quantity of each nutrient as described on the label.

Since December 2014 [59] it was operating the regulation UE 1169/2011 in the whole EC that renewed the legislation on labelling [60] regarding all types of foods: fresh, packaged by manufactures and product by catering services.

This regulation covers all the aspects connected with product presentation and advertising. From a practical point of view [61], since the application of the regulation all food packages are deeply changed regarding its clarity (also font size character has been regulated) and completeness including the whole list of ingredients, the indication on the country of origin, nutritional facts reported as percentage of the RDA (recommended dietary allowance), the list of allergens eventually present in the food, and durability expressed in term of the date of minimum durability that indicate the date until which the food retains its specific properties when properly stored.

This kind of information is directly related to the shelf-life of the product, and it is reported on the package using the form "best before", in which a date (with a format day-month-year) is closely reported or there is an indication on where it can be retrieved. This phrase may be substituted by "best before end" date in some specific cases. The indication on durability of highly perishable food, that represents an actual danger for human health after a short period, must be reported as "use by" date. After the "use by" date a food shall be deemed to be unsafe [62] and so it could not be considered an indication of the shelf-life of the product.

For frozen food, it is mandatory to indicate the freezing date ("frozen on" date) or the date of first freezing if the supply chain provides for more than one freezing operation.

Recently, EFSA's Panel on Biological Hazards (BIOHAZ) provided a series of scientific opinions useful to establish guidelines on date marking and related food information in view of the implementation by food business operators (FBOs) of regulation (EU) No 1169/2011 on food information to consumers as an integral part of their food safety management system (FSMS). Specifically, the guide provides guidance on determining shelf life and storage conditions and identifying factors that affect shelf-life determination [63,64].

4.3. Methods to Improve Food Shelf-Life

Many different methods are known to improve the shelf-life of a food, from ancient times humans had the problem of having a safe food to eat when it was not possible to retrieve it fresh. Therefore, some of the oldest ways to treat food for storage were drying with sun exposure or by osmotic mechanisms, while in recent time cold storage, the use of chemical additives, sterilization by ionizing radiation or the use of engineered packages are normal practices for manufacturers and generally accepted by consumers.

All these methods are not suitable for all kinds of food or preparations so, it is necessary to study a way of reduction or control of the microbial charge for each case.

Food deriving from a cook–chill supply chain belongs to a special class of products for which a shelf-life enhancement would be desirable to reach as complicated to obtain. Freezing, heating or drying procedure procedures can result in significant changes to the organoleptic properties of foods [65,66]. In addition, the use of chemical additives is not very suitable because they may cause a sensory modification of the product and even if not, their presence could make the product less appealing considering today's most prevalent consumer sensitivity [67]. European consumers, for instance, are now used to reading the product label they buy by having more awareness about the food they consume.

The use of high frequency electromagnetic radiation to sterilize cooked food is not generally applicable because the reaction that may occur on the food surface cannot be predicted [68].

The choice of the proper type of package may help to preserve cooked food, so several studies have been carried out to develop new materials and applications taking in due account the needs of consumers, the society, and the manufacturers [69–71]. Consumers are oriented on high quality materials that are more convenient to meet their lifestyle, the society watches on human health safety as well as friendlier products in respect to the environment to meet the needs of public and environmentalists. The manufacturers need better and more cost-effective packaging technologies to satisfy the market and make profits.

To meet the market request, developments in packaging materials have focused four main class of product that are developed in the last decades [72]:

- Sustainable packaging involves environmentally friendly technologies that are socially acceptable and economically advantageous.
- Intelligent packaging involves the use of package integrated devices such as RFID (radio frequencies identification) tags, time or temperature indicators and sensors for tracking activities or sensing the internal or external environment of the package and monitoring the product quality.
- Active packaging utilizes advanced technologies that actively modify the inner atmosphere of the package in order to extend the shelf-life of the stored product.
- Responsive food packaging in which particular materials are able to react against unfavorable stimuli in order to preserve the food quality [73].

In the following section, we will consider the field of application of packages in food technologies that are able to extend the shelf-life of cooked food and research perspectives on these engineered materials.

5. Food Active Packaging

The use of specific packaging could be of help in preserving the nutritional and organoleptic properties of food prepared by using cook–chill technology. In this regard, many new materials have been developed to contain food, constituting a real field of applied research that aims to introduce safe, sustainable and low-cost packaging systems to the food market. The interest in active packaging is confirmed by the increasing number of scientific papers which has more than doubled in the last decade [74,75].

Even if active functions of packaging have gained more visibility than original attributes, such as mechanical strength, barrier performance and thermal stability, every new material must satisfy each of these basic properties until it becomes a potential material for food carrying. In addition, the use and development of new packaging, as well as the use of new additives that provide interesting properties for food preservation, require careful safety and toxicological evaluations due to the potential presence of food contaminants of packaging origin on the quality and safety of fresh food [76–78].

To extend the shelf-life of a foodstuff it is crucial to measure the microbiological count by optimizing some factors including oxygen partial pressure, moisture and water activity, sunlight exposure and initial microbial charge [79].

Packaging with antimicrobial purposes can interact with the food contained in it or with the empty space above it, to reduce, retard or even inhibit the growth of pathogenic microorganisms and food spoilage [80].

A class of material deeply investigated for food packaging is polymers (low-density polyethylene, LDPE, in most cases) which properties such as density and permeability (related with release of small molecules) are tuned looking at the physical and chemical properties of the spread substance [81].

Polymers and composites used as emitting materials can be divided into two main categories:

- Polymers that incorporate organic compounds.
- Polymers that incorporate inorganic compounds.

Summarily, organic compounds that may be used in active packaging material, must have several features to satisfy food technology requirements and improve the shelf-life of the food itself. They must be safe for humans by ingestion, inhalation, and contact, they do not interfere with the organoleptic properties of the food, they should be able to contrast the growth of many microorganisms (bacteria and fungi) that cause food spoilage and they should have some antioxidant activity.

Synthetic additives that may be tailored with all these features have to satisfy many requirements before safety declaration (as expected for drugs). Therefore, organic compounds derived from natural sources that are deeply studied on animal models and humans, are suitable for this application.

Essential oils (EO) are mixture of compounds derived from aromatic plants that have been investigated not only for being natural product, but also because they have used since ancient time for their biological properties as antioxidant, antimicrobial, anti-tumor, analgesic, anti-pest, anti-diabetic, and anti-inflammatory [82–85].

The EOs mechanism of action has been extensively reported in the literature and concerns the breakdown of bacterial cell wall, although the effect on the destruction of enzymes or membrane proteins or the spillage of cellular content after cytoplasmic membrane breakage are also possible [83,86,87].

Our experience in evaluating EOs drug efficacy as antimicrobial agents allows us to promote their use in the engineering of new active packaging [88–92].

Among the class of organic additives, chitosan deserves particular attention: a polymer that shows unique characteristics and potential application.

It is directly derived from chitin, a very cheap natural occurring material which is constituted by a linear polymer of β-1,4-*N*-acetyl-D-glucosamine whose structure is close to the β-1,4-D-glucopyranoside chain of cellulose except for the acetamide group at C-2 position of the monosaccharide unit.

Chitosan and its derivatives show significant antimicrobial activity [93] alone or in combinations with essential oils [94] or other small molecules with antimicrobial properties [95–97]. Different polymers have been studied for food packaging, poly (butyleneadipate-co-terephtalate) (PBAT) [98,99] is used because of its mechanical properties that are comparable with LDPE. It is completely biodegradable and compostable, obtained by total metal-free processes and allowed for food application by FDA.

Cellulose is a very versatile polymer that can be obtained as a nanostructured membrane which may grow in the presence of other polymers and various additives producing bionanocomposites that may be used in food packaging technology (including active packaging) [100].

Other cellulose derivatives (hydroxypropyl-methylcellulose, HPHC, extensively used in drug formulation as cover agent for solid preparations) was recently studied [101] as matrix for controlled release of antioxidants (green tea extract) loaded on polylactic acid nanoparticles.

The release of the active compound from the matrix was studied and results encourage the use of the material as a potential candidate in active food packaging.

Some metals or metal-oxides show antimicrobial properties that several researchers have exploited for food applications. The amount of release of metal particles is the real challenge to meet the legislation of health institutions (EFSA, FDA).

Silver based nanocomposites of poly (3-hydroxybutyrate-co-3-hydroxyvalerate) (PHBV) and silver nanoparticles was studied [102] showing antimicrobial properties against pathogens such as *S. enterica* and *L. monocytogenes* and oxygen permeability in respect to the native polymer. PHBV was used also in composite with ZnO at various particle dimensions [103–105].

A recent work [106] reports an innovative process for deposition of copper-containing hybrid organic-inorganic thin film that improves the antimicrobial activity of the copper at the level of silver.

Some materials for active packaging have already been used in the food industry to preserve highly perishable foods, especially for meat, poultry, seafood, and their derivatives [107–109] but no application has been reported for foods prepared with cook–chill technology.

These packaging materials include substances used against microbial growth, oxidation processes or contain oxygen scavengers, carbon dioxide emitters and absorbers, moisture regulators, flavour releasers and absorbers.

6. Conclusions

Food safety became a topic in scientific publications because of the growing interest of the whole society on nutrition themes and rational exploitation of food resources. By providing an analysis of the literature on food technology we focused our attention on innovative packaging technology and food bio-preservatives aimed to increase the shelf-life of cooked chill foods.

Despite the developments and the strong correlation between the sector of cook–chill procedures and food active packaging research, at the present time there are no reported examples of foods cooked with cook–chill technology and stored in food active packaging. These emerged as new fields of investigation that may lead to commercially viable products to improve sustainable and healthy food system production taking advantage of both manufacturers and consumers.

Since we are certain of the strong correlation between the two sectors, this paper could be a starting point for future investigations aimed to focus light on scientific findings of innovative food technology to raise awareness among food industry managers and stakeholders to the use of advanced food technology able to produce more sustainable and healthy food systems.

Author Contributions: Conceptualization M.L.C.; methodology and validation M.M.; formal analysis, G.F.; data curation, F.C.; writing—original draft preparation, V.F. and M.M.; writing—review and editing, M.M. and V.F.; supervision, F.F.R.C. All authors have read and agreed to the published version of the manuscript.

Funding: This research received no external funding.

Conflicts of Interest: The authors declare no conflict of interest.

References

1. Dai, T.; Yang, Y.; Lee, R.; Fleischer, A.S.; Wemhoff, A.P. Life cycle environmental impacts of food away from home and mitigation strategies—A review. *J. Environ. Manag.* **2020**, *265*, 110471. [CrossRef] [PubMed]
2. Keeble, M.; Adams, J.; Sacks, G.; Vanderlee, L.; White, C.M.; Hammond, D.; Burgoine, T. Use of Online Food Delivery Services to Order Food Prepared Away-From-Home and Associated Sociodemographic Characteristics: A Cross-Sectional, Multi-Country Analysis. *Int. J. Environ. Res. Public Health* **2020**, *17*, 5190. [CrossRef] [PubMed]
3. Manning, P.; Taylor, G.; Hanley, M.E. Bioenergy, Food Production and Biodiversity—An Unlikely alliance? *Glob. Chang. Biol. Bioenergy* **2014**, *7*, 570–576. [CrossRef]
4. Takacs, B.; Borrion, A. The Use of Life Cycle-Based Approaches in the Food Service Sector to Improve Sustainability: A Systematic Review. *Sustainability* **2020**, *12*, 3504. [CrossRef]
5. IEA Report. Available online: https://www.iea.org/reports/world-energy-outlook-2020 (accessed on 31 October 2020).
6. Tilman, D.; Socolow, R.; Foley, J.A.; Hill, J.; Larson, E.; Lynd, L.; Pacala, S.; Reilly, J.; Searchinger, T.; Somerville, C.; et al. Beneficial Biofuels–The Food, Energy, and Environment Trilemma. *Science* **2009**, *325*, 270–271. [CrossRef]
7. Da Rosa, F.S.; Lunkes, R.J.; Spigarelli, F.; Compagnucci, L. Environmental Innovation and the Food, Energy and Water Nexus in the Food Service Industry. *Resour. Conserv. Recycl.* **2021**, *166*, 105350–105359. [CrossRef]
8. Gormley, R.; Tansey, F. Sous Vide and Cook-Chill processing. In *Handbook of Food Safety Engineering*; Sun, D.V., Ed.; Wiley-Blackwel: Hoboken, NJ, USA, 2011; pp. 468–496. ISBN 978−1−444−33334−3. [CrossRef]
9. James, C.; Onarinde, B.A.; James, S.J. The Use and Performance of Household Refrigerators: A Review. *Compr. Rev. Food Sci. Food Saf.* **2017**, *16*, 160–179. [CrossRef] [PubMed]
10. Lund, B.; Baird-Parker, A.C.; Baird-Parker, T.C.; Gould, G.W.; Gould, G.W. (Eds.) *Microbiological Safety and Quality of Food;* Springer Science & Business Media: Berlin, Germany, 2000; Volume 1.
11. Adams, M.R.; Moss, M.O.; McClure, P. *Food Microbiology*, 4th ed.; The Royal Society of Chemistry: Cambridge, UK, 2015.
12. Szymkowiak, A.; Guzik, P.; Kulawik, P.; Zając, M. Attitude-behaviour Dissonance Regarding the Importance of Food Preservation for Customers. *Food Qual. Prefer.* **2020**, *84*, 103935–103943. [CrossRef]
13. Chamchong, M.; Datta, K. Thawing of Foods in a Microwave Oven: I. Effect of Power levels and power cycling. *J. Microwave Power* **1999**, *34*, 9–21. [CrossRef]
14. Vandivabal, R.; Jayas, D.S. Non-uniform Temperature Distribution During Microwave Heating of Food Materials. A Review. *Food Bioprocess Technol.* **2010**, *3*, 161–171. [CrossRef]
15. Lyashchuk, Y.O.; Novak, A.I.; Kostrova, Y.B.; Shibarshina, O.Y.; Evdokimova, O.V.; Kanina, I.V. The study of persistence of microorganisms and parasites in food products. In *IOP Conference Series: Earth and Environmental Science;* IOP Publishing: Bristol, UK, 2021; Volume 640, p. 062002.
16. D'SA, E.M.; Harrison, M.A.; Williams, S.E.; Broccoli, M.H. Effectiveness of Two Cooking Systems in Destroying *Escherichia coli* O157: H7 and *Listeria monocytogenes* in Ground Beef Patties. *J. Food Prot.* **2000**, *63*, 894–899. [CrossRef]
17. Roh, S.H.; Oh, Y.J.; Lee, S.Y.; Kang, J.H.; Min, S.C. Inactivation of *Escherichia coli* O157: H7, *Salmonella, Listeria monocytogenes,* and Tulane virus in Processed Chicken Breast Via Atmospheric In-package Cold Plasma Treatment. *LWT-Food Sci. Technol.* **2020**, *127*, 109429. [CrossRef]
18. Baker, R.C.; Hogarty, S.; Poon, W.; Vadehra, D.V. Survival of *Salmonella typhimurium* and *Staphylococcus aureus* in eggs cooked by different methods. *Poult. Sci.* **1983**, *62*, 1211–1216. [CrossRef]
19. Soni, A.; Smith, J.; Thompson, A.; Brightwell, G. Microwave-induced Thermal Sterilization-A Review on History, Technical Progress, Advantages and Challenges as Compared to the Conventional Methods. *Trends Food Sci. Technol.* **2020**, *97*, 433–442. [CrossRef]
20. Rhee, M.S.; Lee, S.Y.; Hillers, V.N.; McCURDY, S.M.; Kang, D.H. Evaluation of Consumer-style Cooking Methods for Reduction of *Escherichia coli* O157: H7 in Ground Beef. *J. Food Prot.* **2003**, *66*, 1030–1034. [CrossRef]
21. Mattick, K.L.; Bailey, R.A.; Jørgensen, F.; Humphrey, T.J. The Prevalence and Number of *Salmonella* in Sausages and Their Destruction by Frying, Grilling or Barbecuing. *J. Appl. Microbiol.* **2002**, *93*, 541–547. [CrossRef]
22. Avens, T.S.; Albright, S.N.; Morton, A.S.; Prewitt, B.E.; Kendall, P.A.; Sofos, J.N. Destruction of Microorganisms on Chicken Carcasses by Steam and Boiling Water Immersion. *Food Control* **2002**, *13*, 445–450. [CrossRef]
23. Ayadi, M.A.; Makni, I.; Attia, H. Thermal Diffusivities and Influence of Cooking time on Textural, Microbiological and Sensory Characteristics of Turkey Meat Prepared Products. *Food Bioprod. Process.* **2009**, *87*, 327–333. [CrossRef]
24. James, S.J.; James, C. Chilling and Freezing of Food. In *Food Processing: Principles and Application*, 2nd ed.; Clark, S., Jung, S., Lamsal, B., Eds.; John Wiley and Sons Ltd.: New York, NY, USA, 2014; pp. 79–105. ISBN 9781118846315.
25. James, S.J.; James, C. *Chilled Foods. A Comprehensive Guide*, 3rd ed.; Woodhead Publishing Ltd.: Sawston, UK, 2008; Volume 375.
26. Redding, G.P.; Yang, A.; Shim, Y.M.; Olantunji, J.; East, A. A Review of the Use and Design of Produce Simulations for Horticultural Forced Air Cooling Studies. *J. Food Eng.* **2016**, *190*, 80–93. [CrossRef]
27. Tassou, S.A.; Gowreesunker, B.L.; Parpas, D.; Raeisi, A. Modelling Cold Food Chain Processing and Display Environments. In *Modelling Food Processing Operations*; Woodhead Publishing Ltd.: Sawston, UK, 2015; pp. 185–208. ISBN 978−1−78242−284−6.
28. Mondal, A.; Buchanan, R.L.; Lo, Y.M. Computational Fluid Dynamics Approaches in Quality and Hygienic Production of Semisolid Low-Moisture Foods: A Review of Critical Factors. *J. Food Sci.* **2014**, *79*, 1861–1870. [CrossRef] [PubMed]

29. Bojacà, G.R.; Schrevens, E.; Suay, R. Analysis of Air Temperature Distribution Inside a Cold Store by Means of Geostatistical Methods. *Acta Hortic.* **2012**, *94*, 29–37. [CrossRef]
30. Lemus-Mandaca, R.A.; Vega-Gálvez, A.; Moraga, N.O. Computational Simulating and Developments Applied to Food Thermal Processing. *Food Eng. Rev.* **2011**, *3*, 121–135. [CrossRef]
31. Lamanna, D.; Menz, R.J. Prepared Meal System with Lower Hotplate Having a Plurality of Heating Zones and an Upper Radiant Heating Element and Operated in Accordance with Pre-Programmed Cooking Programs Selected by a User. U.S. Patent 5,708,255, 13 January 1998.
32. Hensel, K.J.; Li, Z.K. Toast Oven Having at Least Five Heating Elements. U.S. Patent 8,878,106B2, 3 February 2014.
33. Breunig, M.; Greiner, M. Method of Conducting at Least One Cooking Process. U.S. Patent 20,140,050,826A1, 20 February 2014.
34. Cocchi, G.; Lazzarini, R.; Zaniboni, G. Machine and Method for Producing and Dispensing Liquid or Semi-Liquid Consumer Food Products. U.S. Patent 20140220194A1, 24 June 2014.
35. Pardo, R.J. Cook/Chill Tank. U.S. Patent 5,280,748, 24 February 1992.
36. Kennedy, B.C. Cooker Utilizing a Peltier Device. U.S. Patent 7,174,720, 13 February 2007.
37. Alexander, C. Heated or Cooled Dishware and Drinkware. U.S. Patent 10,188,229B2, 14 November 2017.
38. Alexander, C.; Leith, D.J.; Timperi, M.J.; Wakeham, C.T. Portable Cooler Container with Active Temperature Control. U.S. Patent 10,743,708B2, 13 August 2020.
39. Dreano, C. Cooking Machine. U.S. Patent 5,005,471, 9 April 1991.
40. Polster, L.S. Cooker/Rethermalizer. U.S. Patent 5,539,185, 23 July 1996.
41. Storek, D.; Otillar, R.P.; Sequeira, A.L. Liquid Movement and Control within a Container for Food Preparation. U.S. Patent 20,160,206,136A1, 26 April 2016.
42. Rosalia, J.A.; Chen, H. Modularized Food Preparation Device and Tray Structure for Use Thereof. U.S. Patent 10,674,855B2, 18 June 2015.
43. Svensson, S.A. Rethermalizer with Expansible Rack. U.S. Patent 5,706,718, 13 January 1998.
44. Shei, S.M. Food Warming Apparatus and Method. U.S. Patent 7,105,779B2, 17 November 2005.
45. Veltrop, L.J.; Schroeder, J.; Hartfelder, C.; Guasta, J. Apparatus and Method for Maintaining Cooked Food in a Ready-to-Use Condition. U.S. Patent 8,096,231B2, 28 December 2010.
46. Menges, J.A.; Bachert, E.E.; Schmidt, G.E. Cook and Chill Casing. U.S. Patent 20,040,253,399A1, 16 December 2004.
47. Schif, H.J.; Schiffmann, J.M. Multilayer Planar or Tubular Food Casing or Film. A.U. Patent 2,003,245,922B8, 17 March 2009.
48. Hynes, K.A. Food Container and Method. U.S. Patent 2,0060,034,986A1, 16 February 2006.
49. Lyzenga, D.A.; Weber, J.T. Easy-Opening Flexible Film Packaging Products and Manufacturing Methods. E.S. Patent 2,555,259T3, 30 December 2015.
50. Schumacher, H. EU Patent Device for Distributing Cooled Foods, in Particular for Large Kitchens. EP 2,090,853A2, 19 August 2009.
51. Komiya, H.; Kawabata, K. Out-of -Hospital Cooking Management Method and System Therefor. U.S. Patent 20,030,120,506A1, 26 June 2003.
52. Vilona, M. Pre-Cooked Food Manufacturing System. Patent WO 2,010,091,856A2, 19 August 2010.
53. Fu, B.; Labuza, T.P. Shelf-life prediction: Theory and application. *Food Control* **1993**, *4*, 125–133. [CrossRef]
54. Man, C.M.D.; Jones, A.A. (Eds.) *Shelf-Life Evaluation of Foods*; Springer: Berlin, Germany, 1994.
55. Taormina, P.J.; Hardin, M.D. (Eds.) *Food Safety and Quality-Based Shelf-life of Perishable Foods*; Springer: Berlin, Germany, 2021.
56. *Standard Guide for Sensory Evaluation Methods to Determine the Sensory Shelf-Life of Consumer Products*; ASTM E2454-20 Standard; ASTM: West Conshohocken, PA, USA, 2020.
57. Food Safety and Inspection Service. Available online: https://www.fsis.usda.gov (accessed on 2 October 2019).
58. 9CFR 317.8 9 CFR § 317.8—False or Misleading Labeling or Practices Generally; Specific Prohibitions and Requirements for Labels and Containers. U.S. Code of Federal Regulation. Available online: https://www.govinfo.gov./app/details/CFR-2021-title9-vol2/CFR-2021-title9-vol2-sec317-8/summary (accessed on 3 June 2021).
59. Regulation (EU) No 1169/2011 of the European Parliament and of the Council of 25 October. 2011. Available online: https://eur-lex.europa.eu/LexUriServ/LexUriServ.do?uri=OJ:L:2011:304:0018:0063:en:PDF (accessed on 3 June 2021).
60. Consolidated Text: Directive 2000/13/EC of the European Parliament and of the Council of 20 March 2000 on the Approximation of the Laws of the Member States Relating to the Labelling, Presentation and Advertising of Foodstuffs. Available online: http://data.europa.eu/eli/dir/2000/13/2013-07-01 (accessed on 3 June 2021).
61. Food Safety, European Commission. Available online: https://ec.europa.eu/food/safety/chemical_safety/food_contact_materials/legislation_en (accessed on 30 June 2021).
62. Regulation (EC) No 178/2002 of the European Parliament and of the Council of 28 January 2002, Art. 14. 2002. Available online: https://eur-lex.europa.eu/legal-content/EN/TXT/PDF/?uri=CELEX:02002R0178-20140630&rid=1 (accessed on 3 June 2021).
63. Koutsoumanis, K.; Allende, A.; Alvarez-Ordóñez, A.; Bolton, D.; Bover-Cid, S.; Chemaly, M.; Davies, R.; De Cesare, A.; Herman, L.; Nauta, M.; et al. Guidance on date marking and related food information: Part 1 (date marking). *EFSA J.* **2020**, *18*, 6306–6380. [CrossRef]
64. Koutsoumanis, K.; Allende, A.; Alvarez-Ordóñez, A.; Bolton, D.; Bover-Cid, S.; Chemaly, M.; Davies, R.; De Cesare, A.; Herman, L.; Hilbert, F.; et al. Guidance on date marking and related food information: Part 2 (food information). *EFSA J.* **2021**, *19*, 6510–6555. [CrossRef]

65. Prosapio, V.; Lopez-Quiroga, E. Freeze-Drying Technology in Foods. *Foods* **2020**, *9*, 920. [CrossRef] [PubMed]
66. Li, Y.H.; Wang, W.J.; Zhang, F.; Shao, Z.P.; Guo, L. Formation of the oxidized flavor compounds at different heat treatment and changes in the oxidation stability of milk. *Food Sci. Nutr.* **2018**, *7*, 238–246. [CrossRef]
67. Wu, L.; Zhang, C.; Long, Y.; Chen, Q.; Zhang, W.; Liu, G. Food additives: From functions to analytical methods. *Crit. Rev. Food Sci. Nutr.* **2021**. [CrossRef] [PubMed]
68. Vaclavik, V.A.; Christian, E.W.; Campbell, T. Food PreservationOpen image in new window. In *Essentials of Food Science*; Food Science Text Series; Springer: Cham, Switzerland, 2021. [CrossRef]
69. Brody, A.L.; Strupinsky, E.P.; Kline, L.R. *Active Packaging for Food Applications*; CRC Press: Boca Raton, FL, USA, 2001.
70. Robertson, G.L. *Food Packaging: Principles and Practice*, 3rd ed.; CRC Press: Boca Raton, FL, USA, 2012.
71. Sun, D.W. *Handbook of Frozen Food Processing and Packaging*, 2nd ed.; CRC Press: Boca Raton, FL, USA, 2012.
72. Yam, K.L. *The Wiley Encyclopedia of Packaging Technology*; John Wiley & Sons: New York, NY, USA, 2010.
73. Brandelli, A.; Brum, L.F.W.; Dos Santos, J.H.Z. Nanobiotechnology Methods to Incorporate Bioactive Compounds in Food Packaging. In *Nanoscience in Food and Agriculture 2*; Ranjan, S., Dasgupta, N., Lichtfouse, E., Eds.; Springer: Cham, Switzerland; Berlin, Germany, 2016; Volume 21.
74. Atarés, L.; Chiralt, A. Essential Oils as Additives in Biodegradable Films and Coatings for Active Food Packaging. *Trends Food Sci. Technol.* **2016**, *48*, 51–62. [CrossRef]
75. Firouz, M.S.; Mohi-Alden, K.; Omid, M. A Critical Review on Intelligent and Active Packaging in the Food Industry: Research and Development. *Food Res. Int.* **2021**, *141*, 110113–110137. [CrossRef]
76. Wrona, M.; Nerín, C. Risk Assessment of Plastic Packaging for Food Applications. In *Food Contact Materials Analysis: Mass Spectrometry Techniques*; Royal Society of Chemistry: London, UK, 2019; pp. 163–191. ISBN 978-1-78801-124-2.
77. Wrona, M.; Nerín, C. Polymers/Food Contact and Packaging Materials—Analytical Aspects. In *Reference Module in Chemistry, Molecular Sciences and Chemical Engineering*, 3rd ed.; Elsevier: Amsterdam, The Netherlands, 2018; pp. 350–359. ISBN 9780081019849.
78. Wrona, M.; Nerín, C. Analytical Approaches for Analysis of Safety of Modern Food Packaging: A Review. *Molecules* **2020**, *25*, 752. [CrossRef]
79. Zahra, S.A.; Butt, Y.N.; Nasar, S.; Akram, S.; Fatima, Q.; Ikram, J. Food Packaging in Perspective of Microbial Activity: A Review. *J. Microbiol. Biotechnol. Food Sci.* **2016**, *6*, 752–757. [CrossRef]
80. Otoni, C.G.; Espitia, P.J.P.; Avena-Bustillos, R.J.; McHugh, T.H. Trends in Antimicrobial Food Packaging Systems: Emitting Sachets and Absorbent Pads. *Food Res. Int.* **2016**, *83*, 60–73. [CrossRef]
81. Sahi, S.; Djidjelli, H.; Boukerrou, A. Study of the Properties and Biodegradability of the Native and Plasticized Corn Flour-Filled Low Density Polyethylene Composites for Food Packaging Applications. *Mater. Today Proc.* **2021**, *36*, 67–73. [CrossRef]
82. Ribeiro-Santos, R.; Andrade, M.; de Melo, N.R.; Sanches-Silva, A. Use of Essential Oils in Active Food Packaging: Recent Advances and Future Trends. *Trends Food Sci. Technol.* **2017**, *61*, 132–140. [CrossRef]
83. Langeveld, W.T.; Veldhuizen, E.J.A.; Burt, S.A. Synergy between Essential Oil Components and Antibiotics: A Review. *Crit. Rev. Microbiol.* **2014**, *40*, 76–94. [CrossRef] [PubMed]
84. Aljaafari, M.; Sultan Alhosani, M.; Abushelaibi, A.; Lai, K.S.; Erin Lim, S.H. Essential Oils: Partnering with Antibiotics. In *Essential Oils-Oils Nature*; IntechOpen Limited: London, UK, 2019.
85. Pollini, M.; Sannino, A.; Paladini, F.; Sportelli, M.C.; Picca, R.A.; Cioffi, N.; Fracchiolla, G.; Valentini, A. Chapter 14: Combining Inorganic Antibacterial Nanophases and Essential Oils: Recent Findings and Prospects. In *Essential Oils and Nanotechnology for Treatment of Microbial Diseases*; Rai, M., Zacchino, S., Derita, M.G., Eds.; CRC Press: Boca Raton, FL, USA; Taylor & Francis: Oxfordshire, UK, 2017; ISBN 9781138630727.
86. Zigadlo, J.A.; Zunino, M.P.; Pizzolitto, R.P.; Merlo, C.; Omarini, A.; Dambolena, J.S. Chapter 4: Antibacterial and antibiofilm Activities of Essential Oils and Their Components Including Modes of Action. In *Essential Oils and Nanotechnology for Treatment of Microbial Diseases*; Rai, M., Zacchino, S., Derita, M.G., Eds.; CRC Press: Boca Raton, FL, USA; Taylor & Francis: Oxfordshire, UK, 2017; ISBN 9781138630727.
87. Bueno, J.; Demirci, F.; Baser, K.H.C. Chapter 6: Essential Oils against Microbial Resistance Mechanisms Challenges and Applications in Drug Discovery. In *Essential Oils and Nanotechnology for Treatment of Microbial Diseases*; Rai, M., Zacchino, S., Derita, M.G., Eds.; CRC Press: Boca Raton, FL, USA; Taylor & Francis: Oxfordshire, UK, 2017; ISBN 9781138630727.
88. Rosato, A.; Sblano, S.; Salvagno, L.; Carocci, A.; Clodoveo, M.L.; Corbo, F.; Fracchiolla, G. Anti-biofilm Inhibitory Synergistic Effects of Combinations of Essential Oils and Antibiotics. *Antibiotics* **2020**, *9*, 637. [CrossRef]
89. Salvagno, L.; Sblano, S.; Fracchiolla, G.; Corbo, F.; Clodoveo, M.L.; Rosato, A. Antibiotics—*Mentha Piperita* Essential Oil Synergism Inhibits Mature Bacterial Biofilm. *Chem. Today* **2020**, *38*, 49–52.
90. Rosato, A.; Carocci, A.; Catalano, A.; Clodoveo, M.L.; Franchini, C.; Corbo, F.; Carbonara, G.G.; Carrieri, A.; Fracchiolla, G. Elucidation of the Synergistic Action of *Mentha Piperita* Essential Oil with Common Antimicrobials. *PLoS ONE* **2018**, *13*, e0200902. [CrossRef]
91. Rosato, A.; Maggi, F.; Cianfaglione, K.; Conti, F.; Ciaschetti, G.; Rakotosaona, R.; Fracchiolla, G.; Clodoveo, M.L.; Franchini, C.; Corbo, F. Chemical Composition and Antibacterial Activity of Seven Uncommon Essential Oils. *J. Essent. Oil Res.* **2018**, *30*, 233–243. [CrossRef]

92. Rosato, A.; Piarulli, M.; Corbo, F.; Muraglia, M.; Carone, A.; Vitali, M.E.; Vitali, C. In Vitro Synergistic Action of Certain Combinations of Gentamicin and Essential Oils. *Curr. Med. Chem.* **2010**, *17*, 3289–3299. [CrossRef] [PubMed]
93. Quesada, J.; Sendra, E.; Navarro, C.; Sayas-Barberá, E. Antimicrobial Active Packaging including Chitosan Films with *Thymus vulgaris L.* Essential Oil for Ready-to-Eat Meat. *Foods* **2016**, *5*, 57. [CrossRef]
94. Duran, M.; Aday, M.S.; Demirel Zorba, N.N.; Temizkan, R.; Büyükcan, B.; Caner, C. Potential of Antimicrobial Active Packaging 'containing Natamycin, Nisin, Pomegranate and Grape Seed Extract in Chitosan Coating' To Extend Shelf-life of Fresh Strawberry. *Food Bioprod. Process.* **2016**, *98*, 354–363. [CrossRef]
95. Siripatrawan, U.; Vitchayakitti, W. Improving Functional Properties of Chitosan Film Sas Active Food Packaging by Incorporating with Propolis. *Food Hydrocoll.* **2016**, *61*, 695–702. [CrossRef]
96. Demitri, C.; De Benedictis, V.M.; Madaghiele, M.; Corcione, C.E.; Maffezzoli, A. Nanostructured Active Chitosan-Based Films for Food Packaging Applications: Effect of Graphene Stacks on Mechanical Properties. *Measurement* **2016**, *90*, 418–423. [CrossRef]
97. Zehetmeyer, G.; Meira, S.M.M.; Scheibel, J.M.; de Oliveira, R.V.B.; Brandelli, A.; Soares, R.M.D. Influence of Melt Processing on Biodegradable Nisin-PBAT Films Intended for Active Food Packaging Applications. *J. Appl. Polym. Sci.* **2016**, *133*, 1–10. [CrossRef]
98. Ferreira, F.V.; Cividanes, L.S.; Gouveia, R.F.; Lona, L.M. An Overview on Properties and Applications of poly (butylene adipate-co-terephthalate)–PBAT Based Composites. *Polym. Eng. Sci.* **2019**, *59*, 7–15. [CrossRef]
99. Azeredo, H.M.C.; Rosa, M.F.; Mattoso, L.H.C. Nanocellulose in Bio-based Food Packaging Applications. *Ind. Crops Prod.* **2017**, *97*, 664–671. [CrossRef]
100. Wrona, M.; Cran, M.J.; Nerín, C.; Bigger, S.W. Development and Characterisation of HPMC Films Containing PLA Nanoparticles Loaded with Green Tea Extract for Food Packaging Applications. *Carbohydr. Polym.* **2017**, *156*, 108–117. [CrossRef] [PubMed]
101. Ogiwara, Y.; Roman, M.J.; Decker, E.A.; Goddard, J.M. Iron chelating active packaging: Influence of competing ions and pH value on effectiveness of soluble and immobilized hydroxamate chelators. *Food Chem.* **2016**, *196*, 842–847. [CrossRef] [PubMed]
102. Castro-Mayorga, J.L.; Fabra, M.J.; Lagaron, J.M. Stabilized Nanosilver Based Antimicrobial poly (3-hydroxybutyrate-co-3-hydroxyvalerate) Nanocomposites of Interest in Active Food Packaging. *Innov. Food Sci. Emerg. Technol.* **2016**, *33*, 524–533. [CrossRef]
103. Castro-Mayorga, J.L.; Fabra, M.J.; Pourrahimi, A.M.; Olsson, R.T.; Lagaron, J.M. The Impact of Zinc Oxide Particle Morphology as an Antimicrobial and when Incorporated in poly (3-hydroxybutyrate-co-3-hydroxyvalerate) Films for Food Packaging and Food Contact Surfaces Applications. *Food Bioprod. Process.* **2017**, *101*, 32–44. [CrossRef]
104. Al-Naamani, L.; Dobretsov, S.; Dutta, J. Chitosan-zinc Oxide Nanoparticle Composite Coating for Active Food Packaging Applications. *Innov. Food Sci. Emerg. Technol.* **2016**, *38*, 231–237. [CrossRef]
105. Salarbashi, D.; Mortazavi, S.A.; Noghabi, M.S.; Bazzaz, B.S.F.; Sedaghat, N.; Ramezani, M.; Shahabi-Ghahfarrokhi, I. Development of new Active Packaging Film Made from a Soluble Soybean Polysaccharide Incorporating Zno Nanoparticles. *Carbohydr. Polym.* **2016**, *140*, 220–227. [CrossRef] [PubMed]
106. De Vietro, N.; Conte, A.; Incoronato, A.L.; Del Nobile, M.A.; Fracassi, F. Aerosol-assisted Low Pressure Plasma Deposition of Antimicrobial Hybrid Organic-inorganic Cu-composite Thin Films for Food Packaging Applications. *Innov. Food Sci. Emerg. Technol.* **2017**, *41*, 130–134. [CrossRef]
107. Ahmed, I.; Lin, H.; Zou, L.; Brody, A.L.; Li, Z.; Qazi, I.M.; Pavase, T.R.; Lv, L. A Comprehensive Review on the Application of Active Packaging Technologies to Muscle Foods. *Food Control* **2017**, *82*, 163–178. [CrossRef]
108. Fang, Z.; Zhao, Y.; Warner, R.D.; Johnson, S.K. Active and Intelligent Packaging in Meat Industry. *Trends Food Sci. Technol.* **2017**, *61*, 60–71. [CrossRef]
109. McMillin, K.W. Advancements in Meat Packaging. *Meat Sci.* **2017**, *132*, 153–162. [CrossRef] [PubMed]

 *foods*

Opinion

The Tower of Babel of Pharma-Food Study on Extra Virgin Olive Oil Polyphenols

Maria Lisa Clodoveo [1], Marilena Muraglia [2,*], Pasquale Crupi [1], Rim Hachicha Hbaieb [3], Stefania De Santis [2], Addolorata Desantis [4] and Filomena Corbo [2]

1 Interdisciplinary Department of Medicine, University of Bari "A. Moro", 70124 Bari, Italy; marialisa.clodoveo@uniba.it (M.L.C.); pasquale.crupi@uniba.it (P.C.)
2 Department of Pharmacy-Pharmaceutical Sciences, University of Bari "A. Moro", 70125 Bari, Italy; stefania.desantis@uniba.it (S.D.S.); filomena.corbo@uniba.it (F.C.)
3 Biocatalysis and Industrial Enzymes Group, Laboratory of Microbial Ecology and Technology, Carthage University, National Institute of Applied Sciences and Technology (INSAT), BP 676, Tunis 1080, Tunisia; rimhachicha@yahoo.fr
4 Department of Soil, Plant and Food Sciences (DISPA), University of Bari "A. Moro", 70126 Bari, Italy; addolorata.desantis@uniba.it
* Correspondence: marilena.muraglia@uniba.it

Abstract: Much research has been conducted to reveal the functional properties of extra virgin olive oil polyphenols on human health once EVOO is consumed regularly as part of a balanced diet, as in the Mediterranean lifestyle. Despite the huge variety of research conducted, only one effect of EVOO polyphenols has been formally approved by EFSA as a health claim. This is probably because EFSA's scientific opinion is entrusted to scientific expertise about food and medical sciences, which adopt very different investigative methods and experimental languages, generating a gap in the scientific communication that is essential for the enhancement of the potentially useful effects of EVOO polyphenols on health. Through the model of the Tower of Babel, we propose a challenge for science communication, capable of disrupting the barriers between different scientific areas and building bridges through transparent data analysis from the different investigative methodologies at each stage of health benefits assessment. The goal of this work is the strategic, distinctive, and cost-effective integration of interdisciplinary experiences and technologies into a highly harmonious workflow, organized to build a factual understanding that translates, because of trade, into health benefits for buyers, promoting EVOOs as having certified health benefits, not just as condiments.

Keywords: polyphenols; EVOO; antioxidant and anti-inflammatory properties; health claim; human studies

Citation: Clodoveo, M.L.; Muraglia, M.; Crupi, P.; Hbaieb, R.H.; De Santis, S.; Desantis, A.; Corbo, F. The Tower of Babel of Pharma-Food Study on Extra Virgin Olive Oil Polyphenols. *Foods* **2022**, *11*, 1915. https://doi.org/10.3390/foods11131915

Academic Editor: Marco Poiana

Received: 25 May 2022
Accepted: 24 June 2022
Published: 28 June 2022

1. Introduction

The Tower of Babel Model: A Challenge for Pharma-Food Science Communication on EVOO Polyphenols

Nowadays, there is considerable attention toward the functional properties of extra virgin olive oil (EVOO), the main product obtained from olives, fruits that come from evergreen trees [1,2]. EVOO is a characteristic element of the Mediterranean diet (MD) because of the health-beneficial effects deriving from its chemical composition [3–5] as well as its appreciable taste and usefulness in flavoring a large variety of foods.

The overall literature studies report a wide variety of chemical and enzymatic analyses in vitro, in vivo, and ex vivo to reveal the beneficial effects of EVOO polyphenols on human health. EVOO is one of the foods naturally rich in polyphenols and is the main lipid source in MD. Numerous scientific papers and reviews have shown that the potential health benefits of EVOO are correlated to its high content of functional compounds such as polyphenols, tocopherols, carotenoids, sterols, fatty acids, and squalene. This specific chemical composition helps to prevent the incidence of multiple diseases such as cardiac, vascular, neurodegenerative, metabolic, and inflammatory ones [1,6–9].

In agreement with Visioli et al. [10], there is an urgent need to provide unequivocal scientific evidence that can rationalize EVOO consumption, as an integral and essential part of a balanced diet and healthy lifestyle, to quality of life in terms of prevention and health maintenance. In fact, despite an exorbitant number of papers, only one effect of the EVOO polyphenols has been officially accredited as a health claim by the European Food Safety Authority (EFSA).

The need for new research able to deal systematically with the issue of phenolic compounds present in EVOO and the related actions on human health is confirmed by the fact that many of the requests of authorization to apply health claims on EVOO health effects have received a negative scientific opinion from the EFSA. The latter could be ascribed to different reasons, including the insufficient characterization of food constituent(s), the lack of beneficial physiological effects of the proposed claimed effect, and, most of all, the quality of the studies provided for the scientific substantiation of the claims. Among the numerous olive oil claims that were submitted to EFSA, only one was allowed: "Protection of LDL particles from oxidative damage" [11]. Specifically, the panel based its decision on "a well-conducted and powered study, and two smaller-scale studies that showed dose-dependent and significant effects of olive oil polyphenol consumption (for three weeks) on appropriate markers of LDL peroxidation (oxLDL)".

Moreover, the EFSA Panel on Nutrition, Novel Foods and Food Allergens (NDA) considered that the quantity of olive oil required to obtain the claimed effect is 20 g (two tablespoons) containing at least 5 mg of hydroxytyrosol and its derivatives (e.g., oleuropein complex and tyrosol). This can reasonably be consumed within a balanced diet, and the proposed wording (although the terminology can be considered excessively technical and not fully understandable for most consumers) reflected the scientific evidence, complying with the criteria for the use of claims specified in the EU Regulation 1924/2006 [11,12].

However, no other health claim on EVOO polyphenols has been approved by the competent authorities to date. The critical and objective analysis of this result reveals that most of the scientific investigations so far aimed at highlighting the health value of polyphenols in EVOO have not met the requirements for the authorization of a novel health claim while denouncing the inappropriate choice of investigative tools and the insufficient evidence provided to establish a cause–effect relationship between the daily consumption of polyphenols in EVOO and the claimed beneficial effect.

This evidence sheds new light on the need to bring together skills belonging to different scientific areas, from food sciences to medical sciences, in a context of skills that still do not communicate effectively due to excessively clear boundaries between the different methods of investigation and specific languages of each discipline. To achieve a better understanding of the real potential beneficial effects of EVOO polyphenols on consumers' health and well-being, there is a need to create osmotic communication among different scientific areas, building bridges through a clear evaluation of the information coming out from the various methodologies of investigation in each step of health benefits assessment. The failure rate of applications for the approval of health claims concerning EVOO is enormous compared to the intellectual, human, and financial resources spent to prepare the application reports supporting the cause-and-effect relationship. Unfortunately, science, focusing on bridging information deficits about the health value of EVOO polyphenols, has underestimated the need to rework its communication models to overcome regulatory difficulties in favor of valuing the health claims of EVOOs.

Therefore, this circumstance represents clear evidence of the need to point out some aspects concerning the EVOO phenols, such as the mechanisms of action and the laboratory methods used to identify, quantify, and test the antioxidant and anti-inflammatory properties; these analytical tools need to be critically evaluated to verify their method of defining the mechanisms of action and their practical utility in the transfer of information in the subsequent phases of the cause-dose-effect evaluation, when the tests on humans will have to confirm the studies of drug dynamics, pharmacokinetics, and the potential toxicity, to define the effective dose and limitations of use.

However, each research tool has a different efficacy in the health benefit validation process. Most of them need to be harmonized to accelerate the progression of functional food research through a highly active interconnection with related fields.

The challenge for communication across scientific fields, then, is not to discuss which experimental protocol to adopt to improve the health value of EVOO, but to link scientific knowledge and expertise in a transdisciplinary way by making it objective, meaningful, reproducible, and transferable to public health policy authorities.

Thus, cross-science intellectual fertilization is needed, overcoming the current stratification of the various scientific fields through a fluid and osmotic communicative process among stakeholders. Herein, we purpose the Tower of Babel as a metaphor for a scientific approach made of different disciplines and backgrounds that need to be integrated by the synthesis of methodological procedures which draw upon more than one scientific area, challenging conventional disciplinary approaches to generate the emerging sector of pharma-food research (Figure 1).

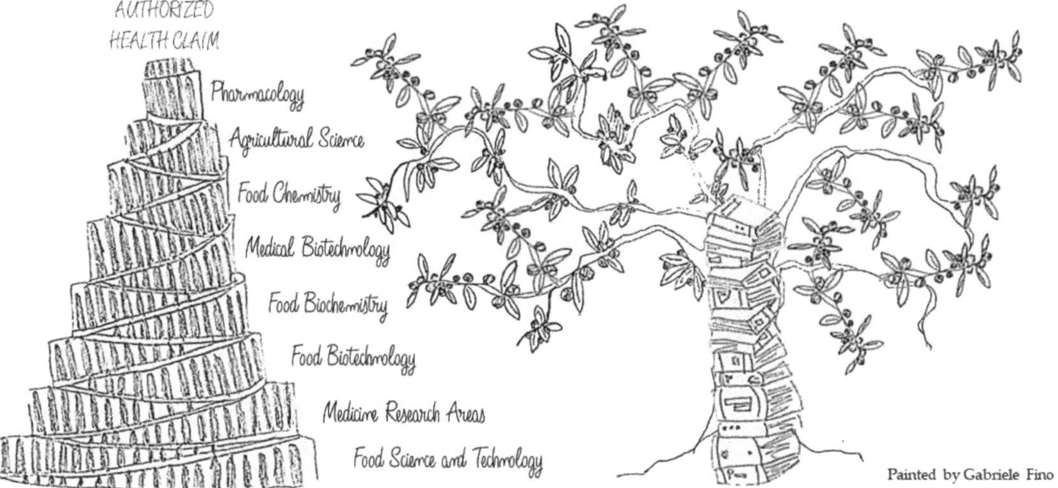

Figure 1. The metaphor of the Tower of Babel of the functional food research sector.

A specialized disciplinary language favors more efficient intra-disciplinary communication but is a tool that excludes outsiders in a particular field. Thus, the Tower of Babel is also a metaphor for different forma mentis. In this regard, a meta-disciplinary awareness is necessary, as the ability to think about the goals, methods, and forms of communication of disciplines harmonizes the focus of disciplinary knowledge and inquiry, and recognizes the roles and constraints imposed by individual disciplines in the goal attainment.

In this work, we report an overview of the main methodologies adopted by the different disciplinary areas involved in characterizing the polyphenolic and antioxidant profile of EVOOs and evaluating the antioxidant and anti-inflammatory properties potentially useful for obtaining a health claim. This paper aims to suggest a strategic, unique, and efficient integration of cross-disciplinary expertise and technology in a harmonized workflow (Figure 2), organized to build factual understanding that can be translated thanks to industry into consumer health benefits, by marketing EVOOs as having certified health benefits, and not only as seasonings.

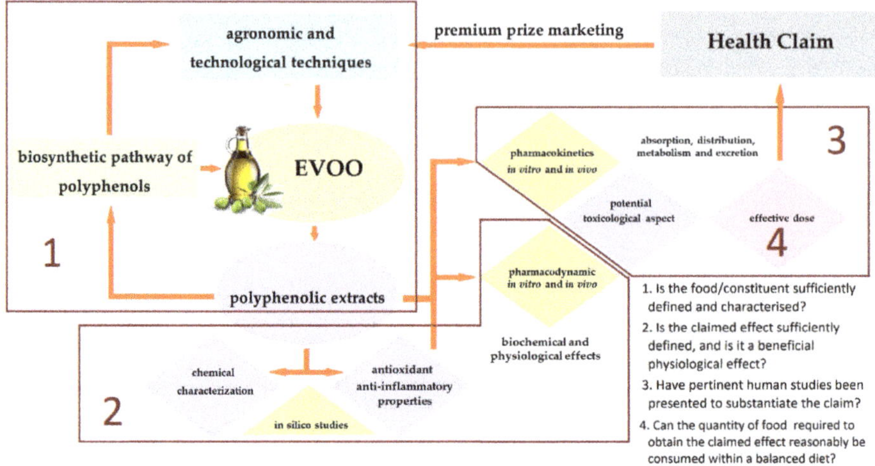

Figure 2. Workflow for pharma-food study on extra virgin olive oil polyphenols.

2. EVOO Polyphenols

2.1. Quali- and Quantitative EVOO Polyphenols Characterization: A Dual Communication?

The most represented chemical classes in *Olea europea* L. trees are mainly classified as nonpolar compounds (present in the lipophilic oil fraction, such as squalene, tocopherols, sterols, and triterpenic compounds) and polar phenolic compounds [1,13]. Among the polyphenolic compounds, the most abundant and studied in olives are tyrosol (TY), hydroxytyrosol (HT), oleuropein (OL), oleocanthal, and verbascoside (Figure 3) [1].

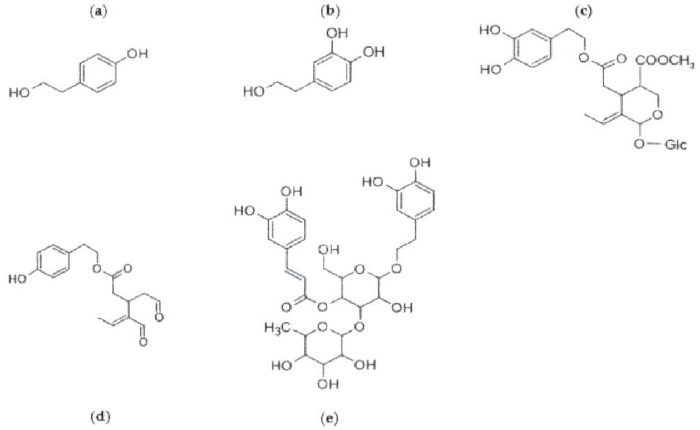

Figure 3. Most abundant polyphenols in olives: (**a**) tyrosol; (**b**) hydroxytyrosol; (**c**) oleuropein; (**d**) oleocanthal; and (**e**) verbascoside.

The secondary metabolites from *Olea europea* L. have high biological value, and they are present in different concentrations in the various parts of the olive plant [1]; as such, many of them are present in the derived EVOOs, but they can also be found in the waste products from the production process. In this regard, it should be noted that knowledge of the biosynthetic pathway of EVOO polyphenols is one of the tools that can meaningfully link distant disciplines such as food science, pharmacology, and medicine and accelerate research progress in the field of functional foods.

Understanding the origin of a bioactive compound and how to control food production phases to modulate the concentration of the biomolecule is a crucial step so that scientific evidence of a benefit to human health and the definition of the effective dose can be used as competitive factors in the industrial sector.

Furthermore, knowledge of the biosynthetic pathway of EVOO polyphenols facilitates the standardization of products and the routine use of health claims on the label. These aspects are important for food biochemistry.

All these considerations underline that the qualitative and quantitative characterization of the pool of bioactive molecules that make up food is a priority requirement for the possible authorization of a health claim.

To ensure the transferability of the health claim in the production world, agronomy and food technology work together with olive oil and EVOO producers to choose select extraction methods that reduce the influence of food matrix variability factors while ensuring the presence of bioactive compounds in an effective concentration in a consumable food dose.

Indeed, it should be emphasized that EVOO mechanically extracted from healthy olive fruits is an important source of phenolic compounds whose profile and content depend on several factors, including endogenous [14,15], agronomic [16,17], and technological factors occurring before, during, or after the EVOO extraction process [15,18–25].

The EVOO production chain can lead to the high variability of polyphenol contents in the final products. However, due to rapid advances in scientific research on the EVOO production process, it is now realistic to develop supply chain models to predict the impact of the supply chain on the polyphenol content of consumed products.

The qualitative and quantitative characterization of polyphenols is essential to classify the different commercial categories of olive oil, knowledge of which could also lead to the informed and responsible use of the product for health purposes.

Many countries have subscribed to the standards of the International Olive Oil Council (IOC), which has a United Nations Charter (UN) that sets criteria for different olive oils, their quality, and purity. The indications of the IOC have been implemented by the European Union in the European Regulation 1308/2013, which lists part of the "Designations and definitions of olive oils and olive-pomace oils" in Part VIII (Figure 4) [26].

A search of bibliographic sources through web collections of scientific literature such as PubMed, Scopus, and Google Scholar using "olive oil" and "health benefits" as keywords yielded more than 200,000 published articles, underscoring the considerable attention that academia is devoting to improving the health of olive oil.

Very often the generic term "olive oil" is used, neglecting the correlation between the different commercial categories of olive oils and the content of bioactive compounds whose chemical composition is based on: the mechanical factors of oil extraction; the climatic factors of the harvest; the geographical area; and the methods of conservation/storage of the final product.

Apart from a general reference to the type of oil used, often no chemical characterization has been carried out, not even of the highly variable saponifiable fraction, which is nevertheless within well-defined ranges (monounsaturated and polyunsaturated fatty acids are the subject of specific health claims for EVOOs), nor of the unsaponifiable fraction (in which both polyphenols and tocopherols can be certified with specific health claims).

This gap is likely due to poor communication between the disciplinary sectors involved in olive oil research.

Consequently, the lack of a precise definition of the commercial category used in the experiments and the complete chemical characterization of the matrix and polyphenolic compounds is both an obstacle to the reproducibility of the experiment and to the use of the research results as evidence to support a health claim. Reproducibility is an important principle of the scientific method. It means that a result obtained by an experiment or observational study should be achieved again with a high degree of consistency when the study is repeated using the same methodology by different researchers. Only after one or more successful repetitions should a result be accepted as scientific knowledge. Unfortunately, many experiments today, although published in reputable journals, do not effectively add to the knowledge of the health benefits of EVOO polyphenols. Some of these reports provided confusing or difficult-to-interpret results; however, in the last 10 years, these gaps have been closed [27].

DESIGNATIONS AND DEFINITIONS OF OLIVE OILS		
Virgin olive oils are oils which are obtained from the fruit of the olive tree (Olea europaea L.) solely by mechanical or other physical means under conditions, particularly thermal conditions, that do not lead to alterations in the oil, and which have not undergone any treatment other than washing, decantation, centrifugation, and filtration. Virgin olive oils shall be classified and designated as follows:		
Extra virgin olive oil (it fit for consumption)	Virgin olive oil which has a free acidity, expressed as oleic acid, of not more than 0.80 grams per 100 grams and the other physico–chemical and organoleptic characteristics of which correspond to those fixed for this category in this standard.	It can contain variable amounts of polyphenols (from 100 to 1000 mg/kg). 250 mg/kg is the fixed level for the health claim on "olive oil polyphenols"
Virgin olive oil (it fit for consumption)	Virgin olive oil which has a free acidity, expressed as oleic acid, of not more than 2.0 grams per 100 grams and the other physico–chemical and organoleptic characteristics of which correspond to those fixed for this category in this standard.	
Lampante virgin olive oil (it must undergo processing prior to consumption)	Virgin olive oil which has a free acidity expressed as oleic acid, of more than 3.3 grams per 100 grams and/or the physico–chemical and organoleptic characteristics of which correspond to those fixed for this category in this standard. It is intended for refining.	It can contain low amounts of polyphenols (from 0 to 50 mg/kg)
Refined olive oil	the acidity of this oil is not less than 3.3 grams per 100 grams and its other physico–chemical and organoleptic characteristics correspond to those fixed for this category in this standard. It is obtained from oils with a rather high level of acidity or with severe organoleptic defects that are subjected to a refining process (deodorization, deacidification and bleaching) through which acidity, organoleptic defects and oxidized substances are eliminated. The refined olive oil obtained is odourless, without antioxidant but still beneficial due to the presence of the unchanged acid profile and it has a very pale straw colour, similar to the colour of common seed oils. It may be sold only in bulk	During refining polyphenols are completely removed, so they are not detectable
Olive oil	Olive oil composed of refined olive oil and virgin olive oils: oil consisting of a blend of refined olive oil and virgin olive oils fit for consumption as they are. It has a free acidity, expressed as oleic acid, of not more than 1.00 gram per 100 grams and its other physico–chemical and organoleptic characteristics correspond to those fixed for this category in this standard	The amount of polyphenols depends on the quality of EVOO/VOO added to the blend, but it is low due to the reduced percentage of added EVOO/VOO.
Olive pomace oil is the oil obtained by treating olive pomace with solvents or other physical treatments, to the exclusion of oils obtained by re-esterification processes and of any mixture with oils of other kinds. It is marketed in accordance with the following designations and definitions		
Crude olive-pomace oil	An olive oil that is obtained from the pomace of olives, that is, the waste substance derived after the extracting process of the olives. In this product, in fact, there is a small portion of olive oil that can be extracted through the use of a solvent: pomace is sent to pomace industry and processed into a shrivelled state; then, the dried substance is mixed with a solvent (the hexane) in which all the oil dissolves. At this point the solid part is separated from the solvent that now contains oil. The solvent is removed by distillation and the residual oil is the crude olive-pomace oil. This oil, finally, is made edible through a refining process from which the refined olive-pomace oil is obtained.	It can contain low amounts of polyphenols (from 0 to 50 mg/kg)
Refined olive-pomace oil	It is obtained by refining crude olive-pomace oil reaching a maximum degree of 0.3% acidity. Being a refined olive oil, it may be sold only in bulk.	During refining polyphenols are completely removed, so they are not detectable
Pomace olive oil	It is obtained from the blend of refined-olive pomace oil with virgin olive oil with a free acidity of not more than 1%	The amount of polyphenols depends on the quality of EVOO/VOO added to the blend, but it is low due to the reduced percentage of added EVOO/VOO

Figure 4. Polyphenol content in the different commercial categories of olive oils and olive-pomace oils [26].

2.2. Characterization of EVOO Polyphenols: The Lack of an Official Methodology

Although the EFSA Food Matrix Characterization Panel's opinion on the approved claim for the ability of polyphenols in olive oil to "protect LDL particles from oxidative damage" is painfully positive, in reality, there is still a gap in terms of the ease of use of the claim on the EVOO label: the lack of an official methodology. After the publication of the approved claim, a great scientific discussion on the subject began. This topic brought together experts in food science and technology, including food chemists and analytical chemists. The HPLC method established by the IOOC for the determination of polyphenols in olive oils has not yet been implemented at the legal level [28].

There is a contrasting need for an analytical method that should be simple, repeatable, and easily adaptable to the conditions of as many laboratories as possible and for the use of high-throughput analytical techniques that are more accurate but also more complicated and that would be mandatory for the identification and quantification of each molecule of the bioactive polar phenols present in each structural form [29,30].

Be that as it may, to date there are no official methods for the determination of total phenols or the most active individual molecules in existing legislation, nor are there any legal limits for their content. As far as the analysis of total phenol content in EVOO is concerned, the most widely used test is the Folin–Ciocalteu colorimetric assay, a well-standardized electron transfer-based method suitable for the routine determination of total phenols in both hydrophilic and lipophilic matrices [31]. It is worth noting that the major limitation of the method is its lack of selectivity, as it leads to the simultaneous determination of all types of phenolic molecules present in the EVOO extract.

In line with previous reports, the EFSA Panel also commented negatively on the usefulness of using the Folin–Ciocalteu assay for the characterization of polyphenols in olive oil. In numerous EFSA opinions, the experts reported that the Folin–Ciocalteu method is not specific for polyphenols because other reducing compounds such as ascorbic acid, sugars, and proteins are also included in the quantification, leading to an overestimation of the actual polyphenol content (see Section 3.1). The total polyphenol content determined with this method is not suitable for the characterization of polyphenols in food.

A lot of elements or compounds (i.e., sugars, proteins, aromatic amines, metals, or other reductones) with high reductive potentials can also interfere with the reagent to give overestimated phenolic concentrations. Finally, even if simple to perform, the Folin–Ciocalteu test is time-consuming (it is necessary to wait for 1.5–2 h before spectrophotometric analysis) [32,33].

In a recent study, however, it was shown how the simple and rapid Folin–Ciocalteu method could give results proportional to the ones obtained by the more laborious and solvent consuming acid hydrolysis-HPLC [33].

In 2009, the International Olive Council (IOC) adopted a high-performance liquid chromatography coupled to the UV/Vis spectrophotometric detector (HPLC-UV) method for the determination of polyphenols in olive oils [34]. Specifically, based on the information about the maximum absorbance values and relative retention times, the method allows up to 27 phenolic molecules to be identified and quantified via converting the sum of the areas of the related chromatographic peaks to tyrosol equivalents (mg of tyrosol/kg of oil). This HPLC-UV determination is certainly characterized by good sensitivity and specificity, but it is also time and solvent consuming [35] and is rather cumbersome due to resolution problems, especially at higher retention times where several isobaric derivatives of secoiridoids appear. Moreover, because of the lack of commercial standards, real external calibration is not performed, which prevents a punctual and reliable identification and quantification [36,37]. To overcome these drawbacks, an analytical method consisting of acidic hydrolysis of bound forms in EVOO methanol/water extract followed by HPLC analysis to indirectly quantify secoiridoids and improve the recognition of lignans, pinoresinol, and acetoxypinoresinol in the total phenolic fraction has been proposed as a reliable alternative thanks to its simplicity and relevance to the bioactive phenolic moieties covered by the claim [38,39]. The method can also allow for assessing the total phenols before hydrolysis according to the IOC official method and can compare their contents before and after the acidic hydrolysis to indirectly investigate the secoiridoidic precursors. However, the issue regarding the correct identification of the phenolic structures persists [12].

To date, reverse-phase (RP) HPLC-MS employing C18 stationary phases has been widely accepted as the main tool in the identification, structural characterization, and quantitative analysis of phenolic compounds in EVOO [32,40]. HPLC-MS/MS methods using triple quadrupole (QqQ) mass analyzers are by far the most proposed ones for the accurate determination of polar phenols in EVOO because of their high sensitivity and selectivity in multiple reaction monitoring (MRM) acquisition modes [6,23]. On the other side, ion-trap mass analyzers (HPLC-MS) can obtain MS spectra which are helpful to establish fragmentation patterns and then elucidate the structure of more complex secoiridoid derivatives.

Even though MS/MS and MS fragmentation are powerful tools for the structural characterization and identification, the low resolution attainable with QqQ and ion-trap instruments sometimes makes the differentiation between isomers (such as open and closed forms of ligstroside and oleuropein aglycones) and isobars (different compounds with the same nominal mass but different elemental compositions, such as oleacinic and oleocanthalic acid) difficult. For these reasons, high-resolution mass spectrometry (HRMS, such as TOF, Q-TOF, Orbitrap, and FTICR), which provides excellent mass accuracy and measurements of the correct isotopic pattern, especially when combined with tandem mass spectrometry experiments, appears as the best choice to achieve definitive characterization and identification of EVOO polyphenols [41–43]. In many cases where mass spectral data are insufficient to establish a definitive structure for these complex phenolic compounds, nuclear magnetic resonance spectroscopy (NMR) is a powerful complementary technique for the structural assignment [44,45]. Further attempts have been used in recent years with a special coupling technique, LC-NMR [46–48].

In recent years, ultra-high-performance LC (UHPLC), either using sub-2 μm particle packed columns or porous-shell columns, has opened new possibilities for improving the analytical methods for complex sample matrices such as EVOO, being able to achieve 5- to 10-fold faster separations than with conventional HPLC, while maintaining or even increasing resolution [49–51]. UHPLC methods can be considered more cost-effective because they typically consume around 80% less organic solvents than conventional HPLC methods. Today, UHPLC coupled to MS and MS/MS is one of the most widely employed techniques in EVOO phenols analysis because it is typically less affected by possible matrix effects [30,52].

Figure 5 graphically summarizes the analytical and colorimetric procedures adopted for the qualitative and quantitative evaluation of EVOO polyphenols.

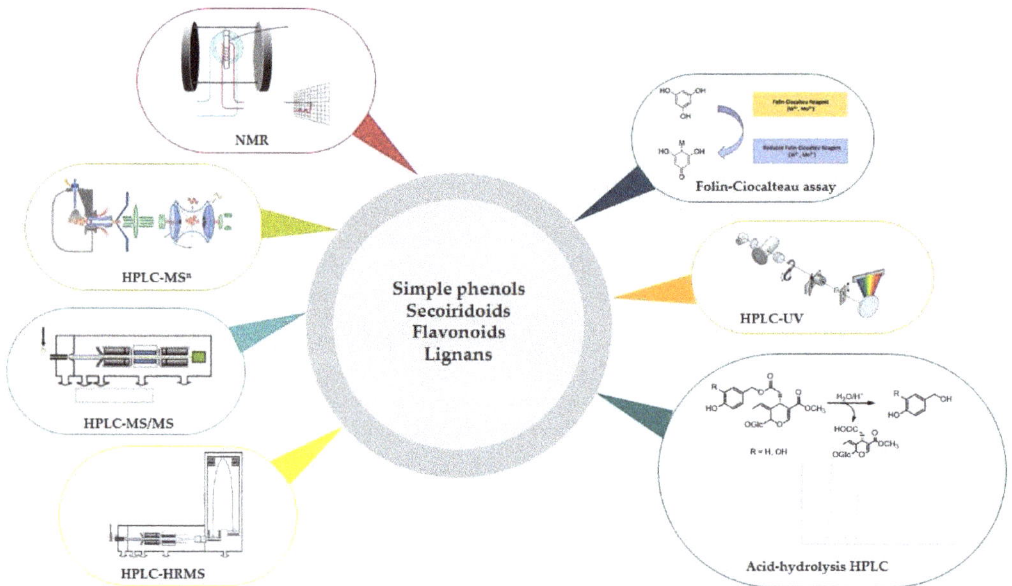

Figure 5. Methods of analysis for polyphenols in EVOO extracts.

3. EVOO Polyphenols Antioxidant and Anti-Inflammatory Properties

3.1. Chemical Antioxidant EVOO Polyphenols Assays: A Preliminary Investigation

Antioxidant assays can be classified concerning different approaches, such as:

- The type of antioxidant measured (e.g., lipophilic, hydrophilic, enzymatic, and non-enzymatic);
- The character of assay medium (e.g., aqueous, and organic solvent, direct or indirect, in situ and ex situ);
- The type of assay reagent (e.g., radical and non-radical initiated reactions);
- The mechanism of action.

Depending upon the mechanism of reactions involved, the antioxidant capacity assays can roughly be classified into two types of reaction-based assays:

1. Single electron transfer (SET);
2. Hydrogen atom transfer (HAT).

The SET mechanism involves a redox reaction with an oxidant which changes color when reduced, as an indicator of reaction endpoint, i.e., when electron transfer has stopped. The degree of color change is correlated with the sample antioxidant concentrations.

SET-based assays include the total phenols assay by:

- Folin–Ciocalteu reagent (FCR);
- Trolox equivalence antioxidant capacity (TEAC);
- 2,2-Azino-bis(3-ethylbenzothiazoline-6-sulfonic acid) assay ($ABTS^+$);
- Ferric ion reducing antioxidant power (FRAP);
- CUPRAC: "total antioxidant potential" assay using a Cu(II) complex as an oxidant;
- 2,2-Diphenyl-1-picrylhydrazyl radical scavenging capacity assay ($DPPH^{\cdot}$);
- Fast Blue BB diazonium salt (FBBB).

These assays provide a relative measure of antioxidant activity, but often the radicals scavenged have little relevance to those present in biological systems [53]. However, these total antioxidant activity assays in test tubes do not necessarily reflect the cellular physiological conditions and do not consider the bioavailability and metabolism issues. In addition, the mechanisms of action of antioxidants go beyond the antioxidant activity of scavenging free radicals in disease prevention and health promotion [54].

Instead, HAT assays involve a synthetic radical generator, an oxidizable probe, and an antioxidant; they apply a competitive reaction scheme, in which the antioxidant and substrate compete for thermally generated peroxyl radicals through the decomposition of azo compounds. These assays include:

- Inhibition of induced low-density lipoprotein autoxidation;
- Oxygen radical absorbance capacity (ORAC);
- Total radical trapping antioxidant parameter (TRAP);
- Crocin bleaching assays [55].

It is worth noting that both SET and HAT reaction-based assays measure the radical scavenging capacity instead of the preventive capacity of EVOO tout court [56,57]. In particular, SET-based assays measure the antioxidant reducing capacity, while HAT-based assays quantify hydrogen atom donating capacity [58]. Usually, a strong relationship between phenolic compounds in the EVOO extracts and the antioxidant capacities is observed, allowing it to concluded that they can act as effective radical chain-breaking antioxidants and that all the methods tested are suitable for determining the antioxidant capacity of phenolic compounds in olive oil [51,57,59].

However, it is necessary to emphasize that the assays described herein are affected by a series of drawbacks, which have opened a serious debate about their reliability and the difficulty of comparing their heterogeneous results to perform quality control for antioxidant products in the food and nutraceutical industry [57,59]; to give an idea, the best method for determining the antioxidant capacity of olive oil is sometimes referred to as ABTS and sometimes as ORAC. At first, these methods are strictly dependent on the experimental conditions, for instance, the pH, which plays a fundamental role in the reducing capacity of antioxidants. Indeed, the antioxidant assays are carried out at acidic (FRAP), neutral (TEAC), or basic (FCR) conditions; at acidic conditions, the reducing capacity may be suppressed due to protonation on antioxidant compounds, whereas at basic conditions, proton dissociation of phenolic compounds would enhance a sample reducing capacity. Another issue regards the substrate involved: HAT-based assays have a mechanistic similarity to lipid peroxidation, but under the assay conditions, the concentration of the substrate (in this case the probe) is often smaller than the concentration of antioxidants (i.e., polar phenols). This is in contradiction with the real situations: in food systems such as EVOO, the antioxidant concentration is much smaller than that of the substrate (e.g., lipid). Then, the antioxidant assays are usually carried out in a controlled manner in a homogeneous solution with an artificial oxidant or radical precursor added to initiate the reaction, whereas in a real food lipid system the reaction occurs without an added radical initiator or oxidant. Instead, the reaction is initiated by light, metal ions, or heat during food processing or storage. Moreover, it is often a heterogeneous mixture (as in food emulsions), and the phase distribution of antioxidants will be critical for its effectiveness [55,60]. Therefore, to have a more realistic assessment, the measurement of the antioxidant capacity of polyphenols that are mixtures, multifunctional, or are acting in complex multiphase systems such as EVOO must be conducted using a combination of a few assays (minimum three) involving different chemical reactions. Of course, this is time-consuming and expensive, since many reagents are needed [61]. A further controversial concern is that antioxidant capacity assays are strictly based on chemical reactions in vitro. They bear no similarity to biological systems and any claims about the bioactivity of EVOO based solely on ORAC, TEAC, FRAP, etc., would be exaggerated, unscientific, and out of context because they do not report any data about whether the measured antioxidants have any biological role [62,63]. Moreover, these assays do not take into account the bioavailability, in vivo stability, retention of antioxidants by tissues, cellular uptake, transport process, and reactivity in situ. Thereby, none of these assays have much value in terms of whether a source of antioxidants provides antioxidant protection in biological systems except in situations where biological fluids, collected before and after EVOO consumption, for example, are applied in the chemical assays, because in this way the gathered data would tell a story about whether antioxidants are assimilated into the bloodstream in a form that can reduce free radical damage. On the other side, the current antioxidant assays have been reported to possess several strengths, such as a simple procedure, rapid analysis time [55,64], cheap reagents, good correlation with bioactive compounds (phenols and flavonoids), reproducibility [65], and the use of simple instrumentation. These strengths are generally good, but are not strong enough to overcome the aforementioned limitations and to support the efficacy and reliability of these assays as they are seriously flawed.

Recent methodologies adopted to study the antioxidant and anti-inflammatory profile, such as membrane-mimicking systems (liposomes, micelles, etc.) and the activation of cellular antioxidant responses, via Nrf2-Keap1 cascades, could provide a valuable model to study the properties of EVOO polyphenols. In particular, the erythroid nuclear factor 2-like transcription (Nrf2) pathway, which can direct the expression of antioxidant and cytoprotective genes, is the focus of worldwide research. Its activity is induced by exposure to oxidative or electrophilic stresses, including so-called indirect antioxidant compounds. Therefore, special clinical attention is being paid to natural compounds that modify Nrf2 activity, whose pharmacological applications in health and disease are being investigated [66,67].

Generally, the widespread use of antioxidant assays as rapid screening tools rather than as chemical reactions to measure kinetics and determine mechanisms are largely responsible for the inconsistencies, inaccuracies, and controversies in the scientific antioxidant literature, in medicine, and in the popular press.

Regardless, even though alternative and more specific methods have recently been proposed [68–71], to date a universal and validated protocol for the determination of antioxidant capacities of polar phenols in EVOO as well as other biological samples is still lacking [55].

3.2. In Vitro Antioxidant and Anti-Inflammatory EVOO Polyphenols Screening

Inflammation constitutes one of the biological responses of body tissues to nocive stimuli, such as pathogens, injuries, and irritants. It is habitually accompanied by pain, redness, heat, and swelling. Two types of inflammation have been distinguished: acute and chronic inflammation [72]. To begin with, acute inflammation is short-term and only takes a few days to disappear. It results from a cut or injury in the skin, bronchitis, or a sore throat. On the contrary, chronic inflammation is long-term, during which the inflammatory response can firstly damage cells, tissues, and organs and thereafter damage DNA and kill tissues. Consequently, it is generally responsible for the development of several diseases such as cancer, cardiovascular and neurodegenerative diseases, type 2 diabetes, obesity, and asthma [73–75]. Non-steroidal anti-inflammatory drugs (NSAIDs) constitute a potent anti-inflammatory drug for the management of inflammatory conditions. However, they are toxic and have secondary effects on human health [76]. For these reasons, the search for natural products with anti-inflammatory activity has been the subject of several studies. Among them, EVOO phenolic compounds have been well proven for their anti-inflammatory activity [7,77–81]. In this regard, in our previous work, we extensively reported and discussed the close correlation between the chemical characterization of EVOO polyphenols, anti-inflammatory potential, and biological activities in human studies [80].

The anti-inflammatory activity has been evaluated using different methods (see Sections 3.2.1 and 3.2.2).

3.2.1. Inhibition of Protein Denaturation Assay

Considering that the denaturation of protein is linked to the occurrence of inflammatory diseases such as diabetes and cancer, measuring the percentage inhibition of protein denaturation by a substance is useful in predicting the anti-inflammatory effect. The substance to be tested prevents the process of denaturing proteins in relation to the dimension of its anti-inflammatory activity. For that, in this method, either egg albumin or bovine serum albumin (BSA) can be used as the protein.

A volume of egg albumin or BSA was mixed with a sample extract to be tested in phosphate-buffered saline. After 15 min of incubation in a 37 °C water bath, the reaction mixture was heated at 70 °C for 5 min to denaturate the proteins. Then, the turbidity was measured spectrophotometrically at 660 nm. Distilled water or phosphate buffer solution was used as the control in place of extracts to be tested. The percentage inhibition of protein denaturation was calculated using the following formula:

$$\% \text{ Inhibition of denaturation} = (1 - A1/A2) \times 100.$$

where A1 is the absorbance of the test sample and A2 is the absorbance of the control sample.

3.2.2. Membrane Stabilization Method

The lysis of the lysosomal membrane with releasing of its constituents is considered one of the inflammation results. Consequently, the anti-inflammatory effect of a sample was estimated by its ability to stabilize the lysosomal membranes and thus prevent the release of lysosomal components such as enzymes.

For testing the anti-inflammatory activity, studies have used human red blood cells (HRBCs) as their membranes resemble those of lysosomes. HRBCs could be hemolyzed either by their treatment in a hypotonic solution or by heat [82].

a. Membrane hemolysis induced by hypotonic solution

Hyposaline solution [83] and distilled water [82] were used as hypotonic solutions to favor HRBC hemolysis. In this method, erythrocyte suspension was firstly prepared from human blood and then homogenized in hypotonic solutions with the sample to test. The mixture was then incubated at 37 °C for 30 min and, after its centrifugation at 3000 rpm for 20 min, the hemoglobin content in the supernatant liquid was determined spectrophotometrically at 560 nm.

For a control test, phosphate buffer solution was used in the place of the sample to test.

The percentage of red blood cell membrane stabilization or protection was calculated by the following equation:

$$\text{Percentage protection} = 100 - (A2/A1) \times 100$$

where A1 = absorption of the control and A2 = absorption of the test sample mixture.

b. Heat-induced hemolysis

In this case, the membranes of HRBCs were lysed due to heat treatment. Briefly, the blood cell suspension and the test sample were homogenized in an isotonic solution, such as phosphate buffer, then maintained in a water bath under shaking for 20 min at 54 °C or 30 min at 60 °C according to Gunathilake's experimental procedure [76]. After centrifugation (2500 rpm for 3 min or 3000 rpm for 5 min), the absorbance of the supernatant was measured spectrophotometrically at 540 nm or 560 nm.

In the control test, a phosphate buffer solution was used. The percentage inhibition of hemolysis was calculated as follows:

$$\text{Percentage inhibition} = (1 - A2/A1) \times 100$$

where A1 = absorption of the control and A2 = absorption of the test sample mixture.

3.2.3. Assay of Cyclooxygenase and 5-Lipooxygenase Inhibition

During an inflammation, arachidonic acid could be metabolized either by the cyclooxygenase (COX) pathway producing prostaglandins and thromboxane A2 or by generating eicosanoids and leukotrienes through the 5-lipoxygenase (5-LOX) pathway [84]. For this reason, anti-inflammatory activity is attained by suppressing the production of prostaglandins and leukotrienes via COX and 5-LOX pathways inactivation.

By focusing on lipoxygenase inhibition assay, this method consists of determining the percentage of LOX activity inhibition in the presence of an anti-inflammatory substance. For that, linoleic acid and soybean lipoxygenase were used as the substrate and enzyme, respectively [76]. Other studies have used lipoxidase and human recombinant lipoxygenase as an enzyme [85].

Briefly, an aliquot of the test sample (10 µL) was homogenized with 20 µL of soybean lipoxygenase solution (167 U/mL) in a sodium phosphate buffer (100 mM; pH 8.0) and then incubated at 25 °C for 10 min. Thereafter, the absorbance was measured spectrophotometrically at 234 nm after the addition of 10 µL of the linoleic acid substrate. In the control sample, the test sample was replaced by a phosphate buffer solution [84].

The percentage of LOX activity inhibition was calculated as follows:

$$\text{Percentage inhibition} = (1 - A2/A1) \times 100$$

where A1 = absorption of the control and A2 = absorption of the test sample mixture.

4. In Vitro EVOO Polyphenols Biological Screening: A Point about the Potential Mechanism of Action

The connection between oxidative stress and inflammation has been laid out by many authors. The pathogenic role of mixed advanced glycoxidation products (AGE) and advanced lipid peroxidation products (ALE) generated during oxidative stress and their adducts with cell biomolecules, such as proteins and nucleic acids, in several chronic inflammatory and autoimmune diseases are well documented [86].

The scientific literature reports a wide range of beneficial effects of EVOO polyphenols which are united by an etiology based on oxidative stress and chronic inflammation. Oxidative stress is viewed as an imbalance between the production of reactive oxygen species (ROS) and their elimination by protective mechanisms, which can lead to chronic inflammation. EVOO polyphenols, based on previous evidence, can be useful as an adjuvant therapy for their potential anti-inflammatory effect, associated with antioxidant activity, and the inhibition of enzymes involved in pathways of inflammatory disorders. EVOO polyphenols provide several high biological properties, such as antioxidant [87–89], anti-inflammatory [90–92], and anti-microbial activities [93,94], which are partially associated with the ability of these natural compounds to scavenge free radicals.

The beneficial effects of EVOO phenolic compounds for human health have been recognized by the American Food and Drug Administration (FDA) [95] and by EFSA, which, as previously reported, advises consumption of about 20 g of EVOO daily. Additionally, in November 2018, the FDA declared the possibility of introducing the "qualified health claim" on EVOO bottle labels.

In this section, we will focus on the antioxidant and anti-inflammatory effects of EVOO polyphenols as they have been the subject of considerable research interest in recent years.

Oxidative stress is the main cause of human diseases [88]. It results from the excessive production of reactive oxygen species (ROS) and/or the low physiological activity of antioxidant defenses against these free radicals. The high production of ROS is responsible for the damage to cell biomolecules such as lipids, DNA, and proteins, thus promoting the appearance of various diseases including cancer, respiratory, cardiovascular, neurodegenerative, and digestive diseases.

To exhibit its antioxidant activity and protect the human body against the previously mentioned diseases, phenolic compounds can act as radical scavengers, chain breakers, or metal chelators. The low redox potentials of polyphenols allow them to reduce highly oxidizing free radicals by chelating metal ions [96]. Moreover, polyphenols can inhibit the activity of enzymes implicated in ROS production and improve the antioxidant defenses system of the cell by acting on gene expression and transduction and enzyme activities [97].

In addition to oxidative stress, the disruption of the balance between pro-inflammatory and anti-inflammatory molecules occurring during an inflammation causes many pathologies [98]. The antioxidant and anti-inflammatory activities of EVOO polyphenols are essentially related to their ability to attenuate reactive oxygen species (ROS), destroy carcinogenic metabolites, and act negatively on the inflammatory processes. Several studies have revealed the antioxidant and anti-inflammatory activities of EVOO polyphenols and their protective effects against several diseases [81,89,99–104]. The antioxidant activity of EVOO is not necessarily related to its high content of total phenols. Each phenolic compound has its antioxidant power which is related essentially to its chemical structure [104].

5. In Silico Studies on EVOO Polyphenols

Molecular docking methods can be considered the first step to investigating the molecular binding properties of EVOO phenolic compounds to bind to, for example, epigenetic enzymes. Molecular docking results suggest that flavonoids, secoiridoids, and glucosides may bind particularly strongly to epigenetic regulators.

The in silico analysis, thanks to public and private molecular databases which contain the primary structure of any gene or protein of biological interest in humans and in various animal species, including three-dimensional structures on which to carry out computer experiments, allows cellular or physiological processes, even complex ones, to be simulated in a static or dynamic way, allowing the kinetics and tissue distribution of a bioactive molecule to be predicted based on a few experimental points. With this type of technique (molecular docking), it is possible, for example, to verify whether the bioactive molecules isolated from food can interact with the biological target. Computer modeling is widely used to compare the chemical characteristics of bioactive compounds on specific molecular targets, identified thanks to screening strategies involving the use of in vitro or in vivo experimental models. These methodologies allow considerable savings in terms of time and work, favoring a more targeted and rational design of biologically active molecules. For these reasons, the nutraceutical industry has been equipped, for several years now, with groups of bioinformaticians and chemists in charge of identifying the biological targets of bioactive molecules.

ROS and reactive nitrogen species (RNS) are the major reactive species causing oxidative damage in the human body. To counteract their assault, living cells have a biological defense system composed of enzymatic antioxidants that convert them to harmless species. In contrast, no enzymatic action is known to scavenge them. Therefore, the burden of defense relies on a variety of nonenzymatic antioxidants such as vitamins and other phytochemicals that, due to their favorable oxidation potential, have the property of scavenging oxidants and free radicals [60]. In this context, polyphenols are compounds of great interest, having antioxidant properties which derive from several potentially synergistic mechanisms such as radical scavenging, hydrogen atom transfer, singlet oxygen quenchers, and metal-chelating activity [105–107]. As radical scavengers or chain breakers, they act by donating hydrogen radicals to alkoxyl and peroxyl radicals formed during the initiation step of lipid oxidation, slowing down the total rate of autoxidation. The presence of catechol moiety (o-diphenols as oleuropein derivatives) stabilizes the phenoxyl radical through an intramolecular hydrogen bond; moreover, they can bind metals, preventing the lipid autoxidation related to EVOO shelf life [40,107].

6. Conclusions

Although the scientific literature reports many studies related to the chemical and biological characterization of polyphenols in EVOOs to reveal the functional properties of polyphenols in extra virgin olive oil on human health, only one effect of EVOO polyphenols has been formally approved by the EFSA as a health claim. The validity of the experimental protocols adopted is indisputable, but the scientific communication of the results is strictly limited to the specific knowledge and skills of the research groups involved in the studies. This generates a stratified and non-osmotic scientific communication among researchers to the detriment of the exploitation of the beneficial effects of EVOO polyphenols on human health. In this context, we suggest the model of the Tower of Babel as an opportunity to challenge scientific communication that can favor scientific contamination between the different fields of investigation involved, building bridges through the transdisciplinary analysis of the data of the different investigative methodologies in each stage of the assessment of health benefits. The main objective of this work is to propose an integrated, strategic, and easily transferable scientific communication between the parties that promote the health value of EVOO polyphenols by applying them to health claims.

Since we are certain that fluid communication between scientific knowledge can generate important results in enhancing the health power of EVOO polyphenols, we hope that the Tower of Babel model can be a starting point for the enhancement of scientific results from research related to the food sector and raise awareness among the authorities involved in favor of the use of healthy foods.

Author Contributions: Conceptualization M.L.C.; methodology and validation M.M.; formal analysis, A.D.; data curation, P.C., S.D.S. and R.H.H.; writing—original draft preparation, M.L.C. and M.M.; writing—review and editing, M.M. and M.L.C.; supervision, F.C. All authors have read and agreed to the published version of the manuscript.

Funding: This research received no external funding.

Acknowledgments: We kindly thank the young artist Gabriele Fino for the realization of Figure 1.

Conflicts of Interest: The authors declare no conflict of interest.

References

1. Mallamaci, R.; Budriesi, R.; Clodoveo, M.L.; Biotti, G.; Micucci, M.; Ragusa, A.; Curci, F.; Muraglia, M.; Corbo, F.; Franchini, C. Olive Tree in Circular Economy as a Source of Secondary Metabolites Active for Human and Animal Health beyond Oxidative Stress and Inflammation. *Molecules* **2021**, *26*, 1072. [CrossRef]
2. Spizzirri, U.G.; Caputo, P.; Oliviero Rossi, C.; Crupi, P.; Muraglia, M.; Rago, V.; Malivindi, R.; Clodoveo, M.L.; Restuccia, D.; Aiello, F. A Tara Gum/Olive Mill Wastewaters Phytochemicals Conjugate as a New Ingredient for the Formulation of an Antioxidant-Enriched Pudding. *Foods* **2022**, *11*, 158. [CrossRef]
3. Boskou, D.; Blekas, G.; Tsimidou, M. History and characteristics of the olive tree. In *Olive Oil Chemistry and Technology*; Boskou, D., Ed.; AOCS Press: Champaign, IL, USA, 1996.
4. Cicerale, S.; Lucas, L.J.; Keast, R.S.J. Antimicrobial, antioxidant and anti-inflammatory phenolic activities in extra virgin olive oil. *Curr. Opin. Biotechnol.* **2012**, *23*, 129. [CrossRef]
5. Marcelino, G.; Hiane, P.A.; de Cássia Freitas, K.; Santana, L.F.; Pott, A.; Donadon, J.R.; de Cássia Avellaneda Guimarães, R. Effects of Olive Oil and Its Minor Components on Cardiovascular Diseases, Inflammation, and Gut Microbiota. *Nutrients* **2019**, *11*, 1826. [CrossRef] [PubMed]
6. De Santis, S.; Liso, M.; Verna, G.; Curci, F.; Milani, G.; Faienza, M.F.; Franchini, C.; Moschetta, A.; Chieppa, M.; Clodoveo, M.L.; et al. Extra Virgin Olive Oil Extracts Modulate the Inflammatory Ability of Murine Dendritic Cells Based on Their Polyphenols Pattern: Correlation between Chemical Composition and Biological Function. *Antioxidants* **2021**, *10*, 1016. [CrossRef]
7. Sarapis, K.; Thomas, C.J.; Hoskin, J.; George, E.S.; Marx, W.; Mayr, H.L.; Kennedy, G.; Pipingas, A.; Willcox, J.C.; Prendergast, L.A.; et al. The Effect of High Polyphenol Extra Virgin Olive Oil on Blood Pressure and Arterial Stiffness in Healthy Australian Adults: A Randomized, Controlled, Cross-Over Study. *Nutrients* **2020**, *12*, 2272. [CrossRef]
8. Marrano, N.; Spagnuolo, R.; Biondi, G.; Cignarelli, A.; Perrini, S.; Vincenti, L.; Laviola, L.; Giorgino, F.; Natalicchio, A. Effects of Extra Virgin Olive Oil Polyphenols on Beta-Cell Function and Survival. *Plants* **2021**, *10*, 286. [CrossRef] [PubMed]
9. Khandouzi, N.; Zahedmehr, A.; Nasrollahzadeh, J. Effect of polyphenol-rich extra-virgin olive oil on lipid profile and inflammatory biomarkers in patients undergoing coronary angiography: A randomised, controlled, clinical trial. *Int. J. Food Sci. Nutr.* **2021**, *72*, 548–558. [CrossRef] [PubMed]
10. Visioli, F.; Franco, M.; Toledo, E.; Luchsinger, J.; Willett, W.C.; Hu, F.B.; Martinez-Gonzalez, M.A. Olive oil and prevention of chronic diseases: Summary of an International conference. *Nutr. Metab. Cardiovasc. Dis.* **2018**, *28*, 649–656. [CrossRef]
11. Agostoni, C.; Bresson, J.L.; Fairweather-Tait, S.; Flynn, A.; Golly, I.; Korhonen, H.; Lagiou, P.; Løvik, M.; Marchelli, R.; Martin, A.; et al. Scientific Opinion on the substantiation of a health claim related to polyphenols in olive and maintenance of normal blood HDL cholesterol concentrations (ID 1639, further assessment) under Article 13(1) of Regulation (EC) No 1924/2006. *EFSA J.* **2012**, *10*, 2848. [CrossRef]
12. Bellumori, M.; Cecchi, L.; Innocenti, M.; Clodoveo, M.L.; Corbo, F.; Mulinacci, N. The EFSA Health Claim on Olive Oil Polyphenols: Acid Hydrolysis Validation and Total Hydroxytyrosol and Tyrosol Determination in Italian Virgin Olive Oils. *Molecules* **2019**, *24*, 2179. [CrossRef] [PubMed]
13. Bucciantini, M.; Leri, M.; Nardiello, P.; Casamenti, F.; Stefani, M. Olive Polyphenols: Antioxidant and Anti-Inflammatory Properties. *Antioxidants* **2021**, *10*, 1044. [CrossRef]
14. García-Rodríguez, R.; Romero-Segura, C.; Sanz, C.; Pérez, A.G. Modulating oxidoreductase activity modifies the phenolic content of virgin olive oil. *Food Chem.* **2015**, *171*, 364–369. [CrossRef] [PubMed]
15. Hbaieb, R.H.; Kotti, F.; García-Rodríguez, R.; Gargouri, M.; Sanz, C.; Pérez, A.G. Monitoring endogenous enzymes during olive fruit ripening and storage: Correlation with virgin olive oil phenolic profiles. *Food Chem.* **2015**, *174*, 240–247. [CrossRef] [PubMed]
16. Hbaieb, R.H.; Kotti, F.; Gargouri, M.; Msallem, M.; Vichi, S. Ripening and storage conditions of Chétoui and Arbequina olives: Part I. Effect on olive oils volatiles profile. *Food Chem.* **2016**, *203*, 548–558. [CrossRef] [PubMed]
17. Hbaieb, R.H.; Kotti, F.; Cortes-Francisco, N.; Caixach, J.; Gargouri, M.; Vichi, S. Ripening and storage conditions of Chétoui and Arbequina olives: Part II. Effect on olive endogenous enzymes and virgin olive oil secoiridoid profile determined by high resolution mass spectrometry. *Food Chem.* **2016**, *210*, 631–639. [CrossRef] [PubMed]
18. Veneziani, G.; Esposto, S.; Taticchi, A.; Urbani, S.; Selvaggini, R.; Sordini, B.; Servili, M. Characterization of phenolic and volatile composition of extra virgin olive oil extracted from six Italian cultivars using a cooling treatment of olive paste. *LWT* **2018**, *87*, 523–528. [CrossRef]
19. Almeida, B.; Valli, E.; Bendini, A.; Gallina Toschi, T. Semi-industrial ultrasound-assisted virgin olive oil extraction: Impact on quality. *Eur. J. Lipid Sci. Technol.* **2017**, *119*, 1600230. [CrossRef]
20. Aymen Bejaoui, M.; Sánchez-Ortiz, A.; Paz Aguilera, M.; Ruiz-Moreno, M.J.; Sánchez, S.; Jiménez, A.; Beltrán, G. High power ultrasound frequency for olive paste conditioning: Effect on the virgin olive oil bioactive compounds and sensorial characteristics. *Innov. Food Sci. Emerg. Technol.* **2018**, *47*, 136–145. [CrossRef]
21. Miho, H.; Moral, J.; López-González, M.A.; Díez, C.M.; Priego-Capote, F. The phenolic profile of virgin olive oil is influenced by malaxation conditions and determines the oxidative stability. *Food Chem.* **2020**, *314*, 126183. [CrossRef]
22. Taticchi, A.; Esposto, S.; Veneziani, G.; Minnocci, A.; Urbani, S.; Selvaggini, R.; Sordini, B.; Daidone, L.; Sebastiani, L.; Servili, M. High vacuum-assisted extraction affects virgin olive oil quality: Impact on phenolic and volatile compounds. *Food Chem.* **2021**, *342*, 128369. [CrossRef]

23. Clodoveo, M.L.; Dipalmo, T.; Crupi, P.; Durante, V.; Pesce, V.; Maiellaro, I.; Lovece, A.; Mercurio, A.; Laghezza, A.; Corbo, F.; et al. Comparison between Different Flavored Olive Oil Production Techniques: Healthy Value and Process Efficiency. *Plant Foods Hum. Nutr.* **2016**, *71*, 81–87. [CrossRef] [PubMed]
24. Clodoveo, M.L.; Moramarco, V.; Paduano, A.; Sacchi, R.; Di Palmo, T.; Crupi, P.; Corbo, F.; Pesce, V.; Distaso, E.; Tamburrano, P.; et al. Engineering design and prototype development of a full scale ultrasound system for virgin olive oil by means of numerical and experimental analysis. *Ultrason. Sonochem.* **2017**, *37*, 169–181. [CrossRef] [PubMed]
25. Abbattista, R.; Losito, I.; Castellaneta, A.; De Ceglie, C.; Calvano, C.D.; Cataldi, T.R.I. Insight into the Storage-Related Oxidative/Hydrolytic Degradation of Olive Oil Secoiridoids by Liquid Chromatography and High-Resolution Fourier Transform Mass Spectrometry. *J. Agric. Food Chem.* **2020**, *68*, 12310. [CrossRef] [PubMed]
26. Official Journal of the European Union. Regulation (EU) No 1308/2013 of the European Parliament and of the Council Establishing a Common Organization of the Markets in Agricultural Products and Repealing Council Official. *J. Eur. Union Bruss. Belg.* **2013**, *L 347/671*, 1. Available online: https://data.europa.eu/eli/reg/2013/1308/oj (accessed on 22 May 2022).
27. Buttriss, J.L.; Benelam, B. Nutrition and health claims: The role of food composition data. *Eur. J. Clin. Nutr.* **2010**, *64* (Suppl. S3), S8–S13. [CrossRef]
28. International Olive Council. Determination of the Content of Waxes, Fatty Acid Methyl Esters and Fatty Acid Ethyl Esters by Capillary Gas Chromatography. 2017. Available online: https://www.internationaloliveoil.org/wp-content/uploads/2019/11/Method-COI-T.20-Doc.-No-28-Rev.2-2017.pdf (accessed on 22 May 2022).
29. Karkoula, E.; Skantzari, A.; Melliou, E.; Magiatis, P. Direct Measurement of Oleocanthal and Oleacein Levels in Olive Oil by Quantitative 1H NMR. Establishment of a New Index for the Characterization of Extra Virgin Olive Oils. *J. Agric. Food Chem.* **2012**, *60*, 11696. [CrossRef]
30. Tsimidou, M.Z.; Boskou, D. The health claim on "olive oil polyphenols" and the need for meaningful terminology and effective analytical protocols. *Eur. J. Lipid Sci. Technol.* **2015**, *117*, 1091–1094. [CrossRef]
31. Dini, I.; Graziani, G.; Gaspari, A.; Fedele, F.L.; Sicari, A.; Vinale, F.; Cavallo, P.; Lorito, M.; Ritieni, A. New Strategies in the Cultivation of Olive Trees and Repercussions on the Nutritional Value of the Extra Virgin Olive Oil. *Molecules* **2020**, *25*, 2345. [CrossRef]
32. Cerretani, L.; Gallina Toschi, T.; Bendini, A. Phenolic fraction of virgin olive oil: An overview on identified compounds and analytical methods for their determination. *Funct. Plant Sci. Biotechnol* **2009**, *3*, 69.
33. Reboredo-Rodríguez, P.; Valli, E.; Bendini, A.; Di Lecce, G.; Simal-Gándara, J.; Gallina Toschi, T. A widely used spectrophotometric assay to quantify olive oil biophenols according to the health claim (EU Reg. 432/2012). *Eur. J. Lipid Sci. Technol.* **2016**, *118*, 1593–1599. [CrossRef]
34. International Olive Council. Determination of Biophenols in Olive Oils by HPLC. 2009 COI/T.20/Doc.No 29/1-8. Available online: https://www.oelea.de/downloads/COI-T20-DOC-29-2009-DETERMINATION-OF-BIOPHENOLS-IN-OLIVE-OILSBY-HPLC.pdf (accessed on 22 May 2022).
35. Romani, A.; Pinelli, P.; Mulinacci, N.; Galardi, C.; Vincieri, F.F.; Liberatore, L.; Cichelli, A. HPLC and HRGC analyses of polyphenols and secoiridoid in olive oil. *Chromatographia* **2001**, *53*, 279–284. [CrossRef]
36. Mastralexi, A.; Nenadis, N.; Tsimidou, M.Z. Addressing analytical requirements to support health claims on "olive oil polyphenols" (EC Regulation 432/2012). *J. Agric. Food Chem.* **2014**, *62*, 2459–2461. [CrossRef] [PubMed]
37. Purcaro, G.; Codony, R.; Pizzale, L.; Mariani, C.; Conte, L. Evaluation of total hydroxytyrosol and tyrosol in extra virgin olive oils. *Eur. J. Lipid Sci. Technol.* **2014**, *116*, 805–811. [CrossRef]
38. European Commission. Regulation EC No. 432/2012 Establishing a List of Permitted Health Claims Made on Foods, Other Than Those Referring to the Reduction of Disease Risk and to Children's Development and Health. *Off. J. Eur. Union* **2012**, *L136*, 1. Available online: https://eur-lex.europa.eu/legal-content/EN/ALL/?uri=celex%3A32012R0432 (accessed on 22 May 2022).
39. Mulinacci, N.; Giaccherini, C.; Ieri, F.; Innocenti, M.; Romani, A.; Vincieri, F.F. Evaluation of lignans and free and linked hydroxy-tyrosol and tyrosol in extra virgin olive oil after hydrolysis processes. *J. Sci. Food Agric.* **2006**, *86*, 757–764. [CrossRef]
40. Ajila, C.M.; Brar, S.K.; Verma, M.; Tyagi, R.D.; Godbout, S.; Valéro, J.R. Extraction and Analysis of Polyphenols: Recent trends. *Crit. Rev. Biotechnol.* **2010**, *31*, 227–249. [CrossRef] [PubMed]
41. Losito, I.; Abbattista, R.; De Ceglie, C.; Castellaneta, A.; Calvano, C.D.; Cataldi, T. Bioactive Secoiridoids in Italian Extra-Virgin Olive Oils: Impact of Olive Plant Cultivars, Cultivation Regions and Processing. *Molecules* **2021**, *26*, 743. [CrossRef]
42. Innocenti, M.; La Marca, G.; Malvagia, S.; Giaccherini, C.; Vincieri, F.F.; Mulinacci, N. Electrospray ionisation tandem mass spectrometric investigation of phenylpropanoids and secoiridoids from solid olive residue. *Rapid Commun. Mass Spectrom.* **2006**, *20*, 2013–2022. [CrossRef]
43. Di Donna, L.; Mazzotti, F.; Napoli, A.; Salerno, R.; Sajjad, A.; Sindona, G. Secondary metabolism of olive secoiridoids. New microcomponents detected in drupes by electrospray ionization and high-resolution tandem mass spectrometry. *Rapid Commun. Mass Spectrom.* **2007**, *21*, 273–278. [CrossRef]
44. Christophoridou, S.; Dais, P. Novel approach to the detection and quantification of phenolic compounds in olive oil based on 31P nuclear magnetic resonance spectroscopy. *J. Agric. Food Chem.* **2006**, *54*, 656–664. [CrossRef]
45. Starec, M.; Calabretti, A.; Berti, F.; Forzato, C. Oleocanthal Quantification Using 1H NMR Spectroscopy and Polyphenols HPLC Analysis of Olive Oil from the Bianchera/Belica Cultivar. *Molecules* **2021**, *26*, 242. [CrossRef] [PubMed]

46. Christophoridou, S.; Dais, P.; Tseng, L.H.; Spraul, M. Separation and identification of phenolic compounds in olive oil by coupling high-performance liquid chromatography with postcolumn solid-phase extraction to nuclear magnetic resonance spectroscopy (LC-SPE-NMR). *J. Agric. Food Chem.* **2005**, *53*, 4667–4679. [CrossRef]
47. Pérez-Trujillo, M.; Gómez-Caravaca, A.M.; Segura-Carretero, A.; Fernández-Gutiérrez, A.; Parella, T. Separation and identification of phenolic compounds of extra virgin olive oil from *Olea europaea* L. by HPLC-DAD-SPE-NMR/MS. Identification of a new diastereoisomer of the aldehydic form of oleuropein aglycone. *J. Agric. Food Chem.* **2010**, *58*, 9129–9136. [CrossRef]
48. Klikarová, J.; Rotondo, A.; Cacciola, F.; Ceslova, L.; Dugo, P.; Mondello, L.; Rignano, F. The Phenolic Fraction of Italian Extra Virgin Olive Oils: Elucidation through Combined Liquid Chromatography and NMR Approaches. *Food Anal. Methods* **2019**, *12*, 1759–1770. [CrossRef]
49. Sánchez de Medina, V.; Calderón-Santiago, M.; El Riachy, M.; Priego-Capote, F.; Luque de Castro, M.D. High-resolution mass spectrometry to evaluate the influence of cross-breeding segregating populations on the phenolic profile of virgin olive oils. *J. Sci. Food Agric.* **2014**, *94*, 3100–3109. [CrossRef] [PubMed]
50. Gosetti, F.; Bolfi, B.; Manfredi, M.; Calabrese, G.; Marengo, E. Determination of eight polyphenols and pantothenic acid in extra-virgin olive oil samples by a simple, fast, high-throughput and sensitive ultra high performance liquid chromatography with tandem mass spectrometry method. *J. Sep. Sci.* **2015**, *38*, 3130–3136. [CrossRef] [PubMed]
51. Negro, C.; Aprile, A.; Luvisi, A.; Nicolì, F.; Nutricati, E.; Vergine, M.; Miceli, A.; Blando, F.; Sabella, E.; De Bellis, L. Phenolic Profile and Antioxidant Activity of Italian Monovarietal Extra Virgin Olive Oils. *Antioxidants* **2019**, *8*, 161. [CrossRef] [PubMed]
52. Blasi, F.; Rocchetti, G.; Montesano, D.; Lucini, L.; Chiodelli, G.; Ghisoni, S.; Baccolo, G.; Simonetti, M.S.; Cossignani, L. Changes in extra-virgin olive oil added with *Lycium barbarum* L. carotenoids during frying: Chemical analyses and metabolomic approach. *Food Res. Int.* **2018**, *105*, 507–516. [CrossRef]
53. Perron, N.R.; Brumaghim, J.L. A review of the antioxidant mechanisms of polyphenol compounds related to iron binding. *Cell Biochem. Biophys.* **2009**, *53*, 75–100. [CrossRef]
54. Rui, H.L. Potential Synergy of Phytochemicals in Cancer Prevention: Mechanism of Action. *J. Nutr.* **2004**, *134* (Suppl. S12), 3479S–3485S. [CrossRef]
55. Bibi Sadeer, N.; Montesano, D.; Albrizio, S.; Zengin, G.; Mahomoodally, M.F. The Versatility of Antioxidant Assays in Food Science and Safety-Chemistry, Applications, Strengths, and Limitations. *Antioxidants* **2020**, *9*, 709. [CrossRef] [PubMed]
56. Augusto Ballus, C.; Dillenburg Meinhart, A.; de Souza Campos, F.A., Jr.; Teixeira Godoy, H. Total Phenolics of Virgin Olive Oils Highly Correlate with the Hydrogen Atom Transfer Mechanism of Antioxidant Capacity. *J. Am. Oil Chem. Soc.* **2015**, *92*, 843–851. [CrossRef]
57. Samaniego Sánchez, C.; Troncoso González, A.M.; García-Parrilla, M.C.; Quesada Granados, J.J.; López García de la Serrana, H.; López Martínez, M.C. Different radical scavenging tests in virgin olive oil and their relation to the total phenol content. *Anal. Chim. Acta* **2007**, *593*, 103–107. [CrossRef] [PubMed]
58. Apak, R.; Güçlü, K.; Özyürek, M.; Esin Celik, S. Mechanism of antioxidant capacity assays and the CUPRAC (cupric ion reducing antioxidant capacity) assay. *Microchim. Acta* **2008**, *160*, 413–419. [CrossRef]
59. Ninfali, P.; Aluigi, G.; Bacchiocca, M.; Magnani, M. Antioxidant capacity of extra-virgin olive oils. *J. Am. Oil Chem. Soc.* **2001**, *78*, 243–247. [CrossRef]
60. Huang, D.; Ou, B.; Prior, R.L. The chemistry behind antioxidant capacity assays. *J. Agric. Food Chem.* **2005**, *53*, 1841–1856. [CrossRef] [PubMed]
61. Opitz, S.E.; Smrke, S.; Goodman, B.A.; Keller, M.; Schenker, S.; Yeretzian, C. Antioxidant Generation during Coffee Roasting: A Comparison and Interpretation from Three Complementary. Assays. *Foods* **2014**, *3*, 586–604. [CrossRef] [PubMed]
62. Granato, D.; Putnik, P.; Kovačević, D.B.; Santos, J.S.; Calado, V.; Rocha, R.S.; Cruz, A.G.D.; Jarvis, B.; Rodionova, O.Y.; Pomerantsev, A. Trends in Chemometrics: Food Authentication, Microbiology, and Effects of Processing. *Compr. Rev. Food Sci. Food Saf.* **2018**, *17*, 663–677. [CrossRef]
63. Magalhães, L.M.; Segundo, M.A.; Reis, S.; Lima, J.L. Methodological aspects about in vitro evaluation of antioxidant properties. *Anal. Chim. Acta* **2008**, *613*, 1–19. [CrossRef]
64. Lee, O.H.; Lee, B.Y.; Lee, J.; Lee, H.B.; Son, Y.Y.; Park, C.S.; Shetty, K.; Kim, Y.C. Assessment of phenolics-enriched extract and fractions of olive leaves and their antioxidant activities. *Bioresour. Technol.* **2009**, *100*, 6107–6113. [CrossRef]
65. Gil, M.I.; Tomás-Barberán, F.A.; Hess-Pierce, B.; Holcroft, D.M.; Kader, A.A. Antioxidant activity of pomegranate juice and its relationship with phenolic composition and processing. *J. Agric. Food Chem.* **2000**, *48*, 4581–4589. [CrossRef]
66. Paunkov, A.; Chartoumpekis, D.V.; Ziros, P.G.; Sykiotis, G.P. A Bibliometric Review of the Keap1/Nrf2 Pathway and its Related Antioxidant Compounds. *Antioxidants* **2019**, *8*, 353. [CrossRef]
67. Zhongyi, J.; Haitao, Z.; Pusen, W.; Weitao, Q.; Lin, Z.; Xiao-Kang, L.; Futian, D. Different subpopulations of regulatory T cells in human autoimmune disease, transplantation, and tumor immunity. *MedComm* **2022**, *3*, e137. [CrossRef]
68. Rodrigues, E.; Mariutti, L.R.; Mercadante, A.Z. Scavenging capacity of marine carotenoids against reactive oxygen and nitrogen species in a membrane-mimicking system. *Mar. Drugs* **2012**, *10*, 1784–1798. [CrossRef]
69. Akanbi, T.O.; Barrow, C.J. Lipase-Produced Hydroxytyrosyl Eicosapentaenoate is an Excellent Antioxidant for the Stabilization of Omega-3 Bulk Oils, Emulsions and Microcapsules. *Molecules* **2018**, *23*, 275. [CrossRef] [PubMed]
70. Condelli, N.; Caruso, M.C.; Galgano, F.; Russo, D.; Milella, L.; Favati, F. Prediction of the antioxidant activity of extra virgin olive oils produced in the Mediterranean area. *Food Chem.* **2015**, *177*, 233–239. [CrossRef]

71. Della Pelle, F.; Compagnone, D. Nanomaterial-Based Sensing and Biosensing of Phenolic Compounds and Related Antioxidant Capacity in Food. *Sensors* **2018**, *18*, 462. [CrossRef] [PubMed]
72. Gambini, J.; Stromsnes, K. Oxidative Stress and Inflammation: From Mechanisms to Therapeutic Approaches. *Biomedicines* **2022**, *10*, 753. [CrossRef] [PubMed]
73. Lopez-Candales, A.; Hernández Burgos, P.M.; Hernandez-Suarez, D.F.; Harris, D. Linking Chronic Inflammation with Cardiovascular Disease: From Normal Aging to the Metabolic Syndrome. *J. Nat. Sci.* **2017**, *3*, e341.
74. Rojas-Morales, P.; León-Contreras, J.C.; Sánchez-Tapia, M.; Silva-Palacios, A.; Cano-Martínez, A.; González-Reyes, S.; Jiménez-Osorio, A.S.; Hernández-Pando, R.; Osorio-Alonso, H.; Sánchez-Lozada, L.G.; et al. A ketogenic diet attenuates acute and chronic ischemic kidney injury and reduces markers of oxidative stress and inflammation. *Life Sci.* **2022**, *289*, 120227. [CrossRef]
75. Uddin, L.Q.; Yeo, B.T.T.; Spreng, R.N. Towards a Universal Taxonomy of Macro-scale Functional Human Brain Networks. *Brain Topogr.* **2019**, *32*, 926–942. [CrossRef] [PubMed]
76. Gunathilake, K.D.P.P.; Ranaweera, K.K.D.S.; Rupasinghe, H.P.V. In Vitro Anti-Inflammatory Properties of Selected Green Leafy Vegetables. *Biomedicines* **2018**, *6*, 107. [CrossRef] [PubMed]
77. Azab, A.; Nassar, A.; Azab, A.N. Anti-Inflammatory Activity of Natural Products. *Molecules* **2016**, *21*, 1321. [CrossRef]
78. Aswad, M.; Rayan, M.; Abu-Lafi, S.; Falah, M.; Raiyn, J.; Abdallah, Z.; Rayan, A. Nature is the best source of anti-inflammatory drugs: Indexing natural products for their anti-inflammatory bioactivity. *Inflamm. Res.* **2018**, *67*, 67–75. [CrossRef]
79. Sim, R.H.; Sirasanagandla, S.R.; Das, S.; Teoh, S.L. Treatment of Glaucoma with Natural Products and Their Mechanism of Action: An Update. *Nutrients* **2022**, *14*, 534. [CrossRef]
80. De Santis, S.; Clodoveo, M.L.; Corbo, F. Correlation between Chemical Characterization and Biological Activity: An Urgent Need for Human Studies Using Extra Virgin Olive Oil. *Antioxidants* **2022**, *11*, 258. [CrossRef]
81. Santangelo, C.; Varì, R.; Scazzocchio, B.; De Sancti, P.; Giovannini, C.; D'Archivio, M.; Masella, R. Anti-inflammatory Activity of Extra Virgin Olive Oil Polyphenols: Which Role in the Prevention and Treatment of Immune-Mediated Inflammatory Diseases? *Endocr. Metab. Immune Disord.-Drug Targets* **2018**, *18*, 36–50. [CrossRef] [PubMed]
82. Anosike, C.A.; Obidoa, O.; Ezeanyika, L.U. Membrane stabilization as a mechanism of the anti-inflammatory activity of methanol extract of garden egg (*Solanum aethiopicum*). *DARU J. Pharm. Sci.* **2012**, *20*, 76. [CrossRef]
83. Labu, Z.K.; Laboni, F.R.; Tarafdar, M.; Howlader, M.S.I.; Rashid, M.H. Membrane stabilization as a mechanism of anti-inflammatory and thrombolytic activities of ethanolic extract of arial parts of *Spondiasis pinanata* (Family: Anacardiaceae). *Health Environ. Res. Online* **2015**, *2*, 44. Available online: https://hero.epa.gov/hero/index.cfm/reference/details/reference_id/8324349 (accessed on 22 May 2022).
84. Giménez-Bastida, J.A.; González-Sarrías, A.; Laparra-Llopis, J.M.; Schneider, C.; Espín, J.C. Targeting Mammalian 5-Lipoxygenase by Dietary Phenolics as an Anti-Inflammatory Mechanism: A Systematic Review. *Int. J. Mol. Sci.* **2021**, *22*, 7937. [CrossRef]
85. Mogana, R.; Teng-Jin, K.; Wiart, C. Anti-Inflammatory, Anticholinesterase, and Antioxidant Potential of Scopoletin Isolated from *Canarium patentinervium* Miq. (*Burseraceae Kunth*). *Evid.-Based Complement. Altern. Med.* **2013**, *2013*, 734824. [CrossRef]
86. Hussain, T.; Tan, B.; Yin; Blachier, F.; Tossou, M.C.; Rahu, N. Oxidative Stress and Inflammation: What Polyphenols Can Do for Us? *Oxid. Med. Cell. Longev.* **2016**, *2016*, 7432797. [CrossRef]
87. Rodríguez-Morató, J.; Boronat, A.; Kotronoulas, A.; Pujadas, M.; Pastor, A.; Olesti, E.; Pérez-Mañá, C.; Khymenets, O.; Fitó, M.; Farré, M.; et al. Metabolic disposition and biological significance of simple phenols of dietary origin: Hydroxytyrosol and tyrosol. *Drug Metab. Rev.* **2016**, *48*, 218. [CrossRef]
88. Silenzi, A.; Giovannini, C.; Scazzocchio, B.; Varì, R.; D'Archivio, M.; Santangelo, C.; Masella, R. Chapter 22—Extra virgin olive oil polyphenols: Biological properties and antioxidant activity. In *Oxidative Stress and Dietary Antioxidants*; Victor, R., Preedy, P., Eds.; Academic Press: Cambridge, MA, USA, 2020; pp. 225–233. ISBN 9780128159729. [CrossRef]
89. Tuberoso, C.I.G.; Jerković, I.; Maldini, M.; Serreli, G. Phenolic Compounds, Antioxidant Activity, and Other Characteristics of Extra Virgin Olive Oils from Italian Autochthonous Varieties *Tonda di Villacidro*, *Tonda di Cagliari*, *Semidana*, and *Bosana*. *J. Chem.* **2016**, *2016*, 8462741. [CrossRef]
90. Abdallah, C.G.; Dutta, A.; Averill, C.L.; McKie, S.; Akiki, T.J.; Averill, L.A.; Deakin, J. Ketamine, but Not the NMDAR Antagonist Lanicemine, Increases Prefrontal Global Connectivity in Depressed Patients. *Chronic Stress* **2018**, *2*, 2470547018796102. [CrossRef]
91. Gambino, C.M.; Accardi, G.; Aiello, A.; Candore, G.; Dara-Guccione, G.; Mirisola, M.; Procopio, A.; Taormina, G.; Caruso, C. Effect of Extra Virgin Olive Oil and Table Olives on the ImmuneInflammatory Responses: Potential Clinical Applications. *Endocr. Metab. Immune Disord.-Drug Targets* **2018**, *18*, 14–22. [CrossRef]
92. Souza, P.A.L.; Marcadenti, A.; Portal, V.L. Effects of Olive Oil Phenolic Compounds on Inflammation in the Prevention and Treatment of Coronary Artery Disease. *Nutrients* **2017**, *9*, 1087. [CrossRef] [PubMed]
93. Nazzaro, F.; Fratianni, F.; Cozzolino, R.; Martignetti, A.; Malorni, L.; De Feo, V.; Cruz, A.G.; d'Acierno, A. Antibacterial Activity of Three Extra Virgin Olive Oils of the Campania Region, Southern Italy, Related to Their Polyphenol Content and Composition. *Microorganisms* **2019**, *7*, 321. [CrossRef] [PubMed]
94. Peng, F.; Md Aslam, A.; Shaoying, G.; Qi, S.; Xue, B.; Sifan, L.; Ling, G. Antimicrobial activity and mechanism of action of olive oil polyphenols extract against *Cronobacter sakazakii*. *Food Control* **2018**, *94*, 289–294. [CrossRef]
95. Vilaplana-Pérez, C.; Auñón, D.; García-Flores, L.A.; Gil-Izquierdo, A. Hydroxytyrosol and potential uses in cardiovascular diseases, cancer, and AIDS. *Front. Nutr.* **2014**, *1*, 18. [CrossRef]

96. Mishra, B.B.; Gautam, S.; Sharma, A. Free Phenolics and Polyphenol Oxidase (PPO): The Factors Affecting Post-Cut Browning in Eggplant (*Solanum melongena*). *Food Chem.* **2013**, *139*, 105–114. [CrossRef]
97. Yahfoufi, N.; Alsadi, N.; Jambi, M.; Matar, C. The Immunomodulatory and Anti-Inflammatory Role of Polyphenols. *Nutrients* **2018**, *10*, 1618. [CrossRef] [PubMed]
98. Cicchese, J.M.; Evans, S.; Hult, C.; Joslyn, L.R.; Wessler, T.; Millar, J.A.; Marino, S.; Cilfone, N.A.; Mattila, J.T.; Linderman, J.J.; et al. Dynamic balance of pro- and anti-inflammatory signals controls disease and limits pathology. *Immunol. Rev.* **2018**, *285*, 147–167. [CrossRef] [PubMed]
99. Bubonja-Šonje, M.; Knežević, S.; Abram, M. Challenges to antimicrobial susceptibility testing of plant-derived polyphenolic compounds. *Arch. Ind. Hyg. Toxicol.* **2020**, *71*, 300–311. [CrossRef] [PubMed]
100. Casas, R.; Castro-Barquero, S.; Estruch, R.; Sacanella, E. Nutrition and Cardiovascular Health. *Int. J. Mol. Sci.* **2020**, *19*, 3988. [CrossRef] [PubMed]
101. Paradiso, V.M.; Flamminii, F.; Pittia, P.; Caponio, F.; Mattia, C.D. Radical Scavenging Activity of Olive Oil Phenolic Antioxidants in Oil or Water Phase during the Oxidation of O/W Emulsions: An Oxidomics Approach. *Antioxidants* **2020**, *9*, 996. [CrossRef]
102. Scoditti, E.; Capurso, C.; Capurso, A.; Massaro, M. Vascular effects of the Mediterranean diet-part II: Role of omega-3 fatty acids and olive oil polyphenols. *Vascul Pharmacol.* **2014**, *63*, 127–134. [CrossRef] [PubMed]
103. Servili, M.; Esposto, S.; Fabiani, R.; Urbani, S.; Taticchi, A.; Mariucci, F.; Selvaggini, R.; Montedoro, G.F. Phenolic compounds in olive oil: Antioxidant, health and organoleptic activities according to their chemical structure. *Inflammopharmacology* **2009**, *17*, 76–84. [CrossRef]
104. Tripoli, E.; Giammanco, M.; Tabacchi, G.; Di Majo, D.; Giammanco, S.; La Guardia, M. The phenolic compounds of olive oil: Structure, biological activity and beneficial effects on human health. *Nutr. Res. Rev.* **2005**, *18*, 98–112. [CrossRef]
105. Salah, N.; Miller, N.J.; Paganga, G.; Tijburg, L.; Bolwell, G.P.; Rice-Evans, C. Polyphenolic flavanols as scavengers of aqueous phase radicals and as chain-breaking antioxidants. *Arch. Biochem. Biophys.* **1995**, *322*, 339–346. [CrossRef]
106. Rice-Evans, C.A.; Miller, N.J.; Paganga, G. Structure-antioxidant activity relationships of flavonoids and phenolic acids. *Free Radic. Biol. Med.* **1996**, *20*, 933–956. [CrossRef]
107. Cerretani, L.; Lerma-García, M.J.; Herrero-Martínez, J.M.; Gallina-Toschi, T.; Simó-Alfonso, E.F. Determination of Tocopherols and Tocotrienols in Vegetable Oils by Nanoliquid Chromatography with Ultraviolet−Visible Detection Using a Silica Monolithic Column. *J. Agric. Food Chem.* **2010**, *58*, 757–761. [CrossRef] [PubMed]

MDPI
St. Alban-Anlage 66
4052 Basel
Switzerland
Tel. +41 61 683 77 34
Fax +41 61 302 89 18
www.mdpi.com

Foods Editorial Office
E-mail: foods@mdpi.com
www.mdpi.com/journal/foods

www.ingramcontent.com/pod-product-compliance
Lightning Source LLC
LaVergne TN
LVHW070111170726
843515LV00007B/1659

* 9 7 8 3 0 3 6 5 5 6 6 3 5 *